LANGUAGE OF THE STARS

ASTROPHYSICS AND SPACE SCIENCE LIBRARY

A SERIES OF BOOKS ON THE RECENT DEVELOPMENTS
OF SPACE SCIENCE AND OF GENERAL GEOPHYSICS AND ASTROPHYSICS
PUBLISHED IN CONNECTION WITH THE JOURNAL
SPACE SCIENCE REVIEWS

VOLUME 77

ZDENĚK KOPAL

Dept. of Astronomy, University of Manchester, England

LANGUAGE OF THE STARS

A Discourse on the Theory of the Light Changes of Eclipsing Variables

D. REIDEL PUBLISHING COMPANY

DORDRECHT : HOLLAND / BOSTON : U.S.A.
LONDON : ENGLAND

Library of Congress Cataloging in Publication Data

Kopal, Zdeněk, 1914–
Language of the stars.

(Astrophysics and space science library; v. 77)
Bibliography: p.
Includes indexes.
1. Eclipsing binaries. I. Title. II. Series.
QB821.K77 523.8′444 79–18901
ISBN 90–277–1001–5
ISBN 90–277–1044–9 pbk.

Published by D. Reidel Publishing Company,
P.O. Box 17, Dordrecht, Holland.

Sold and distributed in the U.S.A., Canada and Mexico
by D. Reidel Publishing Company, Inc.
Lincoln Building, 160 Old Derby Street, Hingham,
Mass. 02043, U.S.A.

Printed in The Netherlands

TABLE OF CONTENTS

APPENDICES

INTRODUCTION

Eclipsing Variables – What They can Tell Us and What We can do with Them

The aim of the present book will be to provide an introduction to the interpretation of the observed light changes of eclipsing binary stars and their analysis for the elements of the respective systems.

Whenever we study the properties of any celestial body – be it a planet or a star – all information we wish to gain can reach us through two different channels: their gravitational attraction, and their light. Gravitational interaction between our Earth and its celestial neighbours is, however, measurable only at distances of the order of the dimensions of our solar system; and the only means of communication with the realm of the stars are their nimble-footed photons reaching us – with appropriate time-lag – across the intervening gaps of space.

As long as a star is single and emits constant light, it does not constitute a very revealing source of information. A spectrometry of its light can disclose, to be sure, the temperature (colour, or ionization) of the star's semi-transparent outer layers, their chemical composition, and prevalent pressure (through Stark effect) or magnetic field (Zeeman effect), it can disclose even some information about its absolute luminosity or rate of spin. It cannot, however, tell us anything about what we should like to know most – namely, the mass or size (i.e., density) of the respective configuration; its absolute dimensions, or its internal structure.

In order to disclose its mass, the star must be made to 'step on the scales' by entering into gravitational partnership with another star to form a 'binary system'. A certain amount of information on masses of the stars can be obtained from observations of nearby 'visual' binaries which are within measurable distances, and close enough to exhibit absolute motions about their common centre of gravity within 1–2 centuries of their observation. The number of such systems is, however, limited by their required proximity; and their available supply is not copious. Binary systems situated at greater distance can be discovered spectroscopically (from periodic variations of Doppler shifts of spectral lines of their components) the more easily, the closer they happen to be; and although their spectroscopic observations can furnish absolute values for the lower bounds of their masses, they can say nothing about (absolute or relative) dimensions of their components.

If, however, the system happens to be sufficiently close for the dimensions of its components to represent not too small a fraction of their separation, and if their orbit is not inclined too much to the line of sight, each component may eclipse – partly or wholly – its mate in the course of each orbit; and the system becomes thus an *eclipsing variable* – recognizable as such not only by fluctuating Doppler shifts in the spectra of each component, but also by the *variation of light* of the system within eclipses. The characteristic variations of light due to this cause represent an even more eloquent (and more easily observable) tell-tale feature of such systems; and their interpretation will constitute the principal objective of this book.

A study of the phenomena exhibited by eclipsing binary systems occupies a very important position in contemporary stellar astronomy for several reasons. First, because of a truly prodigious abundance of the objects of its study. Surveys of stars in our neighbourhood disclose that at least 0.1 per cent (probably more) of them form systems which happen to eclipse; and if stars with masses greater than that of the Sun were considered alone, their percentage would be much higher. If, moreover, the foregoing conservative estimate of their percentage were to apply to our whole galactic system, the total number of eclipsing variables within it should be of the order of 10^9. Only a minute fraction (a few thousands) of these have been identified so far, and their periods determined; but their total number in our Galaxy is beyond the hope of individual discovery. Eclipsing variables are, therefore, manifestly no exceptional or uncommon phenomena!

The significance of eclipsing variables is further emphasized by the fact that they constitute the only class of double stars that can be discovered in more distant parts of the Universe. In the neighbourhood of the Sun – up to distances of the order of 100 parsecs – at least wide binaries can be recognized by their orbital motion (or, for very wide pairs, by common proper motions of their components). Spectroscopic binaries can be discovered as such up to distances of the order of one or two thousand parsecs (depending on their absolute brightness) with the aid of modern powerful reflectors capable of decomposing their light into spectra of sufficient dispersion. Beyond that limit close binaries can, however, be detected if, and only if, they happen to be eclipsing variables. Their characteristic light variations can be measured photometrically almost down to the limit attainable by our telescopes – not only in our Galaxy, but in any system resolvable by them into stars. At present dozens of them are known in external galaxies, down to distances exceeding one million parsecs.

But the significance of eclipsing variables in astronomy is not based only on their ubiquitous presence and enormous numbers; it rests also as much (or more) on the unique nature of the information which they – and they alone – can provide. We mentioned above that spectrographic observations alone can furnish only the minimum values of the masses of the components of close binary systems or of dimensions of their orbits. The missing clue – necessary to convert their lower bounds into actual values – is represented by the inclination of the orbital plane to the line of sight; and it can be obtained from an analysis of the light changes if the respective binary happens to be an eclipsing variable. As will be shown later in this book, an analysis of their observed light curves can specify not only the orbital inclination, but also fractional dimensions of the components – which, on combination with spectroscopic data, can furnish the masses, densities and absolute dimensions of the constituent stars.

Astrophysical data which can be deduced from the observed light changes of eclipsing variables transcend, moreover, information on the absolute dimensions of their components or the characteristics of their orbits; for even their internal constitution may not (under certain conditions) remain concealed from us. Even though the interiors of the stars are concealed from view by enormous opacity of the overlying material, a gravitational field emanates from them which the overlying layers – opaque as they may be – cannot appreciably modify; and the distribution of brightness over the exposed surfaces of the components is governed by the energy flux originating in the deep interior.

With the stars – one is almost tempted to say – it is like with human beings. A solitary individual, watched at a distance, will seldom disclose to a distant observer more than some of his 'boundary conditions', insufficient in general to probe his real nature. If, however, two (or more) such individuals are brought together within hailing distance, their mutual interaction will bring out their 'internal structure' the more, the closer they come to each other. Similarly, as a star in the sky remains single, there is no way of gauging the detailed properties of its gravitational field, or of the distribution of brightness on its surface. Place, however, another star in its proximity to form a pair bound by mutual attraction: many properties of the combined gravitational field can be inferred at once from observable characteristics of the components' motion – just as a distribution of surface brightness can be deduced from analysis of observable light changes. Subsequent parts of this book will bring out many instances of such situations.

These and other possibilities opening up by the studies of eclipsing variables have long attracted due attention on the part of the observers. Largely because of the recurrent nature of the phenomena which they exhibit, eclipsing variables have always been favourites for pioneers of accurate photometry of any kind – visual, photographic or photoelectric – and the total number of observations made in this field runs into several millions. Observations alone are, however, insufficient to disclose to inspection a wealth of information which they contain. To develop this information calls for introduction of systematic methods rooted in physically sound models of the phenomena we observe to decipher what observations have to say. This sets the tasks which we are going to face in this book.

The problem at issue is indeed one of astronomical cryptography: the messages these stars sent out on waves of light are encoded by processes responsible for them; while the task of the analyst is to decode the photometric evidence to yield the information which it contains. To do so requires some knowledge of the code; and to provide it is the task of the theoretician. In the first part of this book – which follows these introductory remarks – such a code will be set up; while the second part of the book will outline the de-coding procedures.

The decoding process constitutes an essentially mathematical problem; but the identification of the code is primarily one for the astrophysicist; and the extent to which this code can be set up determines the gain to be expected from its use. Eclipsing variables do not represent by any means the only sources of encoded information reaching us from the stars: indeed, virtually all stars to the right or left of the Main Sequence in the Hertzsprung–Russell diagram are known to be variable (though not necessarily periodic). The reason, however, why their light variations tell us (at best) only a part of their story is the fact that – because of physical complexities – we lack as yet a code to decipher some (or most) part of it. In a physical sense, the quest for a code to unlock the information contained in the light changes of eclipsing variables will not confront us with unsurmountable problems. However, their gravity should also not be underestimated; for unless the theoretical basis of an analysis of their light changes is well understood – and due care exercised in their interpretation – not only do we fail to do justice to the skill and perseverance of the observers but, worse still, we run the risk that a considerable part of stellar astronomy may rest on inaccurate empirical foundations. The aim of the present book will be to lessen any such possibility; and its subject needs, therefore, scarcely any apology!

It is, incidentally, astonishing to contemplate the range of inspiration which astronomy (and physical science in general) received from the phenomena of eclipses. Not later than in the 4th century B.C., the observed features of the terrestrial shadow cast on the Moon by the Earth enabled Aristotle (384–322 B.C.) to formulate the first scientific proof that the Earth was a sphere; and only somewhat later, the eclipses of the Sun provided Aristarchos (in the early part of the 3rd century B.C.) and Hipparchos (2nd half of the 3rd century) with the geometric means to gauge the distance which separates us from the Sun. In the 17th century A.D., the eclipses of the satellites of Jupiter furnished Roemer (in 1676) with the first experimental proof of the fact that the velocity of light was finite. Total eclipses of the Sun have, in particular, proved a veritable godsend to students of many branches of science as well as humanities – from the historians of ancient times (to whom they enabled to straighten out many an obscure date of early chronology of their subject), to the geophysicists (who used the old eclipse records to detect secular irregularities in the length of the day), or to the chemists who learned (in 1868) of the existence of at least one new element (helium) from the solar flash spectra before it was identified also in the laboratory. Students of the motion of the Moon in the sky found the records of ancient eclipses invaluable for identification of its 'secular acceleration'; and a disappearance ('occultations') of the stars of known positions behind the Moon's limb valuable for accurate location of the instantaneous positions of our satellite; while high-speed photometry of such occultations led in several cases to a more precise determination of the angular diameters of such stars than would (at present) be possible by any other alternative optical method. In recent decades astrophysicists have also taken advantage of the fleeting minutes of total eclipses to measure the extent of light deflection in the gravitational field of the Sun, and thus to study the metric properties of space in the neighbourhood of large masses.

In the face of this array of facts the reader may ask whether a study of stellar eclipses will prove equally rewarding; and he will indeed not be disappointed. Before, however, he will reap full benefits of his labours (partly listed already in the preceding paragraphs), he will have to 'break the code' in which information provided by eclipsing variables is transmitted through space. If, at times, the complexity of this task (as detailed in subsequent chapters of this book) may seem to some exasperating, the reader should keep in mind that eclipses have never yet caused an astronomer to lose his head – at least not since the days of our somewhat legendary predecessors Hi and Ho in old China! Stellar eclipses will, to be sure, have scarcely ever again an opportunity to stop a battle as did (allegedly) the famous solar eclipse of 585 B.C. at the time of Thales. However, the results of their study may well exert a more profound and lasting effect on science than did that abortive skirmish between the Lydians and the Medes on the history of the human race.

The real history of eclipsing binary systems goes back to the year of 1670, when Giovanni Montanari discovered the light changes of the star Algol (Montanari, 1671) – which turned out to be a prototype of one class of eclipsing variables. A discovery of the *periodicity* of Algol's variability was not made till more than a hundred years later by the 18-year old John Goodricke in 1782; and it was Goodricke – that deaf-and-dumb prodigy, who within his short life (he died in 1786 at the age of 22) found time to discover the variability, not only of Algol, but also of β Lyrae – another prototype of eclipsing variables) – who expressed an opinion that "... if it were not perhaps too early to hazard even a conjecture on the cause of this variation, I should imagine it could hardly be accounted for otherwise than ... by the interposition of a large body revolving around Algol" (Goodricke, 1783).

This brilliant conjecture which accurately struck the head of the nail was, however, destined to remain in the realm of hypotheses for another full century. For it was not till H. C. Vogel (1890) recognized Algol as a spectroscopic binary whose conjunctions coincided with the minima of light that the binary nature of Algol (and of other eclipsing variables) was established beyond any doubt. "The wonderful train of consequences that can be drawn from such regular occultations should engage our utmost attention", wrote William Herschel (1783) in a preface to double-star astronomy at the time of discovery of the periodicity of Algol's light changes. On account of the difficulties inherent in such a proposition this train of consequences was, however, rather slow to unroll – in fact, as regards quantitative *interpretation* of the observed light changes, theoreticians trailed behind the observers by another hundred years.

For it was not till in 1880 that Edward C. Pickering made a first attempt to analyze the observed light changes in terms of a simple geometrical model (Pickering, 1880); and the line of research so initiated made at first but very slow progress. Indeed, methods applicable to an interpretation of light changes due to eclipses of any type were not launched till 1912 by Henry Norris Russell (1912), who in collaboration with Harlow Shapley (cf. Russell and Shapley, 1912) developed numerical procedures designed to cope with large numbers of observations of limited accuracy available at that time. However, in order to attain their aims, Russell and Shapley had to base their work on several oversimplified assumptions concerning the form of the two components (spheres, or similar ellipsoids), or the distribution of brightness over their apparent discs – assumptions prompted by the need to make the problem tractable at all by their method rather than their physical reasonableness. In addition, smooth curves drawn by free hand to follow the course of individual observations were substituted for the actual data; and several tables of auxiliary functions which had to be constructed numerically to make such solutions possible at all.*

The original versions of the Russell–Shapley methods, of which certain modifications were subsequently proposed by Vogt (1919), Fetlaar (1923), Stein (1924a), or Sitterly (1930), remained without challenge in their field throughout the first third of this century. An excellent and comprehensive summary of them can be found in the fourth part of the volume by J. Stein on *Die Veränderlichen Sterne* (Freiburg, 1924b);* and, in abridged form, in Part 5 of K. Schiller's *Einführung in das Studium der Veränderlichen Sterne* (Leipzig, 1923); in Volume VI(2) of *Handbuch der Astrophysik* (Chapter 4*h*, Berlin, 1928) by F. C. Henroteau, or in Chapter VII of the book on *The Binary Stars* by R. G. Aitken (New York, 1935). A succinct summary of these early methods can be found also in Chapter III of Kopal's *Introduction to the Study of Eclipsing Variables* (Cambridge, Mass., 1946). Lastly, an up-dated version of these methods was re-stated by Russell and Merrill (1952) without, however, introduction of any new ideas in the problem; but, in the meantime, their gist was rapidly being rendered obsolescent under the impact of alternative approaches.

The aspects in which the methods proposed in 1912 by Russell and Shapley (and their subsequent modifications) were found to be wanting go back to their use of the observational data: or, more specifically, to their replacement of the actual observations – consisting of the sets of discrete measurements of the star's brightness at particular times – by the continuous light curves drawn by free hand to follow the course of individual observations. There is no doubt that, at the time when such a strategy was adopted and when the actual observations consisted largely of estimates of brightness (in the sky or on photographic plates), this process constituted a shrewd move whose time-saving features may have outweighed its intrinsic disadvantages. However, with the advent and increasing use of photoelectric photometry, the time was bound to be reached when the initial assets of this approach to the problem turned into a liability.

For once we replace the actual observations by graphical inferences in the form of free-hand curves, and limit ourselves to their use as a basis for subsequent analysis, we not only give up any possibility of specifying the *uncertainty* within which the results of this analysis are defined by the data on hand; but – worse still – we have no means to ascertain the *uniqueness* of the solution, or to investigate the extent to which the elements of the system are defined by given data for different types of eclipses. Russell's approach to the problem – even in the simplest cases – could lead to any solution for the elements

* Tables provided for this purpose by Russell and Shapley (1912) (based as they were largely on graphical quadratures) were soon found insufficiently accurate (cf. Kopal, 1948b) to do justice to the observations of improving quality. Forty years later, a more accurate version of these tables was produced by Merrill (1950–1953); however, advances scored in the meantime have rendered them largely obsolescent to deal with tasks confronting the analyst by that time.

* Stein's book contains also a comprehensive historical account of all work in this field preceding Russell's, commencing with Pickering's.

of the eclipse only if certain data (such as the location of 'fixed points' of the light curve; cf. Section V-I) can be read off the free-hand curves without any error. It is well known that indeterminate systems of algebraic equations can furnish unique solutions if the offending degrees of freedom are eliminated. The adoption of fixed location of certain points of the light curve is, in effect, tantamount to a reduction in the number of the degrees of freedom of the respective solution; and the consequences of such a reduction cannot, in general, be anticipated.

To re-state this matter in different terms – no point on a light curve drawn by free hand can, of course, be expected to be localized exactly; but its position will be subject to a finite error (say, ϵ): this is true as much of those points adopted as pivots in the Russell–Shapley procedures as of any other. How will, however, the effects of such an error ϵ propagate through subsequent steps of the reduction? If the errors caused by ϵ will be diminished in the course of subsequent operations, the solution should be stable; but if the opposite were the case, it should become unstable. However, if we disregard ϵ to begin with, we cannot obviously expect to learn which one of the two alternatives may apply.

It should be stressed that an agreement of the theoretical light curve (i.e., one computed from the elements resulting from an analysis) with the observations constitutes a necessary, but *not* sufficient, condition for the correctness of the elements in question; for should the solution border on indeterminacy, nothing should be as easy as to represent the observed light changes by a right combination of wrong elements! Unfortunately for much of the work performed in the first third of the 20th century, Russell's method by its very nature was incapable of serving as a basis for an appropriate error analysis, so that any question as to the determinacy of the results obtained with its aid have remained unanswered. To be sure, if we remove possible indeterminacy by an adoption of fixed points (the positions of which are read off a free-hand curve) Russell's procedure will always furnish *some* results; but how reliable these results may be is another matter. We should – above all – emphasize that, in the absence of any independent check of the correctness of our results, *any confidence in the significance of the results obtained must be based on the justification and internal consistency of the reduction procedures* employed to obtain the solution. In this respect, the methods developed in 1912 by Russell and Shapley can be regarded only as historically the first steps towards our ultimate goal; and as we shall demonstrate in subsequent parts of this book, not necessarily in the right direction.

As will be demonstrated in Chapter 1 of this book which follows these introductory remarks, the principal reason of the difficulties encountered in the approach to the problem in the first third of this century is inherent in the mathematical structure of the underlying phenomena. The relation between the observed brightness of the system and the elements of the eclipse at any particular phase is not only highly non-linear, but transcendental; and, as a result, its mathematical solution can be approached only by successive approximations by *iteration*. Direct procedures à la Russell constitute, in fact, no solution in the mathematical sense at all, for a part of it (i.e., the location of 'fixed points') had to be anticipated in advance; and no subsequent effort could verify its correctness. On the other hand, iterative methods which should permit us to adjust all unknowns simultaneously were initiated by the writer of this book (Kopal, 1941), and subsequently developed (cf. Kopal, 1946, 1948a) in collaboration with Piotrowski (1947, 1948) to furnish the first solution of the underlying problem in the mathematical sense; in which the uncertainty of all elements, and the stability of the solution itself, come out as by-products of the iterative process. These methods have been used extensively in practice (cf. Huffer and Kopal, 1951; Kopal and Shapley, 1956), and written up in a number of sources – in particular, in Kopal's *Introduction to the Study of Eclipsing Variables* (1946), or *Computation of the Elements of Eclipsing Binary Systems* (1950), as well as in Chapter VI of his *Close Binary Systems* (1959). A good summary of these methods can also be found in Chapter 3 of a treatise on *Zatmennye Peremennye Zvezdy* (*Eclipsing Variable Stars*, ed. by V. P. Tsesevich; Moscow, 1971), which appeared in English translation (John Wiley and Sons, New York) in 1973.

A succinct summary of these iterative methods will be given in Part II (Chapter IV) of this volume. In looking back over their initial development between 1941–1948 one cannot, however, fail to note that they appeared on the scene somewhat ahead of their time – almost immediately after the appearance of the tables of the $p(k, \alpha)$-functions (for their definition, see Chapter IV) for different degrees of (linear) limb-darkening by Tsesevich (1940) – which remain still the best set of tables of these functions in existence today – without which an application of the iterative methods to practical cases would have been unthinkable. To carry out such iterations proved much more time-consuming than a relatively simple algebra involved in Russell's methods – well suited to slide-rule age. In the 1940's the best aids to computation were still desk-type electrical models that could multiply 10-digit numbers in approximately 10 seconds of real time; and to carry out an iterative solution with their aid may have occupied the investigator for several hours. Even this time was, to be sure, short in comparison with the time and the effort needed to secure the underlying observations; but with the advent of the era of automatic computers in the 1950's the iterative solutions have been programmed for machine use, e.g., by Huffer and Collins (1962), Jurkevich (1970), Linnell and Proctor (1970, 1971), Budding (1973), Söderhjelm (1974) or Look *et al.* (1978); and the time of machine solutions cut from hours to minutes.

The success with which the numerical burden of the iterative process can now be relegated to automatic computers should not, however, make us lose sight of the *limitations* inherent in *all* methods mentioned so far – direct as well as iterative – which are inherent in their roots and are not affected by a different strategy of approach to a solution of the respective equations of the problem. All methods proposed for their solution between 1880–1960 have been based on the use of the actual observations of the star's brightness made at different phases of the eclipse – direct use in iterative methods, or indirect (through free-hand light curves drawn to follow the course of individual observations) in Russell's method. Each approach aims basically at the same end: namely, *to interpret the observed light changes in the time-domain.* Such an interpretation does not, however, represent by any means the only avenue of approach to the solution of the underlying problem. The alternative approach – equally valid in principle, is to interpret the observed light changes in the *frequency domain* – i.e., to subject to an appropriate analysis, not the light curve of eclipsing variables as a plot of brightness versus time, but its *Fourier transform.*

The legitimacy of such a procedure is obvious. The Fourier (or, indeed, any other integral) transform of the light changes of eclipsing variables contains exactly the same amount of information as their original light curves in the time-domain; and the latter can be synthesized from the former in every respect – indeed, a 'spectral analysis' and 'synthesis' of any function constitute complementary operations covered by the Fourier theorem. In which respect can an approach to the solution of our problem via the frequency-domain offer advantage over that through the time-domain? A more detailed answer to this question will be given in Chapters V and VI of this book; here we wish only to summarize the salient features.

As we already mentioned (and shall rigorously demonstrate in Chapters I–III of Part I), the relation between the elements of the eclipses and their observed characteristics are *transcendental,* and not capable of any type of analytic inversion. In the frequency domain, however, these relations become *algebraic* (albeit nonlinear); and can be inverted – at least in certain cases (cf. Chapters V and VI). This difference entails, in turn, a different extent of insight into the nature of our problem, and permits us to obtain a deeper understanding of the relations between the observational data and the unknowns of the problem.

A much more fundamental difference is, moreover, noted in the behaviour of the photometric *proximity effects* in the time- and frequency-domains. In the time-domain, the light changes of close eclipsing systems (and, in view of the probability of their discovery, a large majority of known eclipsing variables are bound to constitute close systems) consist of a superposition of photometric proximity effects (due to mutual distortion as well as irradiation of both components), which are *continuous* in time, upon *discontinuous* changes arising from the eclipses.

In the time-domain – if we plot the observed brightness of the system against the time (or the phase angle), the two phenomena are difficult to separate by inspection, and may merge indistinguishably with each other for really close systems. In the frequency-domain, on the other hand, *the two causes of light variation are clearly distinguished by their frequency-spectra* – the proximity effect being characterized by *discrete* spectra; while those arising from the eclipses, by *continuous* spectra of the fundamental frequency range.

In the time-domain, the customary procedure in the past had been to 'rectify' the light curve – i.e., attempt to remove the photometric proximity effect algebraically from the light curve *before* solving the latter for the elements of the eclipse; and to do so with the aid of the 'rectification constants', deduced from the light changes between eclipses. Apart from practical difficulties inherent in such a scheme (which become the greater, the closer the system; for the range of the light curve which is to furnish these 'rectification constants' shrinks as the proximity effects grow in magnitude), the procedure itself (which goes back to Russell, 1912) can be theoretically justified only under severely simplified assumptions which are unlikely to be met in reality. In order to fulfil these assumptions, the constituent components must be similar ellipsoids, seen in projection as stars with isophotes constantly symmetrical around the centre of the apparent discs. The latter part can be fulfilled only if the discs in question were either uniformly bright; or for combinations of limb- and gravity-darkening so unlikely on physical grounds as to be of no more than geometrical significance. And if any one of these conditions is not fulfilled, the 'rectified' light curve within in minima does *not* become equivalent to one arising from eclipses of circular discs (to which standard methods of solution could then be applied), but becomes an artifact whose interpretation would be even more complicated than that of the light changes directly observed.

These facts were eventually realized by Russell (1942, 1945); but a more general treatment of these phenomena by Kopal (1945; cf. also 1959, Section V.1) showed that as soon as spherical-harmonic distortion of order higher than the second is taken into account, a symmetry of the isophotes required for rectification is irretrievably lost for *any* distribution of brightness over the apparent discs of such stars; and no amount of limb- or gravity-darkening can help to restore it.

In the face of this situation – now known for more than thirty years – the reaction of the investigators was two-fold. An increasing fraction of more critical students of close eclipsing systems preferred to abstain from any solution which would have to be based on manifestly shaky premises, and limited themselves to publish their observations as they stood – leaving the task of their analysis to the future. Others – less patient to wait – persisted in the use of discredited techniques (described aptly by a contemporary expert* as 'organized cheating') because nothing better could be put in its place; and in the hope that errors committed by the use of 'rectified' light curves as a basis for their solution for the elements of the eclipses (as if the latter were caused by mutual eclipses of spherical stars) may not have too serious an effect on the outcome.

This obviously unsatisfactory situation continued to obtain as long as our operations remained restricted to the time-domain. In the frequency-domain, however, no 'rectification' of old style is needed for decoding of the light curve at all; for – as will be shown in Chapter VI – the photometric proximity effects (due to ellipticity or reflection) can be removed from the observed light changes by a suitable 'modulation' of the light curves. It should also be mentioned, in this connection, that some of our less patient contemporary confrères have been inclined to 'jump' the first part of the Fourier theorem (its 'spectral analysis') and confined (in effect) their efforts to the second – a 'synthesis' of the light curve from an estimated model. Major effort, backed by adequate computing machinery necessary for this task, has been mounted in recent years to construct such 'synthetic light curves' – an effort profitable mainly for those who sell computing

* Professor P. B. Fellgett, in a letter to the author of this book, dated June 6th, 1975.

time at a profit; but scarcely to those aiming to advance the cause of double-star astronomy.

The reason why this is so rests in the basic indeterminacy of such an effort. Even given a sound model, the number of parameters necessary to specify the principal photometric properties of close eclipsing systems is considerable (generally not less than seven); and their adjustment by trial and error – a time-consuming procedure – to match the observed light changes still means but little unless one can prove that the solution obtained is also *unique*. Unfortunately, the 'synthesis' part of the Fourier theorem alone cannot ensure the uniqueness of the outcome: in point of fact, the less determinate the solution, the easier it becomes to match the observations by a right combination of wrong elements – as more than one investigator learned the hard way to his sorrow. The only part of the Fourier theorem which *can* ensure uniqueness (*and* indicate its uncertainty) is the 'spectral analysis'; and for this reason we shall confine our attention to it almost exclusively in this book.

On this background, the general plan of this volume and the structure of its individual chapters becomes self-evident. The problem confronting us is, indeed, one of 'celestial cryptography' – of decoding messages reaching us from this class of celestial objects on the waves of light, and recorded to a different degree of precision by our photometers. A necessary prerequisite for such a task is, of course, an access to the 'code' which is embodied in the general physical characteristics of the systems under consideration. Naturally we do not know these exactly; but in Part I (Chapters I–III) of this volume we shall construct a model of such a code which should go a long way towards unlocking the meaning of their message.

The code developed in the chapters just referred to, and based on the application of the equilibrium theory of tides, is not quite new: in fact, the essential features of it have been known to us for at least one generation. It is, moreover, neither perfect nor all-inclusive; and certain recurrent features exhibited by many close systems (asymmetry of the light changes!) have so far remained outside its domain. In the face of such a situation some readers may, therefore, be apt to ask: if our knowledge of the code of solution is still lacking in precision to this extent, is it justifiable to analyze, on its basis, the available observations to the degree of accuracy attainable by the methods developed in Chapters V–VII? The answer to this question is manifestly in the affirmative; for the secondary phenomena whose cause is not yet securely within the grasp of our understanding cannot obviously be isolated and brought out in the open until the major effects – which are bound to be operative – have been satisfactorily dealt with first. It is the nature of the residuals by which the observed light curves depart from the best fit to a simplified model which should guide us towards a refinement of our code. Nevertheless, even as it is, this code can unlock a truly astonishing amount of information of high astrophysical value – much more extensive as well as accurate than those we currently possess to decipher the secrets of other classes of variable stars.

While Part I of this book will be devoted to the task of setting up a model ('code') of close binary systems based on the equilibrium theory of the tides, in Part II (Chapters II–VII) we shall turn to the task of decoding of the messages in the time- and frequency-domain. The setting-up of the code constitutes an essentially physical problem, but the problem of decoding is primarily a mathematical one. In reflecting on its contents, many a reader may wonder why it took astronomers so long to reach the stage reflected in Chapters V and VI. The answer is probably the fact that for too long the astronomers – that small band of rare creatures minding usually their own business – tried to 'go it alone', and failed to realize (or take advantage of the fact) that the problem of an analysis of the light changes of eclipsing variables was only a particular case of the 'theory of information', developed already to considerable lengths in the service of communication engineering and other branches of human endeavour. The fact that their information may reach us through other channels than light (acoustic, electrical, etc.) does not preclude application of techniques originally developed to other ends; and vice versa. Indeed, any such information can be regarded as a 'language' of the source; and what matters is its internal logic – rather than technical aspects of its transmission.

In order to illustrate this situation on particular instances, let it be stressed that the situation facing the investigator of the phenomena exhibited by eclipsing variables can be compared with the function of a television camera which optically scans a three-dimensional image and transforms what it sees sequentially, at high speed, into a series of linear elements which are reassembled by the receiver into a two-dimensional picture. An eclipsing variable represents, in principle, an analogous source of information; for as one component proceeds to eclipse its mate, a scanning takes place (by the eclipsing limb) which gives rise to characteristic changes of light that can be observed at a distance. Our eclipsing variable can, indeed, be regarded as an elementary one-channel television system – transmitting continuous light signals which our detector (i.e., photometer) records sequentially; and the aim of the receiver must be to reconstruct, from time-variations of the intensity of one picture-point, a time-independent two-dimensional picture of the system. Needless to say, the simplicity of the receiver is bound to add to the complexity of the interpretation of its messages: the task is more complicated, but it can be done.

Or – to use another parallel – in the preceding part of this Introduction we stressed that, in close binary systems, the photometric proximity effects between eclipses are caused by a superposition of partial tides ('diurnal', 'semidiurnal', etc.) which sweep around in front of our view with discrete characteristic frequencies. A spectroscopist would say that the variations of light arising from this cause are 'multiplexed' by Nature, and reach us along the same (optical) channel – i.e., our photometer – with different frequencies. The task of the analyst is to separate ('unscramble') these frequencies by a suitable (mathematical) procedure, in order to reconstruct their message.

The technique of 'modulation' of the light changes for the removal of photometric proximity effects – so familiar in the field of communication engineering – was mentioned already earlier in this introduction. Let us add to it, in this place, a comment that the observational errors – which are responsible for the dispersion of individual observations around the mean light curve in the time-domain – will manifest themselves in the frequency-domain by affecting the high-frequency tail of their (empirical) Fourier transforms in a way which a communication engineer would describe as 'noise'. And while, in the time-domain, we minimize the cumulative effects of observational errors by least-squares solutions of the respective equations of condition (cf. Chapter IV), in the frequency-domain this noise will be filtered out by a disregard of the high-frequency end of the respective transform (to which a Fourier series terminating after a certain number of terms will provide a suitable approximation).

A sequential receipt and registration of all information contained in periodically varying light signals from the stars is forced upon us by a very low frequency at which these signals are emitted. Since the orbital periods of close binary systems are, on the average, of the order of 10^5 seconds, periodic information from them on the wave of light will reach us at frequencies generally of the order of 10^{-5} Hertz. The messages which these systems keep transmitting with such insistence are, therefore, played in a very low key – in fact, some 23–24 octaves below that of audible sound. Moreover, the photometric eclipse phenomena translated into the frequency-domain produce a dissonant cacophony (characterized as they are by continuous frequency spectra); while – in contrast – the 'signature tune' of the proximity effects consist of harmonic chords of individual tones whose wavelengths bear integral ratios to each other. It is the combined frequencies of all these tones that add up to (*sit venia verbo*) the 'music of the spheres' with which we shall be concerned in this book.

In all these instances an analogy between astronomical and terrestrial phenomena encountered in other branches of science or engineering is manifest; and so should be among their treatment. As it happened so often in the history of science in the past, a dissolution of inter-disciplinary barriers – and a transfer of methods in one branch to bear on problems of another – is almost invariably found to be beneficial to both. The extent to which this may prove to be true in our case will be for the reader of judge – now and in the future.

Habent sua fata libelli! The roots of the present book – which probably represents its author's last major contribution to the study of eclipsing variables – go back almost half a century in the past, and to his first contributions (Kopal, 1932a, b) to an analysis of their light changes by Russell's method, an acquaintance with which (being unfamiliar with the English language at that time) I had to acquire from secondary sources (Schiller, 1923: Stein, 1924). In concluding now a work which has been on and off (never really absent from) my mind for so many years I wish to use this last opportunity to express once more my gratitude to my masters whose own activities and contributions to the subject made mine possible. At the inception, my debt to Professor Vincent Nechvíle (1890–1964) of the Charles University in Prague was already repaid partly (one never can repay in full) by a dedication to him of my *Close Binary Systems* in 1959; and my debt to Professor E. Finlay–Freundlich (1884–1964) for inspiration received during his stay in Prague between 1936–1938 comes second only because it was so brief.

The seeds planted during those years in the young man's mind concerned first a quest for the models ('code') of close binary systems (the urge to 'decode' by new methods did not come till somewhat later); and none proved more pregnant with consequences than those which followed (belatedly) from an acquaintance I was privileged to make of Professor Shin-ichiro Takeda (1901–1939) of the University of Kyoto during my visit to Japan in 1936. Fate cut Takeda's life short soon thereafter, but vouchsafed him time to leave behind two papers (Takeda, 1934, 1937) which became of historical importance for further development of our subject; and their influence can be felt throughout the first part of this book. Shin-ichiro Takeda was indeed a man to whom an epitaph could be applied with which Isaac Newton immortalized the memory of Roger Cotes (1682–1716): "had he lived longer, we would have learned something".

A year later (1938) it became my fate to come to the United States (whose adopted son I eventually became) and work for several years in close proximity of Henry Norris Russell (1877–1957) and Harlow Shapley (1885–1972) – a team who did so much in 1912 to advance the knowledge of our subject. As it happened, a reprint of Takeda's 1937 paper reached Princeton around that time; and Russell passed it on to me at Harvard with a note to "try to make some sense out of it". I did; and one of my early contributions of more permanent nature (Kopal, 1942) grew out of this posthumous encounter with a man I had the privilege to meet in Japan a few years before his death.

In the meantime, the time has come to turn to the 'decoding' aspects of our problem. The first ideas that not all was well with the foundations of the methods – then reigning supreme – which Russell and Shapley developed to this end in 1912 go back to my early years at the university; and although my late teachers did their best to discourage me from such an iconoclasy (they should certainly be exonerated from any complicity!), these ideas began to gain urgency since Tsesevich (1939, 1940) provided new extensive and accurate sets of basic auxiliary tables (in particular, of the p-functions appropriate for intermediate degrees of limb-darkening, not previously available), suitable to serve as a basis for renewed work in this field. First copies of the 1940 tables (the more important set of the two) reached the Harvard Observatory in the summer of that year, and Dr. Shapley promptly placed them in my hands: the outcome proved to be the birth of the 'iterative methods' (Kopal, 1941) described in Chapter IV of this book. A brief summary of them appeared already in Chapter III of my *Introduction to the Study of Eclipsing Variables* (1946) – a little volume appropriately dedicated to Henry Norris Russell and Harlow Shapley.

Although – truth be said – the emergence of these new methods was greeted by Russell with somewhat less than enthusiasm (cf. Russell, 1942), their subsequent development received more encouragement on the part of Harlow Shapley who, although he never really returned to the subject of his own doctoral thesis in 1912, retained a lasting interest in it. It was, moreover, not till 1947 – during a year of happy collaboration with Dr. S. L. Piotrowski (then at Harvard, and since 1953 Professor of Astronomy at the University of Warsaw), who applied to our problem his mastery of the least-squares methods acquired from Professor Banachiewicz, that the subject matter outlined in the latter parts of Chapter IV of this book assumed its present form; and by that time Professor Russell was already 70 years old – still interested in the subject, but no longer in position to contribute new ideas to its further outgrowth.

In retrospect, one can perhaps wonder why it took so long for the iterative methods to be more widely accepted. The reasons were several – partly historical, and partly technical. By 1940 the Russell–Shapley methods (with roots going back to 1912) attained a canonical age; and were reduced to a series of 'precepts for the computer' (human, not machine) enjoining him to "do this, do that; and what you get is this". To follow such precepts – backed by the reputation which their originators earned for themselves since 1912 by work in other parts of our vineyard – called for only a limited amount of computation which could be carried out with a slide-rule – Russell's favourite computing instrument – and even less thinking. Who would wonder that, under these circumstances, the users of time-honoured precepts were reluctant to abandon them for alternative approaches to the solution of our problem, requiring (at least initially) hours of computation with desk-type machines – unless their reluctance could be overcome by the attraction of a basically different quality of the outcome; and unless its knowledge was appreciated enough to justify the increased effort to obtain it? At first not many answered the call; and yet how few were aware at that time that, in not many years to come, astronomers (as well as others) would come in possession of automatic computers working 'faster than thought', and rendering practicable feats which would have staggered the imagination of our predecessors in the first half of this century?

We may add that while the new iterative methods were slowly taking root among the younger and more dedicated students of the subject, the earlier (1912) methods gained a new lease on life by the publication of new sets of auxiliary tables by Russell and Merrill in 1952. With their aid, old methods were kept alive by a dwindling band of faithful epigons, whose devotion to what they learned when they were young helped to prolong the lifetime of obsolescent ideas well beyond their natural term. Merrill's tables represented certainly a beautiful piece of printing. However, whether or not their ready availability contributed positively to further development of our subject is frankly questionable.

First, their existence facilitated easy production of second-rate results which have continued to dilute the professional literature up to the present time. Secondly – and worse – they helped to perpetuate a false impression that little or nothing of significance remained to be added to methods already known – an impression which those willing to accept it must have found discouraging of further efforts. This was particularly true of many young investigators who, as a result, turned their creative abilities to other fields. However, doubts continued to grow in the minds of inveterate iconoclasts on whether we may not be on the wrong track altogether by insisting to seek solutions – by *any* method – in the *time-domain* – doubts arising mainly from our inability to deal effectively with the photometric proximity effects in close eclipsing systems. This time the present writer must accept full responsibility for the consequences of such an inconoclasm – unmitigated by the complicity of any predecessors – and should alone be blamed for any troubles in which the adoption of this new approach may land us, as they transpire in the latter parts of this book.

The shortcomings of the old methods have – we repeat – been due not only to their 'original sins' already listed (such as their reliance on fixed points, and consequent inability to provide any basis for error analysis), but also to the imperviousness of their practitioners to these 'blind spots' – merely because what was once convenient became sacrosanct to those of a conformist state of mind. As is well known also from other branches of science, 'vested interests' of this type are seldom prepared to welcome the advent of new ideas; but eventually go under by being unable to withstand the challenge of evidence with which they cannot contend; and (in our case) this proved to be the photometric 'proximity effects'.

Thoughtful readers of Chapter VI of the writer's 1959 treatise on *Close Binary Systems* have no doubt noticed that some sections of it (in particular, VI.11 and 12) were written veritably with 'tongue in cheek' – as the author was losing faith in the professed validity of the 'rectification' procedures then practised, which stood as an obstacle to further progress, and which seemed impossible to bypass in the time-domain. That treatise was written in 1957. However, shortly thereafter it dawned upon its author that the proper way of escape from these difficulties was to transpose the entire problem from the time- to the frequency-domain, in which a separation of the eclipse and proximity effects can be performed in parallel rather than serially.

The first open statement to this effect was made at the 10th General Assembly of the International Astronomical Union in Moscow (August 1958); and re-iterated at its 11th General Assembly in Berkeley, August 1961 (cf. Kopal, 1962). In more detail, the process was outlined at the Congress for Commemoration of the 50th Anniversary of the Death of G. V. Schiaparelli and published in its Proceedings (Kopal, 1960). Unscheduled preoccupations with problems of the solar system between 1961–1972 made, however, the writer to postpone any further development of this work for more than a decade; and he all but forgot about it till 1973. His return to the problem at that time owed much to the gentle insistence and friendly persuasion of Drs. Igor Jurkevich and Alan F. Petty of the U.S. Naval Research Laboratory in Washington; for soon after the initial re-start the individual pieces of the puzzle began to fall quickly into a more complete picture. The material presented later in Paper I (Kopal, 1975a) of the series of papers on 'Fourier Analysis of the Light Curves of Eclipsing Variables' (1975–1978) was virtually completed in one day (August 1, 1973) during the writer's vacations in the Swiss Alps; though it was not written up till some time after completion of Parts II and III of the same series (Kopal, 1975b, c).

The next major methodical advance was made in Paper XI (Kopal, 1977b), when the fractional loss of light $\alpha(k, p)$ of an arbitrarily darkened star (as well as other special functions of the theory of the light changes introduced in Chapter III of this book) were recognized as particular types of *Hankel transforms*. It was not till then that our subject became at last a firmly established part of physical optics; and that its treatment could be extended – e.g., to cases in which the eclipsing body may be semi-transparent (atmospheric eclipses!) with arbitrary distribution of opacity.

If so, however, the question can once more be asked: why was not the underlying problem transferred to the frequency-domain at an earlier date? The answer is similar to that already given concerning the advent of the iterative methods thirty years before. If methods to be expounded in Chapters IV–VI of this book had been developed more than 10 years ago, they would have remained at best of only theoretical interest; and their practical impact would have been very small; for the technical means of automatic computation of the scale required to make these new methods practicable have not yet arrived, or were only around the corner.

We mentioned above that, in the early 1940's, multiplication of ten-digit numbers on desk-type machines then available required approximately 10 seconds; and this was adequate – but only just – to meet the needs of the iterative methods. Since that time, continuing advances in computer engineering have increased the speed of arithmetical operations on electronic machines 10^6 to 10^7 times – an achievement which offered an entirely new dimension for scientific computation – including analysis of the light curves of eclipsing variables. The need to perform a large number of arithmetical operations ceased to be its bottleneck; and emphasis has shifted to their automation – i.e., to the programming of the tasks on hand. In this sense, the earlier methods connected with the names of Russell, Shapley or Merrill rapidly became obsolete; for they did *not* lend themselves to effective automation (since the 'geometrical depth' $p(k, \alpha)$ of the eclipses, underlying their approach, cannot be automated by any practicable programme). On the other hand – as we shall see – an outstanding feature of the frequency-domain approach is the fact that *all operations involved can be written out explicitly in algebraic form*; and *this* makes them intelligible to an automatic computer.

As is well known, such computers experience considerable difficulties in digesting large amounts of auxiliary tabular data, from which other entries are to be obtained by multi-dimensional interpolation; their memory stores can be used to better ends. Electronic computers prefer to evaluate their current needs internally from step to step, and they can do so almost unimaginably fast (performing arithmetical operations on the time-scale of microseconds). However, in order to capitalize on this ability, they require their instructions to be spelled out in algebraic form, and be so programmed that large parts of the work can be carried out automatically – without interruptions for extraneous scrutiny to slow down automatic progress of the work. As we shall demonstrate in subsequent parts of this book, our problem lends itself to this task quite conveniently in the frequency-domain, but *not* in the time-domain; and therein rests the principal difference between these two different ways of approach.

In the present state of research, the emphasis is on the *programming* of the work for use on automatic computers; the memory requirements are but moderate; while arithmetical operations entailed in the process can be performed with incredible speed. Any solution for the elements of the eclipsing variables by the methods presented in Part II of this book can be performed on an automatic computer of the class of (say) CDC 7600 in less than *one minute* of real time – in fact, less than it would take to write down the results by hand if we knew them. On smaller computers this time can be prolonged to several minutes – but still remains incomparably shorter than that taken by the slide-rule methods of bygone generations.

This is, in brief, the state in which our subject finds itself at the present time; and its more recent developments – coupled with the breath-taking speed of electronic calculation now attainable – can be greeted by investigators of the generation of the present writer with truly Simeon's joy. Although these developments have enrolled quite suddenly in the recent past, they do not by any means constitute the work of one man: in fact, the material presented in Chapters VI and VII of this book would not have reached the present state if it had not been for contributions made also by an able group of young scholars from different parts of the world who came in the past several years to do research in this field at the University of Manchester under the writer's supervision; and among whom the contributions of Dr. Mehmet Kurutac (Ph.D., Manchester, 1976), Dr. Helen J. Livaniou Rovithis (Ph.D., Manchester, 1976), Dr. Panayotis G. Niarchos (Ph.D., Manchester, 1977), Dr. Hamid M. K. Al-Naimiy (Ph.D., Manchester, 1977), Dr. S. A. H. Smith (Ph.D., Manchester, 1977), Dr. T. Edalati (Ph.D., Manchester, 1978) and – above all – Dr. Osman Demircan (Ph.D., Manchester, 1978) – should be especially mentioned; the results of their work contained in their theses as well as in subsequently published papers will be frequently referred to in many parts of this book.

Last but not least, the writer's sincere thanks are due – as always – to Mrs. Ellen B. Carling-Finlay, who saw one more branch of new astronomy to develop under her eyes, and who contributed to its development by her faithful and talented assistance to the best of her abilities.

On the Day of Epiphany, A.D. 1979. ZDENĚK KOPAL

PART I
THE CODE OF THE LANGUAGE

In the introductory chapter of this book we outlined already the range of information – much of it unique – concealed in the light changes of eclipsing binary systems, which it should be possible to extract from them by appropriate analysis. The observable aspects of their messages (transmitted continuously for those who care to listen) has many facets – such as are reflected in variations of the intensity of light exhibited by such systems, its coherence, polarization, spectral composition, etc. – all these can be recorded to a different degree of precision; and all contain information unique of its kind. However, this information is not sent out 'en clair', but encoded in a language to which we must first find a clue before we can reap full benefits of it. In order to obtain such a clue, in the first part of this book we shall set out to construct a theoretical model of systems which could serve as a 'code' to the language we shall try to decipher; while in the second part, we shall detail the decoding procedures based upon it.

In doing so, we shall hereafter restrict our attention to the interpretation of mainly one facet of available observational evidence bearing on the nature of eclipsing binary systems: namely, to the *intensity* of their measured energy output, and its variation with the phase – not only because this represents the most important channel of potential information, but also because the amount of photometric data secured since this field began attracting attention of the observers is truly prodigious. For many generations now eclipsing variables have been favourite objects for pioneers of every type of starlight photometry – visual, photographic, photoelectric – and the number of observations accumulated by them goes into several millions (published or unpublished). However (as we already mentioned), methods for unlocking information concealed in them have consistently lagged behind observational advances up to the present time – to the discouragement of the observers and detriment of stellar astronomy. It will be the principal aim of this volume to strive to redress this imbalance; and, in doing so, to help extending the present frontiers of double-star astronomy.

CHAPTER I

LIGHT CHANGES OF ECLIPSING BINARY SYSTEMS: SPHERICAL STARS

In order to embark on the task just set forth, our first aim will be to outline a geometrical theory of eclipsing binary systems consisting of components which can be regarded as spheres. At the first sight, the problem so restricted might seem to suffer from serious limitations: for how can two stars finding themselves in the proximity of each other retain a spherical shape? This objection is true in principle; and a discussion of its limitations will be taken up in Chapter II.

Nevertheless, for a large number of known eclipsing variables, the dimensions of the components expressed in terms of their separation are sufficiently small to justify the retention of spherical model for their shape – if not as an exact representation of reality, at least as an approximation to it which may be sufficient within the limits of observational errors. In such cases – and many belong to this category – the geometry of eclipses, and an analysis of the light changes produced by them, not only becomes simplest, but offers also the most suitable avenue of approach to the treatment of a more realistic problem, in which the shape of each star – being the resultant of actual forces (centrifugal, tidal) prevailing in each particular system – is allowed to depart from a sphere to an extent specified by the fractional dimensions, masses and internal structure of its components. Therefore, in the present chapter we shall carry out a preliminary analysis of our problem strictly on the basis of an assumption of spherical geometry of both components; and its extension to incorporate the effects of distortion will be postponed to Chapter II.

I-1. Equations of the Problem

In accordance with the restrictions just imposed, let us consider a system consisting of two stars of luminosities L_1 and L_2, revolving around a common centre of gravity in an orbit inclined to a plane tangent to the celestial sphere (at a point intersected by the line of sight) by an angle i. The two components – regarded as spherical – should appear in projection on the celestial sphere as circular discs; and if their dimensions are expressed in terms of their separation, we shall hereafter refer to them as fractional radii $r_{1,2}$ of the two stars.

If the sum $r_1 + r_2 < \cos i$, both components of the system would remain continuously in sight of a distant observer; and the sum $L_1 + L_2$ of their luminosities would remain constant: such a binary would emit constant light; and nothing obvious would distinguish it from the neighbouring field stars. Should, however, $r_1 + r_2 > \cos i$, the light of such a system could no longer remain constant at all times; for twice in the course of each orbital cycle one component would step out in front of its mate to *eclipse* (partly or wholly) the apparent disc of the star behind – a phenomenon whose alternation would, to a distant observer, disclose its nature by a characteristic *variation of light* of the system as a whole.

In more specific terms, the light of a binary system consisting of spherical stars – constant between eclipses – would exhibit two minima of light as the components eclipse alternately each other at the time of conjunctions. If, moreover, the orbits of the components are circular (not otherwise; cf. Section IV-4) both minima will be symmetrical, of equal duration, and separated in time by exactly half the period of the orbit (see Figure 1-1). At any moment during a minimum one component will eclipse a certain area (and thus cut off a certain amount of light) of the apparent disc of the other. Half a revolution later – at the corresponding phase during the other minimum – the geometrical relations of the two discs will be exactly the same, except that now the other star is in front and eclipses an equal area – though not necessarily an equal proportion – of the disc of the first star. Moreover, the deeper minimum of the two usually (though not always) corresponds to the star of greater surface brightness; though whether this star is the smaller or the larger of the two cannot be decided by inspection and remains to be established by further investigation.

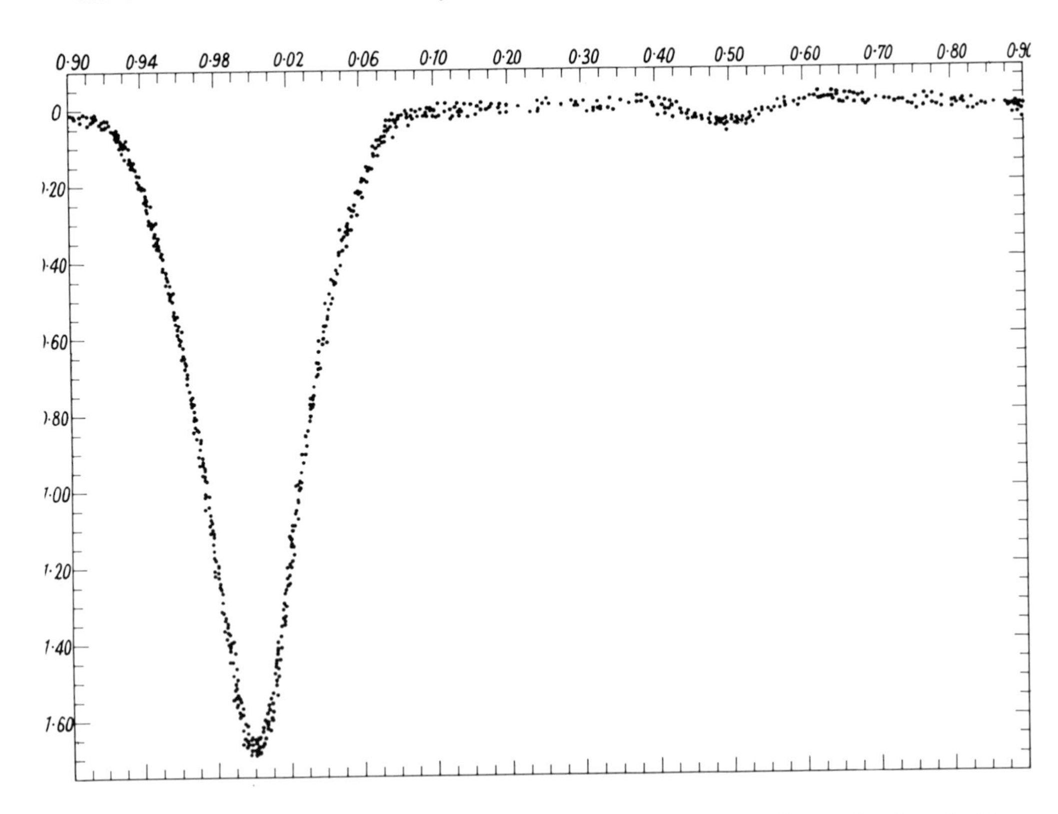

Fig. 1-1. Light changes of the eclipsing system of RZ Cassiopeiae according to the photoelectric observations by C. M. Huffer (*Astrophys. J.*, **114**, 297, 1951). Abscissae: the relative brightness of the system in stellar magnitudes; ordinates: the phase in fractions of the orbital cycle. The time-scale outside primary minimum has been contracted for convenience of presentation.

The *loss of light* $\Delta\mathfrak{L}$ suffered at any phase of eclipse can generally be expressed as

$$\Delta\mathfrak{L} = \int_S J \cos\gamma \, \mathrm{d}\sigma, \tag{1.1}$$

where J represents the distribution of brightness over the apparent disc of the star undergoing eclipse, of surface element $\mathrm{d}\sigma$; and γ, the angle of foreshortening; while the range S of integration is to be extended over the entire eclipsed area (where $\gamma \leqslant 90°$).

The distribution of brightness J will, in general, be a function of the cosine of the angle of foreshortening; and can be specified from the appropriate equations of radiative transfer in stellar atmospheres. For spherical stars, it is generally legitimate to approximate their solution in the limit of zero optical depth by a finite expansion of the form

$$J = H(1 - u_1 - u_2 - \cdots - u_\Lambda + u_1 \cos\gamma + u_2 \cos^2\gamma + \cdots + u_\Lambda \cos^\Lambda \gamma), \tag{1.2}$$

where H stands for the intensity of radiation emerging normally to the surface, and the u_j's are the 'coefficients of limb-darkening' of j-th degree ($j = 1, 2, 3, \ldots, \Lambda$). The approximation represented by Equation (1.2) can be made the more accurate, the higher the value of Λ.

Numerical values of the u_j's have been listed by Kopal (1949) for $\Lambda = 1(1)4$ in the case of plane-parallel gray atmospheres, and the quality of such approximations is diagramatically shown on Figure 5-4. In more general situations the coefficients u_j should, however, be regarded as additional unknowns, to be determined from an analysis of the light curve simultaneously with all other characteristics of the respective eclipsing system.

I-2. Loss of Light During Eclipses

In order to evaluate the loss of light $\Delta\mathfrak{L}$ as defined by Equation (1.1), we find it convenient to express the geometry of our problem in terms of rectangular xy-coordinates defining a plane perpendicular to the line of sight; with the origin at the centre of the disc undergoing eclipse, and the positive x-axis oriented constantly in the direction of the projected centre of the eclipsing star.

Let – in what follows and throughout this book – r_1 be the fractional radius of the star undergoing eclipse; and r_2, that of the eclipsing body (regardless of whether $r_1 \gtrless r_2$). If so, the boundary of the eclipsed disc will, in our xy-coordinates defined above, be given by the equation

$$x^2 + y^2 = r_1^2, \tag{2.1}$$

while the intersection of the shadow cylinder with the xy-plane will be given by

$$(\delta - x)^2 + y^2 = r_2^2 \tag{2.2}$$

where δ stands for the apparent (projected) separation of the centres of the two discs (see Figure 1-2).

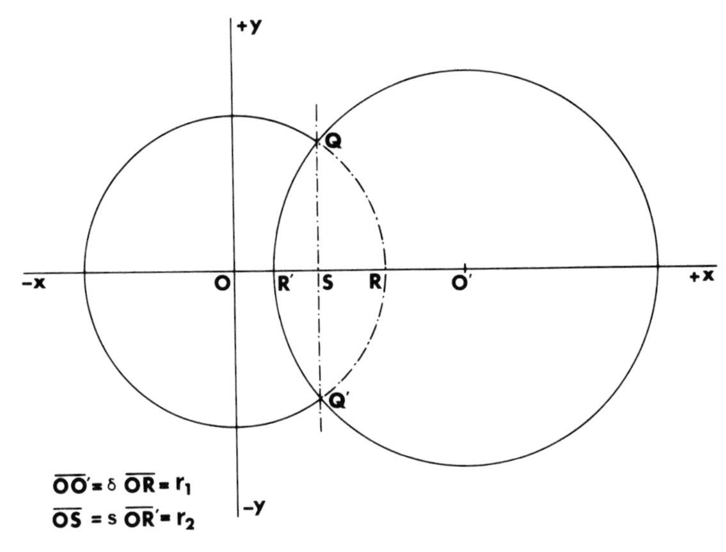

Fig. 1-2. Geometry of the eclipses of spherical stars.

Outside eclipses, the total luminosity L_1 of the component of radius r_1, and limb-darkened to n-th degree in accordance with Equation (1.2), will be given by

$$L_1 = 2\pi \int_0^{r_1} J \sin^{-1}(r/r_1) r \, dr$$

$$= \pi r_1^2 H \left\{ 1 - \sum_{l=0}^{\Lambda} \frac{l u_l}{l+2} \right\}; \tag{2.3}$$

but to establish the fractional loss of light $\delta\mathfrak{L}_1/L_1$ within eclipses we must first specify the limits S of integration on the r.h.s. of Equation (1.1).

Provided that $\delta < r_1 + r_2$, the circles defined by Equations (2.1) and (2.2) will intersect (cf. again Figure 1-2) at points $Q(s, \pm r_1^2 - s^2)$, where

$$s = \frac{r_1^2 - r_2^2 + \delta^2}{2\delta}, \tag{2.4}$$

constrained to vary between $+r_1$ at the moment of first contact of the eclipse to $-r_1$ at the commencement of totality. Since, furthermore,

$$r_1 \cos \gamma = \sqrt{r_1^2 - x^2 - y^2} \tag{2.5}$$

and

$$\cos \gamma \, d\sigma = dx \, dy, \tag{2.6}$$

the fractional loss of light during the eclipse of a star whose surface brightness falls off from centre to limb in accordance with a law of Λ-th degree of the form (1.2) can be expressed as

$$\alpha \equiv \frac{\delta\mathfrak{L}_1}{L_1} = \sum_{l=0}^{\Lambda} C^{(l)} \alpha_l^0, \tag{2.7}$$

where

$$C^{(0)} = \frac{1 - u_1 - u_2 - \cdots - u_\Lambda}{1 - \sum_{l=1}^{\Lambda} \frac{l u_l}{l+2}}; \tag{2.8}$$

while, for $l > 0$,

$$C^{(l)} = \frac{u_l}{1 - \sum_{l=1}^{\Lambda} \frac{l u_l}{l+2}}; \tag{2.9}$$

and

$$\pi r_1^{l+2} \alpha_l^0 = \left\{ \int_s^{r_2} \int_{-\sqrt{r_1^2 - x^2}}^{\sqrt{r_1^2 - x^2}} + \int_{\delta - r_2}^{s} \int_{-\sqrt{r_2^2 - (\delta - x)^2}}^{\sqrt{r_2^2 - (\delta - x)^2}} \right\} z^l \, dx \, dy \tag{2.10}$$

are the so-called 'associated α-functions' of order l and index zero, where

$$z^2 = r_1^2 - x^2 - y^2 \tag{2.11}$$

represents the limb-darkening factor. In Equation (2.10) we meet for the first time an example of a class of functions with which we shall have much to do throughout this book, and in terms of which much of the more advanced parts of a theory of the light changes of eclipsing binary systems can be formulated.

The foregoing equation (2.10) holds good throughout *partial* phases of the eclipse, irrespective of whether the eclipsed star of fractional radius r_1 is the smaller or the larger of the two. In the former case – when $r_1 < r_2$ – we shall hereafter refer (for obvious analogy) to such an eclipse as an *occultation*; while if $r_1 > r_2$, we shall call it a *transit*. If in the course of an occultation $\delta < r_2 - r_1$ (so that $s = -r_1$), the eclipse of the star of radius r_1 becomes *total* – in which case Equation (2.10) discloses that, during totality,

$$\alpha_l^0 = \frac{2}{l+2}. \tag{2.12}$$

If, however, $\delta < r_1 - r_2$ during a transit, the eclipse becomes *annular* – in which case Equation (2.10) should be replaced by

$$\pi r_1^{l+2} \alpha_l^0 = \int_{\delta - r_2}^{\delta + r_2} \int_{-\sqrt{r_2^2 - (\delta - x)^2}}^{\sqrt{r_2^2 - (\delta - x)^2}} z^l \, dx \, dy. \tag{2.13}$$

A. EVALUATION OF THE INTEGRALS

The procedures for literal evaluation of integrals on the right-hand sides of Equations (2.10) or (2.13) will be fully developed in Section III-1; while the Fourier approximations of arbitrary precision to the desired results will be obtained in the next two sections of this chapter. Anticipating the outcome, we find that if l is zero or an even integer, the α_l^0-functions corresponding to any phase of the eclipse can be literally evaluated in terms of elementary functions; but if l is odd, to establish literal expressions calls for the use of elliptic integrals (both complete and incomplete) of the first and second kind.

To give examples, if $l = 0$ (corresponding to the case when the disc undergoing eclipse happens to be uniformly bright) Equation (2.10) discloses that

$$\begin{aligned} \pi r_1^2 \alpha_0^0 &= r_1^2 \cos^{-1}\frac{s}{r_1} + r_2^2 \cos^{-1}\frac{\delta - s}{r_2} - \delta\sqrt{r_1^2 - s^2} \\ &= \tfrac{1}{2} r_1^2 (2\phi_1 - \sin 2\phi_1) + \tfrac{1}{2} r_2^2 (2\phi_2 - \sin 2\phi_2), \end{aligned} \tag{2.14}$$

where (cf. Figure 1-3), the angles $\phi_{1,2}$ are defined by the equation

$$\phi_{1,2} = \cos^{-1}\frac{\delta^2 \pm r_1^2 \mp r_2^2}{2\delta r_{1,2}}. \tag{2.15}$$

On the other hand, if $l = 1$ (corresponding to the case of linear limb-darkening) Equation (2.10) yields

$$\begin{aligned} \pi r_1^3 \alpha_1^0 &= \tfrac{2}{3}\pi r_1^3 - \tfrac{4}{3} r_1^3 \{[(\tfrac{1}{2}\pi, \kappa) - F(\tfrac{1}{2}\pi, \kappa)]F(\phi_2, \kappa) + F(\tfrac{1}{2}\pi, \kappa)E(\phi_2, \kappa')\} \\ &\quad + \tfrac{2}{9} r_2 \sqrt{\frac{\delta}{r_2}} (7r_2^2 - 4r_1^2 + \delta^2)\{2E(\tfrac{1}{2}\pi, \kappa) - F(\tfrac{1}{2}\pi, \kappa)\} \\ &\quad - \frac{2}{9\delta}\sqrt{\frac{\delta}{r_2}} \{5\delta^2 r_2^2 + 3(r_1^2 - r_2^2)^2 - 3\delta r_1^3\} F(\tfrac{1}{2}\pi, \kappa) , \end{aligned} \tag{2.16}$$

where $F \equiv F(\frac{1}{2}\pi, \kappa)$ and $E \equiv E(\frac{1}{2}\pi, \kappa)$ denote the Legendre standard forms of complete elliptic integrals of the first and second kind, with the modulus

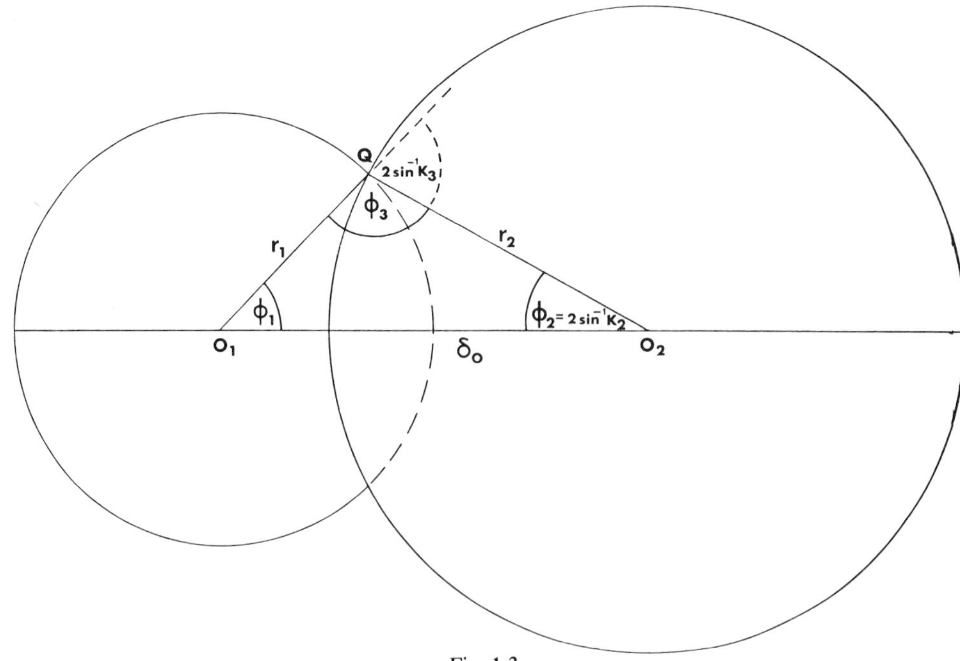

Fig. 1-3

$$\kappa^2 = \frac{1}{2}\left\{1 - \frac{\delta - s}{r_2}\right\} = \frac{r_1^2 - (\delta - r_2)^2}{4\delta r_2}; \tag{2.17}$$

while $F' \equiv F(\phi, \kappa')$ and $E' \equiv E(\phi, \kappa')$ stand for incomplete elliptic integrals of the same kinds, with a complementary modulus $\kappa' = \sqrt{1-\kappa^2}$ and the amplitude

$$\phi = \sin^{-1}\left(\frac{2s}{r_1 + r_2 + \delta}\right)^{1/2}. \tag{2.18}$$

The foregoing equations (2.16)–(2.18) hold good again only during *partial* phases of the eclipse – regardless of whether it is an occultation or transit. Should, however, the latter be the case and $\delta < r_1 - r_2$, the eclipse becomes *annular*. If so, Equation (2.16) based on the integration of Equation (2.10) is then to be replaced by

$$\begin{aligned} \pi r_1^3 \alpha_1^0 = \tfrac{4}{3} r_1^3 \Big\{ \Big[E\left(\frac{\pi}{2}, \frac{1}{\kappa}\right) - F\left(\frac{\pi}{2}, \frac{1}{\kappa}\right) \Big] & F(\sqrt{1-\kappa^{-2}}, \phi) \\ & + F\left(\frac{\pi}{2}, \frac{1}{\kappa}\right) E(\sqrt{1-\kappa^{-2}}, \phi) \Big\} \\ & + \tfrac{2}{9}(7r_2^2 - 4r_1^2 + \delta^2)\sqrt{2\delta(r_2 - \delta + s)}\, E\left(\frac{\pi}{2}, \frac{1}{\kappa}\right) \\ & + \tfrac{2}{9}\{(\delta^2 - r_2^2)^2 - r_1^2(2r_1^2 - r_2^2) - 5\delta^2 r_1^2 + 6\delta r_1^3\} \frac{F\left(\frac{\pi}{2}, \frac{1}{\kappa}\right)}{\sqrt{2\delta(r_2 - \delta + s)}}, \end{aligned} \tag{2.19}$$

deduced similarly from Equation (2.13), where the modulus of complete elliptic integrals becomes κ^{-1}; and the amplitude of the incomplete integrals

$$\phi = \sin^{-1}\left(\frac{r_1 + r_2 - \delta}{r_1 + r_2 + \delta}\right)^{1/2}. \tag{2.20}$$

The results obtained above hold good for $0 \leqslant r_2/r_1 \leqslant \infty$. If $r_2/r_1 = 0$, the eclipsing component dwindles to a point and the loss of light caused by its transit becomes infinitesimal. On the other hand, the condition $r_2/r_1 = \infty$ implies that the curvature of the eclipsing component becomes zero, and the latter acts as a straight occulating edge. In such a case, the second integral on the r.h.s. of Equation (2.10) vanishes; while the first one yields

$$\pi\alpha_0^0 = \cos^{-1}\frac{s}{r_1} - \frac{s}{r_1}\sqrt{1 - \left(\frac{s}{r_1}\right)^2} \tag{2.21}$$

and

$$6\alpha_1^0 = 2 - 3\frac{s}{r_1} + \left(\frac{s}{r_1}\right)^3. \tag{2.22}$$

At the moment of the first contact of the eclipse ($\delta = r_1 + r_2$) all α_l^0's are identically zero. As δ diminishes, the loss of light increases until the inner contact is reached. Should the eclipse be an occultation and the inner contact (at which $\delta = r_2 - r_1$) mark the commencement of totality, the values of α_l^0 become constant and equal to Equation (2.12). Should, however, $r_1 > r_2$ and the internal tangency of both discs mark the beginning of the annular phase, at $\delta = r_1 - r_2$ Equations (2.14) and (2.19) reduce to

$$\alpha_0^0 = k^2 \tag{2.23}$$

and

$$\alpha_1^0 = \frac{4}{3\pi}\{\sin^{-1}\sqrt{k} + \tfrac{1}{3}(4k-3)(2k+1)\sqrt{k(1-k)}\}, \tag{2.24}$$

where we have abbreviated $k \equiv r_2/r_1$. For $\delta < r_1 - r_2$, the loss of light during annular eclipse of a limb-darkened star continues to increase, until a maximum is reached at the moment of *central* eclipse, when

$$\alpha_1^0 = \tfrac{2}{3}\{1 - (1-k^2)^{3/2}\}. \tag{2.25}$$

For an extension of the foregoing results to the α_2^0's for $l > 1$, cf. Section III-1 or Appendix I.

I-3. Loss of Light as Aperture Cross-Correlation

In the preceding section of this chapter we have expressed the fractional loss of light α accompanying eclipses of spherical stars arbitrarily darkened at the limb in terms of elementary geometry of areas eclipsed. However, this definition can be greatly generalized (and expressed in more symmetrical form) if we re-formulate this fractional loss of light as a *cross-correlation of two apertures* – one representing the star undergoing eclipse, and the other the eclipsing disc. In particular, we shall be able to relate the loss of light arising by mutual stellar eclipses with the 'diffraction patterns' of the 'apertures' representing these stars – with arbitrary distribution of light within the 'aperture', or arbitrary transparency of the occulting discs; and by explicit introduction of diffraction integrals in our work we shall connect our subject more closely with the relevant parts of physical optics than has been the case so far.

A. CROSS-CORRELATION: A DEFINITION

In order to approach our subject in this manner, let us revert (temporarily) to the cosmology of ancient Greeks, and consider the stars to represent 'apertures' in the firmament of the Heavens, through which we can see the 'eternal fire'; of brightness yet to be specified. Let x, y be the rectangular coordinates in the plane of the plane tangent to the celestial sphere, with the origin at the centre of our aperture; and let, moreover, the function $f(x, y)$ represent the distribution of brightness within this aperture. If so, the two-dimensional Fourier transform $F(u, v)$ of $f(x, y)$ is known to be given by

$$F(u, v) = \int\int_{-\infty}^{+\infty} f(x, y)\, e^{-2\pi i(xu+yv)}\, dx\, dy, \tag{3.1}$$

where $i \equiv \sqrt{-1}$ denotes the imaginary unit. If, moreover, the distribution of brightness $f(x, y)$ within the aperture is radially-symmetrical – so that

$$f(x, y) \equiv f(r) \tag{3.2}$$

where

$$x + iy = r\,\mathrm{e}^{i\theta}, \tag{3.3}$$

it follows from Equation (3.1) that

$$F(q, \phi) = \int_0^\infty \int_{-\pi}^{\pi} f(r)\,\mathrm{e}^{-2\pi iqr\cos(\theta-\phi)} r\,\mathrm{d}r\,\mathrm{d}\theta, \tag{3.4}$$

where

$$u + iv = q\,\mathrm{e}^{i\phi}. \tag{3.5}$$

In order to evaluate the integrals on the right-hand side of Equation (3.4), let us invoke the use of Jacobi's well-known expansion theorem (cf., e.g., Watson, 1945; pp. 22 or 368) which permits us to assert that

$$\begin{aligned}\mathrm{e}^{-2\pi iqr\cos(\theta-\phi)} &= \mathrm{e}^{-2\pi iqr\sin(1/2\pi+\theta-\phi)}\\ &= J_0(2\pi qr) + 2\sum_{n=1}^{\infty} J_{2n}(2\pi qr)\cos\{n\pi + 2n\theta - 2n\phi\}\\ &\quad - 2i\sum_{n=0}^{\infty} J_{2n+1}(2\pi qr)\sin\{n\pi + \tfrac{1}{2}\pi + (2n+1)(\theta-\phi)\},\end{aligned} \tag{3.6}$$

where the symbols $J_n(x)$ denote the respective Bessel functions of the first kind, with real arguments.

Since, however, for $n > 0$

$$\int_{-\pi}^{\pi} \cos 2n(\tfrac{1}{2}\pi + \theta - \phi)\,\mathrm{d}\theta = 0 \tag{3.7}$$

while

$$\int_{-\pi}^{\pi} \sin(2n+1)(\tfrac{1}{2}\pi + \theta - \phi)\,\mathrm{d}\theta = 0 \tag{3.8}$$

for any value of n including zero, it follows that

$$\int_{-\pi}^{\pi} \mathrm{e}^{-2\pi iqr\cos(\theta-\phi)}\,\mathrm{d}\theta = 2\pi J_0(2\pi qr); \tag{3.9}$$

and, accordingly,

$$F(q) = 2\pi \int_0^\infty f(r) J_0(2\pi qr) r\,\mathrm{d}r, \tag{3.10}$$

where, by (3.5), $q^2 = u^2 + v^2$. If, moreover, the 'aperture function' $f(r)$ vanishes for $r > r_1$, the foregoing equation can be rewritten as

$$F(q) = 2\pi \int_0^{r_1} f(r) J_0(2\pi q r) r \, dr; \tag{3.11}$$

i.e., as a Hankel transform of $f(r)$ of zero order, which represents the well-known Airy diffraction pattern of a finite circular aperture of radius r_1 for a distribution of brightness $f(r)$ within that aperture (and zero outside of it).

Let us assume now that (in accordance with Equation (1.2) of Section 1) this aperture function is given by a law of limb-darkening of the form

$$f(r) = f(0) \left\{ 1 - u_1 - u_2 - u_3 - \cdots + \sum_{l=1}^{\Lambda} u_l \cos^l \gamma \right\}, \tag{3.12}$$

where $u_1, u_2, u_3, \ldots$ denote the respective coefficients of limb-darkening, and

$$\cos \gamma = \frac{\sqrt{r_1^2 - r^2}}{r_1} \tag{3.13}$$

represents the cosine of the angle of foreshortening (such that $r = r_1 \sin \gamma$). Since for any value of $\nu > 0$ (integral or fractional)

$$\int_0^{r_1} (r_1^2 - r^2)^{\nu - 1} J_0(2\pi q r) r \, dr = 2^{\nu - 1} \Gamma(\nu) \frac{J_\nu(2\pi q r_1)}{(2\pi q r_1)^\nu} r_1^{2\nu}, \tag{3.14}$$

the diffraction pattern (3.11) of our limb-darkened circular 'aperture' can be expressed as

$$F(q) = L_1 \sum_{l=0}^{\infty} C^{(l)} 2^\nu \Gamma(\nu) \frac{J_\nu(2\pi q r_1)}{(2\pi q r_1)^\nu}, \tag{3.15}$$

where the luminosity L_1 of the aperture and the coefficients $C^{(l)}$ dependent on its limb-darkening have been defined by Equations (2.8)–(2.9) of Chapter I, and where we have abbreviated

$$\nu = \frac{l + 2}{2}. \tag{3.16}$$

Let us now (cf. Figure 1-4) turn our attention to the off-centre aperture which represents the eclipsing component, situated on the x-axis at a distance δ from the origin of coordinates. If so, the Fourier transform of this latter aperture should – by analogy with (3.1) – be defined by

$$G(u, v) = \iint_{-\infty}^{+\infty} g(\xi, \eta) \, e^{-2\pi i [(\delta + \xi) u + \eta v]} \, d\xi \, d\eta, \tag{3.17}$$

where

$$x \equiv \xi + \delta \quad \text{and} \quad y \equiv \eta. \tag{3.18}$$

Let us, moreover, introduce again a set of plane polar coordinates ρ, ζ related with the rectangular coordinates ξ, η by

$$\xi + i\eta = \rho \, e^{i\zeta}; \tag{3.19}$$

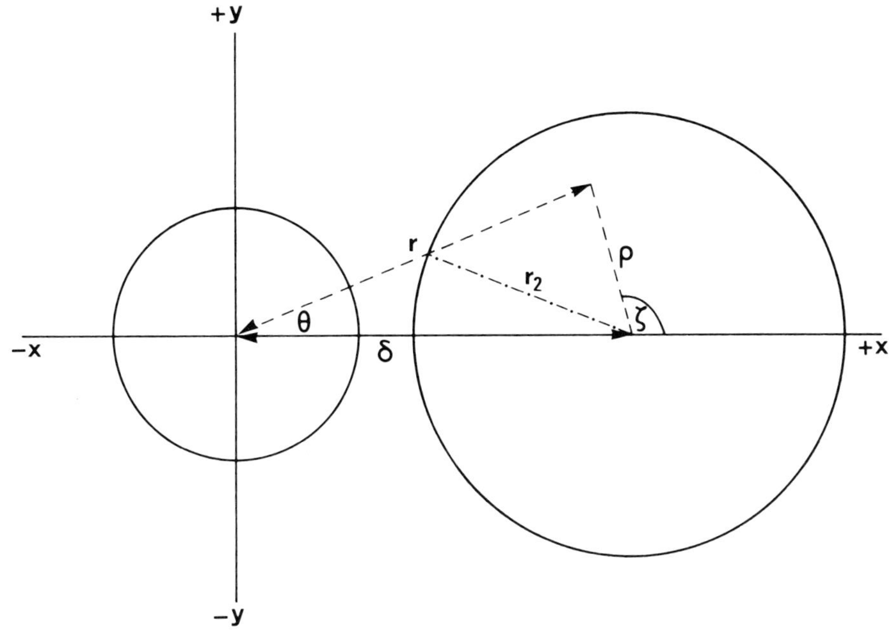

Fig. 1-4

so that

$$dx\,dy = d\xi\,d\eta = \rho\,d\rho\,d\zeta. \tag{3.20}$$

If so, then obviously

$$G(u, v) = e^{-2\pi i\delta u} \int_0^{\infty} \int_{-\pi}^{\pi} g(\rho, \zeta)\, e^{-2\pi i q\rho \cos(\zeta - \phi)} \rho\, d\rho\, d\zeta, \tag{3.21}$$

where $g(\rho, \zeta)$ denotes the 'transparency function' of the second aperture. If, in particular, the latter represents a circular disc which is wholly opaque for $\rho \leqslant r_2$ and wholly transparent for $\rho > r_2$, it would follow that

$$g(\rho, \zeta) = \begin{cases} 1 & \text{for} \quad \rho \leqslant r_2, \\ 0 & \text{for} \quad \rho > r_2; \end{cases} \tag{3.22}$$

in which case

$$G(u, v) = e^{-2\pi i\delta u} \int_0^{r_2} \rho\, d\rho \int_{-\pi}^{\pi} e^{-2\pi i q\rho \cos(\zeta - \xi)}\, d\zeta. \tag{3.23}$$

Both integrals on the right-hand side of the foregoing equation can be evaluated in a closed form: for a resort to Equations (3.6)–(3.8) discloses readily that, since

$$2\pi q\rho J_0(2\pi q\rho) = \frac{d}{d\rho}\{\rho J_1(2\pi q\rho)\} \tag{3.24}$$

and $J_1(0) = 0$,

$$G(u, v) = 2\pi\, e^{-2\pi i \delta u} \int_0^{r_2} J_0(2\pi q\rho)\rho\, d\rho$$

$$= 2\pi r_2^2\, e^{-2\pi i \delta u} \frac{J_1(2\pi q r_2)}{2\pi q r_2}. \tag{3.25}$$

This latter expression holds good, to be sure, only for the transparency function $g(\rho, \zeta)$ as given by Equation (3.22) and characteristic of an opaque circular disc. In the case of its partial transparency (obtaining, for example, if the aperture of radius r_1 undergoes eclipse by a semi-transparent atmosphere of its mate) the Fourier transform $G(u, v)$ of the second aperture will continue to be given by Equation (3.5) with an appropriate form of the function $g(\rho, \zeta)$; only its integration will be somewhat more complicated.

Before we follow this subject in further detail later in this section, let us stop to grasp the fact that *the loss of light arising if the aperture centred at* $(\delta, 0)$ *begins to obscure that centred at* $(0, 0)$ *can be defined as a cross-correlation between the two discs* – which is obviously zero if these do not overlap, and increases with an increase of their common intercept, weighted in accordance with the relative brightness of each element occulted (or the transparency of the respective area of the occulting star). Moreover, the measure of this cross-correlation is expressed by the *convolution integral*

$$L_1\alpha(r_1, r_2, \delta) = \int\int_{-\infty}^{+\infty} F(u, v)G(u, v)\, du\, dv, \tag{3.26}$$

where, by Equations (3.15) and (3.23), the integrand

$$F(u, v)G(u, v) = 2\pi r_2^2 L_1\, e^{-2\pi i \delta u} \frac{J_1(2\pi q r_2)}{2\pi q r_2} \sum_{l=0}^{\infty} C^{(l)} 2^{\nu} \Gamma(\nu) \frac{J_\nu(2\pi q r_1)}{(2\pi q r_1)^\nu}. \tag{3.27}$$

If, moreover, we remember that, in accordance with Equation (3.5),

$$\left.\begin{aligned} u = q \cos\phi, \quad u = q \sin\phi \\ du\, dv = q\, dq\, d\phi, \end{aligned}\right\} \tag{3.28}$$

it follows by use of (3.6)–(3.8) that

$$\int_{-\pi}^{\pi} e^{-2\pi i \delta q \cos\phi}\, d\phi = 2\pi J_0(2\pi q\delta). \tag{3.29}$$

Accordingly,

$$L_1\alpha(r_1, r_2, \delta) = \int_0^{\infty} \int_{-\pi}^{\pi} F(u, v)G(u, v) q\, dq\, d\phi$$

$$= L_1(2\pi r_2)^2 \sum_{l=0}^{\infty} C^{(l)} 2^\nu \Gamma(\nu) \int_0^\infty \frac{J_\nu(2\pi q r_1)}{(2\pi q r_1)^\nu} \frac{J_1(2\pi q r_2)}{2\pi q r_2} \times$$

$$\times J_0(2\pi q \delta) q \, dq; \tag{3.30}$$

and if we set, in accordance with Equation (2.7),

$$\alpha = \sum_{l=0}^{\infty} C^{(l)} \alpha_l^0, \tag{3.31}$$

if follows by a comparison of Equation (3.30) with Equation (3.31) that the 'associated α-functions' α_l^0 of order l are expressible as

$$\alpha_l^0(r_1, r_2, \delta) = (2\pi r_2)^2 2^\nu \Gamma(\nu) \int_0^\infty \left\{ \frac{J_\nu(2\pi q r_1)}{(2\pi q r_1)^\nu} \frac{J_1(2\pi q r_2)}{2\pi q r_2} \right\} J_0(2\pi q \delta) q \, dq. \tag{3.32}$$

This equation represents the associated alpha-functions of order of index $m = 0$ and order $n = 1$ as *Hankel transforms* of zero order of the products of two Bessel functions of orders 1 and ν in curly brackets on the right-hand side of Equation (3.32). This definition of α_l^0 is *more general* than that given by Equations (2.10) or (2.13); for while that latter represents real quantities only if $\delta \leqslant r_1 + r_2$, and their explicit forms are, moreover, different for different types of eclipses (partial, total, annular), the definition (3.32) holds good for *any* type of eclipse and any value of r_1, r_2 or δ. In other words, *the* α_l^0*'s as given by Equation* (3.32) *represent real and continuous non-negative functions of* δ *between minima as well as within eclipses of any type*; such that $\alpha_l^0 > 0$ for $\delta < r_1 + r_2$, and $\alpha_l^0 = 0$ for $\delta \geqslant r_1 + r_2$; the moments of first or second contacts of the eclipse (when $\delta = |r_1 \pm r_2|$) mark merely the points at which the derivatives of $\alpha_l^0(\delta)$ – but not the function itself – may become discontinuous. For a proof of the fact that the definite integral on the right-hand side of Equation (3.32) vanishes when the arguments of the three Bessel functions behind the integral sign obey the inequality $2\pi q r_1 + 2\pi q r_2 < 2\pi q \delta$ the reader is referred, e.g., to Bailey (1936) or Rice (1935).

B. HANKEL TRANSFORMS

The right-hand side of Equation (3.32) contains three parameters as multiplicative factors of the arguments of the Bessel functions occurring in it: namely, r_1, r_2 and δ. It is, however, easy to see that the functions α_l^0 depend on these parameters only through two *ratios* which can be formed of them. For if we set, say,

$$2\pi q r_2 \equiv x, \tag{3.33}$$

Equation (3.32) can be readily rewritten as

$$\alpha_l^0 = 2^\nu \Gamma(\nu) \int_0^\infty (kx)^{-\nu} J_\nu(kx) J_1(x) J_0(hx) \, dx, \tag{3.34}$$

where we have abbreviated

$$h = \frac{\delta}{r_2} \quad \text{and} \quad k = \frac{r_1}{r_2}. \tag{3.35}$$

It may be added that, for odd values of l (i.e., half-integral values of ν), the spherical Bessel functions $J_\nu(kx)$ become expressible in terms of trigonometric functions of kx.

The foregoing equations (3.32) or (3.34) define the associated α-functions of zero index and arbitrary order, regarded as convolution integrals of two apertures representing the two components of an eclipsing binary system – one bright, the other dark. This result can, moreover, be readily generalized to a case in which the light changes arise from occultations by discs which are *semi-transparent*.

Suppose, for the sake of argument, that the transparency of the occulting disc increases with the angle of foreshortening in the same manner as the limb-darkening of the eclipsed star – i.e., that the transparency function $g(\rho, \zeta)$ in Equation (3.21) varies as

$$g(\rho, \zeta) = \begin{cases} [1 - (\rho/r_1)^2]^\lambda, & \rho \leq r_2 \\ 0, & \rho > r_2; \end{cases} \tag{3.36}$$

replacing Equation (3.22) appropriate for opaque circular discs. If so, an exactly the same type of analysis as followed earlier in this section discloses that, in such a case, Equation (3.25) for $G(u, v)$ is to be replaced by

$$G(u, v) = 2^{\lambda+1}\pi r_2^2 \Gamma(\lambda + 1) \frac{J_{\lambda+1}(2\pi q r_2)}{(2\pi q r_2)^{\lambda+1}} \mathrm{e}^{-2\pi i \delta u}; \tag{3.37}$$

and Equation (3.34) for α_l^0, by

$$\alpha_l^0 = 2^{\lambda+\nu}\Gamma(\lambda + 1)\Gamma(\nu)k^{-\nu} \int_0^\infty x^{-\lambda-\nu} J_\nu(kx) J_{\lambda+1}(x) J_0(hx)\, \mathrm{d}x, \tag{3.38}$$

which is of the same form as Equation (3.34), except that the order of the second Bessel function behind the integral sign has been augmented by λ.

The associated α-functions as given by Equation (3.34) are identical with those defined by Equations (2.10) or (2.13) earlier in this chapter. Unlike in that earlier formulation, however, the parameters h and k specifying the geometry of eclipses in Equation (3.34) do not occur in the limits of integration, but solely in the integrand; and in such a way that either h or k appears as a factor in the argument of only one of the three Bessel functions constituting it. This represents a fact of which advantage can be taken in several respects.

First, let us utilize this circumstance to expand the expression (3.34) for $\alpha_l^0(h, k)$ in a Fourier cosine series. By setting (cf. Equation (3.17))

$$\delta^2 = r_1^2 - 2r_1 r_2 \cos\phi + r_2^2, \tag{3.39}$$

and making appeal to Neumann's addition theorem (cf. Watson, 1945; p. 358) for Bessel functions of the first kind, we find that

$$\begin{aligned} J_0(2\pi q\delta) &\equiv J_0(\sqrt{x^2 - 2xx'\cos\phi + x'^2}) \\ &= J_0(x)J_0(x') + 2\sum_{n=1}^{\infty} J_n(x)J_n(x')\cos n\phi, \end{aligned} \tag{3.40}$$

where, as before,

$$x = 2\pi q r_2 \quad \text{while} \quad x' = 2\pi q r_1 \equiv kx. \tag{3.41}$$

On insertion of the expansion (3.40) in the right-hand side of Equation (3.34) we can rewrite the latter in the form of a Fourier series

$$\alpha_l^0(h, k) = K_0^{(l)}(k) + 2\sum_{n=1}^{\infty} K_n^{(l)}(k)\cos n\phi \tag{3.42}$$

in cosines of the multiples of the angle

$$\phi = 2\sin^{-1}\sqrt{\frac{(r_1 + r_2)^2 - \delta^2}{4r_1r_2}} = 2\sin^{-1}\sqrt{\frac{(1+k)^2 - h^2}{4k}} \tag{3.43}$$

following from (3.39), which alone depends on δ (or h); the coefficients

$$K_n^{(l)}(k) \equiv 2^\nu\Gamma(\nu)\int_0^\infty \frac{J_\nu(kx)}{(kx)^\nu} J_n(kx)J_1(x)J_n(x)\,\mathrm{d}x \tag{3.44}$$

being functions of k only.

It may be added that (by Kummer's formula)

$$\cos n\phi = {}_2F_1(-\tfrac{1}{2}n, \tfrac{1}{2}n; \tfrac{1}{2}; \sin^2\phi) = {}_2F_1(-n, n; \tfrac{1}{2}; \sin^2\tfrac{1}{2}\phi), \tag{3.45}$$

where ${}_2F_1$ stands for an ordinary hypergeometric series (see Appendix III), which for integral values of n reduces to the Jacobi polynomials $G_n(0, \frac{1}{2}, \sin^2\frac{1}{2}\phi)$. Moreover, its argument $\sin^2\frac{1}{2}\phi$ can, in accordance with Equations (3.43), be rewritten in terms of the elements of the eclipse as

$$\sin^2\tfrac{1}{2}\phi = \frac{\sin^2\theta_1 - \sin^2\theta}{4r_1r_2\csc^2 i}. \tag{3.46}$$

Next, let us observe the reciprocal relations obtaining between the associated α-functions of the type α_l^0 and the integrand on the right-hand side of Equation (3.34). If we abbreviate

$$\mathfrak{A}_\nu(x) \equiv 2^\nu\Gamma(\nu)\frac{J_\nu(kx)}{(kx)^\nu}\frac{J_1(x)}{x}, \tag{3.47}$$

Equation (2.34) can be rewritten as

$$\alpha_\nu^0(h) = \int_0^\infty \mathfrak{A}_\nu(x)J_0(hx)x\,\mathrm{d}x; \tag{3.48}$$

from which it follows that

$$\mathfrak{A}_\nu(x) = \int_0^\infty \alpha_\nu^0(h) J_0(hx) h \, \mathrm{d}h, \tag{3.49}$$

where the upper limit of integration with respect to h on the right-hand side can be restricted from ∞ to $1+k$ – since, for $h \geqslant 1+k$, the eclipse function $\alpha_\nu^0(h, k)$ vanishes.

The foregoing Equations (3.48) and (3.49) constitute a pair of reciprocal Hankel transforms of zero order. We note that their both sides are even functions of x and can be so expanded, with coefficients of equal powers of x being equal on both sides. In terms of our original notations Equation (3.49) can be rewritten to read

$$2^\nu \Gamma(\nu) \left\{ \frac{J_\nu(2\pi q r_1)}{(2\pi q r_1)^\nu} \frac{J_1(2\pi q r_2)}{2\pi q r_2} \right\} r_2^2 = \int_0^{\delta_1} \alpha_l^0(\delta) J_0(2\pi q \delta) \delta \, \mathrm{d}\delta ; \tag{3.50}$$

which on insertion of the well-known series for

$$J_0(2\pi q \delta) = \sum_{s=0}^\infty \frac{(-1)^s (\pi q \delta)^{2s}}{(s!)^2} \tag{3.51}$$

leads to a generating function

$$2^{\nu+1} \Gamma(\nu) \left\{ \frac{J_\nu(2\pi q r_1)}{(2\pi q r_1)^\nu} \frac{J_1(2\pi q r_2)}{2\pi q r_2} \right\} r_2^2 = \sum_{s=0}^\infty \frac{(-\pi^2)^s a_l^{(s+1)}}{s!(s+1)!} q^{2s} \tag{3.52}$$

of the coefficients

$$a_l^{(s)} = \int_0^{\delta_1} \alpha_l^0 \, \mathrm{d}\delta^{2s} = - \int_0^{\delta_1} \delta^{2s} \frac{\partial \alpha_l^0}{\partial \delta} \mathrm{d}\delta, \tag{3.53}$$

where δ_1 stands for the values of δ at the moments of the first contact of the eclipse. The coefficients $a_l^{(s)}$ as defined by the foregoing equation have been found (cf. Kopal, 1975b, c, d) to play an important role in the formation of the 'moments of the light curves', to be discussed in the second part of this book.

C. EVALUATION OF THE HANKEL TRANSFORMS

Next, let us turn to the actual *evaluation* of the associated α-functions of the form α_l^0 as defined by the Hankel transform (3.34) in terms of the elements of eclipse. In order to do so, we find it convenient to change over on the r.h.s. of Equation (3.34) to a new variable y of integration, related with x by

$$x = \frac{r_2 y}{r_1 + r_2}, \tag{3.54}$$

so that

$$\alpha_l^0 = 2^\nu \Gamma(\nu) b \int_0^\infty (ay)^{-\nu} J_\nu(ay) J_1(by) J_0(cy) \, \mathrm{d}y, \tag{3.55}$$

where

$$a = \frac{r_1}{r_1 + r_2}, \tag{3.56}$$

$$b = \frac{r_2}{r_1 + r_2} = 1 - a, \tag{3.57}$$

$$c = \frac{\delta}{r_1 + r_2}; \tag{3.58}$$

such that, for any type of eclipse, $0 \leqslant a \leqslant 1$, $0 \leqslant b \leqslant 1$ – such that $a + b = 1$; while $1 \geqslant c \geqslant 0$ between the moment of first contact (when $\delta = r_1 + r_2$) and that of central eclipse ($\delta = 0$).

In order to evaluate the integrals on the r.h.s. of Equation (3.55) we can avail ourselves of a theorem by Bailey (1936), asserting that

$$\int_0^\infty t^{\rho-1} J_\kappa(\alpha t) J_\lambda(\beta t) J_\mu(\gamma t)\,dt$$

$$= \frac{2^{\rho-1}\alpha^\kappa\beta^\lambda\Gamma\left(\frac{\kappa+\lambda+\mu+\rho}{2}\right)}{\gamma^{\kappa+\lambda+\rho}\Gamma(\kappa+1)\Gamma(\lambda+1)\Gamma\left(1 - \frac{\kappa+\lambda-\mu+\rho}{2}\right)}$$

$$\times F^{(4)}\left(\frac{\kappa+\lambda-\mu+\rho}{2}, \frac{\kappa+\lambda+\mu+\rho}{2}; \kappa+1, \lambda+1; \frac{\alpha^2}{\gamma^2}, \frac{\beta^2}{\gamma^2}\right), \tag{3.59}$$

where α, β, γ are positive quantities,

$$\rho < \tfrac{5}{2}, \quad \kappa + \lambda + \mu + \rho > 0; \tag{3.60}$$

and the Appell generalized hypergeometric series (cf. Appendix III) of the form

$$F^{(4)}(\mathfrak{a}, \mathfrak{b}; \mathfrak{c}, \mathfrak{c}'; x, y) \equiv \sum_{m=0}^{\infty}\sum_{n=0}^{\infty} \frac{(\mathfrak{a})_{m+n}(\mathfrak{b})_{m+n}}{m!\,n!\,(\mathfrak{c})_m(\mathfrak{c}')_n} x^m y^n \tag{3.61}$$

on the r.h.s. of Equation (3.59) converges whenever

$$\gamma \geqslant \alpha + \beta. \tag{3.62}$$

Consider first the case in which

$$\alpha = a, \quad \beta = b, \quad \gamma = c. \tag{3.63}$$

If so, the condition (3.62) will be satisfied by Equations (3.56)–(3.58) if

$$\delta \geqslant r_1 + r_2 \tag{3.64}$$

i.e., *outside eclipses* (the equality sign corresponding to the moment of first contact). If so, the constants

$$\kappa \equiv \nu, \quad \lambda \equiv 1, \quad \mu \equiv 0, \quad \rho \equiv 1 - \nu \tag{3.65}$$

satisfy the conditions

$$\kappa+\lambda+\mu+\rho=2>0 \tag{3.66}$$

and

$$\rho\equiv 1-\nu=-\tfrac{1}{2}l<\tfrac{5}{2}; \tag{3.67}$$

while

$$\Gamma\left(1-\frac{\kappa+\lambda-\mu+\rho}{2}\right)=\Gamma(0)=\infty. \tag{3.68}$$

Since the Appell series $F^{(4)}$ on the r.h.s. of Equation (3.59) converges for condition (3.64), it follows from Equations (3.55) and (3.59) that, outside eclipses,

$$\alpha_l^0(h\geqslant 1+k, k)=0; \tag{3.69}$$

not perhaps an unexpected result, but one which proves that the definition of α_l^0 in the form (3.34) or (3.55) holds good outside minima as well.

Consider now what happens when the eclipses become *total* or *annular.* The former occur whenever

$$r_2\geqslant\delta+r_1; \tag{3.70}$$

and if so, the condition (3.62) will be satisfied with

$$\alpha=a,\quad \beta=c,\quad \gamma=b \tag{3.71}$$

and

$$\kappa\equiv\nu,\quad \lambda\equiv 0,\quad \mu\equiv 1,\quad \rho\equiv 1-\nu. \tag{3.72}$$

If so, Equations (3.55) and (3.59) disclose that

$$\alpha_l^0(1-k, k)=\frac{1}{\nu}=\frac{2}{l+2}, \tag{3.73}$$

again in agreement with the expectations.

Should, however, the eclipse become annular,

$$r_1\geqslant\delta+r_2, \tag{3.74}$$

which conforms again to the condition (3.62) if

$$a=b,\quad \beta=c,\quad \gamma=a \tag{3.75}$$

and

$$\kappa\equiv 1,\quad \lambda\equiv 0,\quad \mu\equiv\nu,\quad \rho\equiv 1-\nu. \tag{3.76}$$

In such a case, Equations (3.55)–(3.59) yield

$$\alpha_l^0=(r_2/r_1)^2F^{(4)}(1-\nu, 1; 2, 1; r_2^2/r_1^2, \delta^2/r_1^2). \tag{3.77}$$

For uniformly bright discs ($\nu=1$) the r.h.s. of the foregoing equation reduces to $(r_2/r_1)^2$; and for even values of l (i.e., integral values of ν) it will become a polynomial; though for odd l's (i.e., half-integral ν's) the corresponding $F^{(4)}$ will remain an infinite series. If $\delta=0$ (corresponding to the moment of central eclipse)

$$F^{(4)}(1-\nu, 1; 2, 1; r_2^2/r_1^2, 0) = {}_2F_1(1-\nu, 1, 2; r_2^2/r_1^2); \tag{3.78}$$

and, therefore,

$$\begin{aligned}\alpha_l^0(0, r_2/r_1) &= (r_2/r_1)^2 {}_2F_1(1-\nu, 1; 2; r_2^2/r_1^2)\\ &= \frac{1}{\nu}\left\{1-\left(1-\frac{r_2^2}{r_1^2}\right)^{\nu}\right\},\end{aligned} \tag{3.79}$$

which for $l=0$ and 1 reduces to Equations (2.23) and (2.25). Similarly, at the moment of internal tangency (when $\delta = r_1 - r_2$) Equation (3.77) reduces to

$$\begin{aligned}\alpha_l^0\left(\frac{r_1-r_2}{r_1}, \frac{r_2}{r_1}\right) &= \left(\frac{r_2}{r_1}\right)^2 F^{(4)}\left(1-\nu, 1; 2, 1; \frac{r_2^2}{r_1^2}, \frac{(r_1-r_2)^2}{r_1^2}\right)\\ &= \frac{(2\nu)!}{\Gamma(\nu+1)\Gamma(\nu+2)}\left(\frac{r_2}{r_1}\right)^{\nu+1} {}_2F_1(1-\nu, \nu+1; \nu+2; r_2/r_1),\end{aligned} \tag{3.80}$$

which for $l=0$ and 1 reduce to Equations (2.23) and (2.24).

Should, however, the eclipse become partial, none of the three quantities r_1, r_2, δ (or a, b, c) can be equal to, or larger than, the sum of the two others. In consequence, the condition (3.60) can no longer be met, Bailey's theorem (3.59) ceases to be applicable. In order to evaluate the integral on the r.h.s. of Equation (3.55) under these conditions, use can be made of an expansion due to Bateman (1905) of the form

$$\begin{aligned}J_\mu(y\cos\phi\cos\Phi)J_\nu(y\sin\phi\sin\Phi) &= (2/y)(\cos\phi\cos\Phi)\\ &\times(\sin\phi\sin\Phi)^\nu\sum_{n=0}^{\infty}(-1)^n(\mu+\nu+2n+1)\\ &\times\frac{\Gamma(\mu+\nu+n+1)\Gamma(\nu+n+1)}{n!\,\Gamma(\mu+n+1)\{\Gamma(\nu+1)\}^2}\\ &\times{}_2F_1(-n, \mu+\nu+n+1; \nu+1; \sin^2\phi)\\ &\times{}_2F_1(-n, \mu+\nu+n+1; \nu+1; \sin^2\Phi)J_{\mu+\nu+2n+1}(y),\end{aligned} \tag{3.81}$$

valid for all values of μ and ν other than those of negative integers.

Let us, in what follows, set $\mu \equiv 1$ and $\nu \equiv \nu$; while

$$\sin\phi\sin\Phi = a \quad \text{and} \quad \cos\phi\cos\Phi = b \tag{3.82}$$

where a and b continue to be given by Equations (3.56) and (3.57). A solution of Equations (3.82) yields

$$\sin^2\phi = \sin^2\Phi = a, \tag{3.83}$$

$$\cos^2\phi = \cos^2\Phi = b; \tag{3.84}$$

so that the product of the first two Bessel functions on the r.h.s. of Equation (3.55) can be expressed with the aid of Equation (3.81) in the form of a summation

$$J_\nu(ay)J_1(by) = 2\frac{a^\nu b}{y}\sum_{n=0}^{\infty}(-1)^n(\nu+2n+2)\frac{\Gamma(\nu+n+1)\Gamma(\nu+n+2)}{n!(n+1)!} \times\left\{\frac{{}_2F_1(-n, n+\nu+2; \nu+1; a)}{\Gamma(\nu+1)}\right\}^2 J_{\nu+2n+2}(y), \tag{3.85}$$

where, it may be recalled, $b = 1 - a$. If, furthermore, we note (cf., e.g., Erdélyi *et al.*, 1954; p. 349 (1)) that

$$\int_0^\infty y^{-\nu-1}J_{\nu+2n+2}(y)J_0(cy)\,\mathrm{d}y = \frac{2^{-\nu-1}n!}{\Gamma(\nu+n+1)}\,{}_2F_1(-\nu-n-1, n+1; 1; c^2), \tag{3.86}$$

– which converges absolutely for $c^2 \leqslant 1$ for ν integral or fractional, a combination of Equations (3.55), (3.85) and (3.86) discloses that

$$\alpha_l^0 = (1-a)^2\sum_{n=0}^{\infty}(-1)^n\frac{(\nu+2n+2)\Gamma(\nu+n+1)}{(n+1)!\nu\Gamma(n+1)} \times\{{}_2F_1(-n, n+\nu+2; \nu+1; a\}^2{}_2F_1(-\nu-n-1, n+1; 1; c^2). \tag{3.87}$$

We may note also that

$$(1-a)\,{}_2F_1(-n, n+\nu+2; \nu+1; a) \equiv {}_2F_1(-n-1, n+\nu+1; \nu+1; a), \tag{3.88}$$

and

$${}_2F_1(-\nu-n-1, n+1; 1; c^2) \equiv (1-c^2)^{\nu+1}{}_2F_1(-n, \nu+n+2; 1; c^2). \tag{3.89}$$

Therefore, by introducing the Jacobi polynomials

$${}_2F_1(-n, n+\alpha; \gamma; x) \equiv G_n(\alpha, \gamma, x) \tag{3.90}$$

we can rewrite Equation (3.87) as

$$\alpha_l^0(a, c) = \frac{(1-c^2)^{\nu+1}}{\nu\Gamma(\nu+1)}\sum_{n=0}^{\infty}(-1)^n(\nu+2n+1)\frac{\Gamma(\nu+n+1)}{(n+1)!} \times\{G_{n+1}(\nu, \nu+1, a)\}^2 G_n(\nu+2, 1, c^2). \tag{3.91}$$

The series on the r.h.s. of Equation (3.91), obtained by an evaluation of the Hankel transform (3.55) for the associated α-functions of zero index, *represents the most general expansion for* α_l^0 *in algebraic form, valid for every type of eclipse, and for any arbitrary degree l of the adopted law of limb-darkening.* Our previous equations (3.73) or (3.77) and (3.79) obtained by direct use of Bailey's theorem (3.59), represent but particular forms of Equation (3.91), to which the latter reduces for total or annular eclipses; and the reader can easily verify that, in these cases, they indeed yield identical results.

Secondly, the reader may note that the r.h.s. of Equation (3.91) constitutes a rapidly convergent series of terms in which the effects exerted by the two parameters a and c are *factorized* in the form $f(a)g(c)$ – a feat not attained in any previous formulation of the problem; and a simple structure of the factors

$f(a) \equiv \{G_{n+1}(\nu, \nu+1, a)\}^2$ and $g(c) \equiv (1-c^2)^{\nu+1} G_n(\nu+1, 1, c^2)$ lends itself conveniently for automatic computation.

Equation (3.91) holds good, moreover, equally for occultation or transit eclipses – regardless of whether $r_1 \gtrless r_2$. Should the eclipse in question be an occultation ($r_1 < r_2$), it follows from Equation (3.56) that, in such a case, $0 \leqslant a \leqslant \frac{1}{2}$. Should, on the other hand, the eclipse be a transit ($r_1 > r_2$), $\frac{1}{2} \leqslant a \leqslant 1$; in which case it may be of advantage to use in Equation (3.87) the identity (3.88). If $a = 0$, the eclipse corresponds to one by a straight edge; while if $a = 1$, the eclipsing disc dwindles to a point and, as a result, $\alpha_l^0(1, c) = 0$.

The parameter c as defined by Equation (3.58) varies from 1 at the moment of first contact of the eclipse (when $\delta = r_1 + r_2$) to 0 if the eclipse becomes central (i.e., $\delta_0 = 0$). As was to be expected, for $c = 1$, $\alpha_l^0(a, 1) = 0$. At the inner contact of an occultation eclipse (when $\delta_2 = r_2 - r_1$), the r.h.s. of Equation (3.91) adds up to ν^{-1} in accordance with Equation (3.73); while at the moment of internal tangency of a transit eclipse (when $\delta_2 = r_1 - r_2$), Equation (3.91) reduces to (3.80). Lastly, at the moment of central eclipse ($\delta = 0$), Equation (3.91) reduces to (3.79).

It may also be added that an expression for α_l^0 alternative to Equation (3.91) can be obtained by a resort to the summation theorem for Bessel functions (cf., e.g., Magnus and Oberhettinger, 1948; p. 21) in the form

$$J_0(cy) = \left(\frac{2}{y}\right)^{1/2} \sum_{n=0}^{\infty} (2n+\tfrac{1}{2}) \frac{\Gamma(n+\frac{1}{2})}{\Gamma(n+1)} P_{2n}(\sqrt{1-c^2}) J_{2n+1/2}(y), \tag{3.92}$$

where the P_{2n}'s denote the Legendre polynomials of even orders, of the argument $\sqrt{1-c^2}$.

On inserting the expansion (3.92) on the r.h.s. of Equation (3.55) we find the latter to assume the form

$$\alpha_l^0 = 2^{\nu+1/2} \Gamma(\nu) a^{-\nu} b \sum_{n=0}^{\infty} (2n+\tfrac{1}{2}) \frac{\Gamma(n+\frac{1}{2})}{\Gamma(n+1)} P_{2n}(\sqrt{1-c^2}) \times \int_0^{\infty} y^{-\nu-1/2} J_\nu(ay) J_1(by) J_{2n+1/2}(y)\, dy. \tag{3.93}$$

Since, in accordance with Equations (3.56)–(3.57), $a + b = 1$, Bailey's theorem (3.59) can help once more to evaluate the integral on the r.h.s. of Equation (3.93); with the outcome disclosing that

$$\alpha_l^0 = \frac{b^2}{\nu} \sum_{n=0}^{\infty} (2n+\tfrac{1}{2}) F^{(4)}(-n+\tfrac{1}{2}, n+1; \nu+1, 2; a^2, b^2) P_{2n}(\sqrt{1-c^2}), \tag{3.94}$$

where

$$P_{2n}(\sqrt{1-c^2}) = \frac{(-1)^n (2n)!}{2^{2n}(n!)^2}\, {}_2F_1(-n, n+\tfrac{1}{2}; \tfrac{1}{2}; 1-c^2) = P_{2n}(0) G_n(\tfrac{1}{2}, \tfrac{1}{2}, 1-c^2). \tag{3.95}$$

The expression on the r.h.s. of the preceding Equation (3.94) is identical with that on the r.h.s. of Equation (3.91); but (unlike the latter) it constitutes a doubly-infinite summation, because the $F^{(4)}$-series in (3.94) itself never terminates. It should also be noted that whereas, for $c = 1$ (i.e., $\delta = r_1 + r_2$) the series on the r.h.s. of Equation (3.91) is factored by zero, so that the r.h.s. of (3.54) only adds up to zero by virtue of the fact that

$$P_{2n}(0) = (-1)^n \frac{(2n)!}{(n!2^n)^2} \tag{3.96}$$

renders it an alternating series.

Equations (3.91) or (3.94) do not, by any means, exhaust the forms of the expansions by which the function $\alpha_l^0(a, c)$ can be expressed in terms of its parameters. Thus if account is taken of the identity (cf. again Watson, 1945; p. 370) that

$$\begin{aligned} &(-1)^n \{G_{n+1}(\nu, \nu+1, a)\}^2 \\ &= \frac{(n+1)!\Gamma(\nu+1)}{\Gamma(\nu+n+1)} b^2 F^{(4)}(-n, n+\nu+2; 2, \nu+1; b, a), \end{aligned} \tag{3.97}$$

an insertion of Equation (3.97) in (3.91) discloses that, alternatively,

$$\begin{aligned} \nu\alpha_l^0 = b^2(1-c^2)^{\nu+1} &\sum_{n=0}^{\infty} (\nu+2n+2) \\ &\times F^{(4)}(-n, n+\nu+2; 2, \nu+1; a, b) G_n(\nu+2, 1; c^2), \end{aligned} \tag{3.98}$$

where for integral values of n the series $F^{(4)}$ reduces likewise to a polynomial (cf. Appendix III). Moreover, Demircan (1978d) proved that, even more simply,

$$\nu\alpha_l^0 = 2b^2 \sum_{n=1}^{\infty} F^{(4)}(1-n, 1+n; 1+\nu, 2; a^2, b^2) G_n(0, 1; c^2), \tag{3.99}$$

where the Jacobi polynomial on the right-hand side no longer depends on ν. None of the Fourier series on the r.h. sides of Equations (3.91), (3.98) or (3.99) can be matched term-by-term with each other, but all should converge to the same result; only their speed of convergence (or, rather, asymptotic properties) may differ in different domains of the range $0 \leqslant a \leqslant 1$ $(b = 1 - a)$ and $0 \leqslant c \leqslant 1$ for which they are valid.

Should, moreover, $a = 0$ (i.e., the occulting limb to become a straight edge), the known asymptotic properties of Bessel functions enable us to express the α_l^0's in a more elementary form. In order to do so, let us return to Equation (3.32) to consider what happens when the values of both r_2 and δ tend to infinity, but in such a way that

$$\delta = r_2 + r_1 p = r_2(1 + kp), \tag{3.100}$$

where

$$k \ll 1 \quad \text{and} \quad -1 \leqslant p \leqslant 1 \tag{3.101}$$

during eclipse.

As the arguments δ and r_2 become very large, it is legitimate to resort to asymptotic expressions for Bessel functions of large arguments, asserting that, under these circumstances,

$$J_0(2\pi q\delta)\to\frac{\cos(2\pi q\delta-45°)}{\pi\sqrt{q\delta}} \tag{3.102}$$

and

$$J_1(2\pi qr_2)\to\frac{\sin(2\pi qr_2-45°)}{\pi\sqrt{qr_2}}, \tag{3.103}$$

so that

$$\frac{J_2(2\pi qr_2)}{2\pi qr_2}J_0(2\pi q\delta)\to\frac{\cos 2\pi q(\delta+r_2)+\sin 2\pi qr_1p}{\pi(2\pi qr_2)^2}. \tag{3.104}$$

If so, however, Equation (3.32) can be rewritten as

$$\pi\alpha_l^0=-2^{\nu}\Gamma(\nu)\int_0^{\infty}y^{-\nu-1}\{\cos\Pi y+\sin py\}J_{\nu}(y)\,\mathrm{d}y, \tag{3.105}$$

where

$$y\equiv 2\pi qr_1=kx, \tag{3.106}$$

and where we have abbreviated (cf. Equation (3.100))

$$\Pi=\frac{\delta+r_2}{r_1}\quad\text{and}\quad p=\frac{\delta-r_2}{r_1}. \tag{3.107}$$

Of the two parts of the integral on the right-hand side of Equation (3.105), it can be shown that

$$\lim\int_0^{\infty}y^{-\nu-1}J_{\nu}(y)\cos\Pi y\,\mathrm{d}y=0,\quad \Pi\to\infty, \tag{3.108}$$

so that

$$\pi\alpha_l^0=-2^{\nu}\Gamma(\nu)\int_0^{\infty}y^{-\nu-1}J_{\nu}(y)\sin py\,\mathrm{d}y; \tag{3.109}$$

rendering $\alpha_l^0(0,p)$ a Hankel transform, of ν-th order, of the function $y^{-\nu-1}\sin py$.

A repeated differentiation of Equation (3.109) with respect to p discloses that, for eclipses by a straight edge,

$$\begin{aligned}\frac{\partial^2\alpha_l^0}{\partial p^2}&=2^{\nu}\frac{\Gamma(\nu)}{\pi}\int_0^{\infty}y^{1-\nu}J_{\nu}(y)\sin py\,\mathrm{d}y\\&=\frac{2}{\sqrt{\pi}}\frac{\Gamma(\nu)}{\Gamma(\nu-\frac{1}{2})}p(1-p^2)^{\nu-3/2},\end{aligned} \tag{3.110}$$

which on successive integration discloses that

$$\frac{\partial \alpha_l^0}{\partial p} = -\frac{\Gamma(\nu)}{\sqrt{\pi}\,\Gamma(\nu+\frac{1}{2})}(1-p^2)^{\nu-1/2} \tag{3.111}$$

and, therefore,

$$\alpha_l^0 = \frac{1}{2\nu} - \frac{\Gamma(\nu)}{\sqrt{\pi}\,\Gamma(\nu+\frac{1}{2})}\int_0^p (1-p^2)^{\nu-1/2}\,\mathrm{d}p\,; \tag{3.112}$$

where the constant of integration has been adjusted so that, at the beginning of the eclipse, $\alpha_l^0(0, 1) = 0$. The reader may verify that, for $l = 0$ and 1 (i.e., $\nu = 1$ and $\frac{3}{2}$), the foregoing equation (3.112) reduces to (2.21) and (2.22) in which $s/r_1 \equiv p$.

I-4. Fourier Transform of the Light Changes

Having established, in the preceding section, the Fourier approximations to the fractional loss of light suffered in the course of mutual eclipses of spherical stars appearing in projection as arbitrarily darkened circular discs, we wish now to turn our attention to another fundamental aspect of our problem of eminent theoretical as well as practical importance: namely, to a specification of their *Fourier transforms*.

In order to do so, let us depart from the fundamental equations of the Fourier theorem in the form

$$F(\nu) = \int_{-c}^{c} f(\theta)\,\mathrm{e}^{-2\pi i\nu\theta}\,\mathrm{d}\theta, \tag{4.1}$$

$$f(\theta) = \int_{-\infty}^{\infty} F(\nu)\,\mathrm{e}^{2\pi i\nu\theta}\,\mathrm{d}\nu, \tag{4.2}$$

where $f(\theta)$ denotes an 'input function' – such as represented, e.g., by the light changes of an eclipsing variable as functions of the phase angle θ – and $F(\nu)$, its Fourier transform regarded as a function of frequency ν. The limits $\pm c$ of integration on the right-hand side of Equation (4.1) can be finite or infinite – depending (in effect) on the range of θ within which the function $f(\theta)$ is different from zero. For example, if $f(\theta)$ represents a loss of light arising from the eclipses of spherical stars, the quantities $\pm c$ may be identified with the moments of the first and last contact θ_1 of the eclipses; for since $f(\theta) = 0$ for $\theta > c$, the value of the integral on the r.h.s. of Equation (4.1) would clearly be unaffected by any further extension of its limits.

As is well known, even for real functions $f(\theta)$ their transform $F(\nu)$ will generally be a complex quantity. In order to separate its real and imaginary form, let us consider the identity

$$f(\theta) \equiv \tfrac{1}{2}[f(\theta) + f(-\theta)] + \tfrac{1}{2}[f(\theta) - f(-\theta)], \tag{4.3}$$

by virtue of which

$$\int_{-c}^{c} \tfrac{1}{2}[f(\theta)+f(-\theta)]\,\mathrm{e}^{ih\theta}\,\mathrm{d}\theta = \tfrac{1}{2}\int_{0}^{c} [f(\theta)+f(-\theta)](\mathrm{e}^{ih\theta}+\mathrm{e}^{-ih\theta})\,\mathrm{d}\theta$$

$$= \int_{0}^{c} [f(\theta)+f(-\theta)]\cos h\theta\,\mathrm{d}\theta; \tag{4.4}$$

and, similarly,

$$\int_{0}^{c} \tfrac{1}{2}[f(\theta)-f(-\theta)]\,\mathrm{e}^{-ih\theta}\,\mathrm{d}\theta = \tfrac{1}{2}\int_{0}^{c} [f(\theta)-f(-\theta)](\mathrm{e}^{-ih\theta}-\mathrm{e}^{ih\theta})\,\mathrm{d}\theta$$

$$= -i\int_{0}^{c} [f(\theta)-f(-\theta)]\sin h\theta\,\mathrm{d}\theta, \tag{4.5}$$

where we have abbreviated

$$h = 2\pi\nu. \tag{4.6}$$

In consequence, the Fourier transform $F(\nu)$ as defined by Equation (4.1) can obviously be expressed as

$$F(\nu) = F_1(\nu) - iF_2(\nu), \tag{4.7}$$

where

$$F_1(\nu) = \int_{0}^{c} [f(\theta)+f(-\theta)]\cos(h\theta)\,\mathrm{d}\theta \tag{4.8}$$

and

$$F_2(\nu) = \int_{0}^{c} [f(\theta)-f(-\theta)]\sin(h\theta)\,\mathrm{d}\theta \tag{4.9}$$

are – for real functions $f(\theta)$ – both real quantities.

The Fourier transform $F(\nu)$ represents a de-composition of the input function $f(\theta)$ into a *continuous* spectrum of its fundamental frequencies ν. The second half of Equation (4.2) of the Fourier theorem discloses, however, that the original input function $f(\theta)$ can be *synthesized* again from its transform $F(\nu)$ by

$$\begin{aligned} f(\theta) &= \int_{-\infty}^{\infty} [F_1(\nu) - iF_2(\nu)]\,\mathrm{e}^{2\pi i\nu\theta}\,\mathrm{d}\nu \\ &= \int_{0}^{\infty} F_1(\nu)(\mathrm{e}^{2\pi i\nu\theta}+\mathrm{e}^{-2\pi i\nu\theta})\,\mathrm{d}\nu - i\int_{0}^{\infty} F_2(\nu)(\mathrm{e}^{2\pi i\nu\theta}-\mathrm{e}^{-2\pi i\nu\theta})\,\mathrm{d}\nu \\ &= 2\int_{0}^{\infty} [F_1(\nu)\cos 2\pi\nu\theta + F_2(\nu)\sin 2\pi\nu\theta]\,\mathrm{d}\nu \end{aligned} \tag{4.10}$$

$$= 2\int_0^\infty \sqrt{F_1^2 + F_2^2}\cos\left(2\pi\nu\theta - \tan^{-1}\frac{F_2}{F_1}\right)\mathrm{d}\nu.$$

Moreover, if the limits $\pm c$ of the Fourier integral (4.1) are *finite* – as they happen to be in our case – the original input function can be alternatively expressed (cf., e.g., Sneddon, 1951; Section 13.2) by use of the Euler–Maclaurin summation formula exactly as

$$\begin{aligned} f(\theta) &= \sum_{n=-\infty}^{\infty} \frac{\mathrm{e}^{\pi i n\theta/c}}{2c} F\left(\frac{n}{2c}\right) \\ &= \frac{F_1(0)}{2c} + \frac{1}{c}\sum_{n=1}^{\infty}\left\{F_1\left(\frac{n}{2c}\right)\cos\frac{n\pi\theta}{c} + F_2\left(\frac{n}{2c}\right)\sin\frac{n\pi\theta}{c}\right\} \\ &= \tfrac{1}{2}a_0 + \sum_{n=1}^{\infty}\left\{a_n\cos\frac{n\pi\theta}{c} + b_n\sin\frac{n\pi\theta}{c}\right\}, \end{aligned} \tag{4.11}$$

where

$$a_n = \frac{1}{c}F_1\left(\frac{n}{2c}\right) \quad \text{and} \quad b_n = \frac{1}{c}F_2\left(\frac{n}{2c}\right). \tag{4.12}$$

Moreover, the functions $f(\theta)$ which are *symmetrical* with respect to $\theta = 0$ – and, therefore, for which

$$f(\theta) = f(-\theta), \tag{4.13}$$

the imaginary part of $F(\nu)$ as given by Equation (4.9) will clearly vanish, rendering the Fourier transform (4.7) real; also, since the coefficients b_n in Equation (4.12) are now identically zero, Equation (4.11) reduces to a Fourier cosine series of the form

$$f(\theta) = \frac{F_1(0)}{2c} + \frac{1}{c}\sum_{n=1}^{\infty} F_1\left(\frac{n}{2c}\right)\cos\frac{n\pi\theta}{c}, \tag{4.14}$$

representing an expansion of $f(\theta)$ in terms of the functions $\cos(n\pi\theta/c)$ of the phase, the coefficients of which are identical with local values of the Fourier cosine transform $F_1(\nu)$ of $f(\theta)$.

A. EVALUATION OF FOURIER TRANSFORMS

In order to evaluate the transform $F_1(\nu)$, let us return to Equations (4.8) and (4.13), yielding

$$F_1(\nu) = 2\int_0^{\theta_1} f(\theta)\cos h\theta\,\mathrm{d}\theta, \tag{4.15}$$

where

$$\cos h\theta = \frac{\sin\pi h}{\pi h}\left\{1 + 2\sum_{m=1}^{\infty}\frac{(-1)^m h^2}{h^2 - m^2}\cos m\theta\right\}, \tag{4.16}$$

valid for $-\pi < \theta < \pi$ and any value of $h \geqslant 0$ (whether integral or not); in which, for integral values of m,

$$\cos m\theta = {}_2F_1\left(-\frac{m}{2}; \frac{m}{2}; \frac{1}{2}; \sin^2\theta\right). \tag{4.17}$$

Let, as before,

$$f(\theta) \equiv 1 - l = L_1\alpha = L_1 \sum_{l=0}^{\infty} C^{(l)}\alpha_l^0, \tag{4.18}$$

where the coefficients $C^{(l)}$ continue to be given by Equations (2.8)–(2.9), and in which the functions $\alpha_l^0(a, c)$ continue to be given by Equations (3.91)–(3.99). Let us, furthermore, introduce at this stage a new variable u, defined by

$$u = \left(\frac{\sin\theta}{\sin\theta_1}\right)^2, \tag{4.19}$$

by virtue of which the limits $(0, \theta_1)$ of integration with respect to θ on the r.h.s. of Equation (4.15) become normalized to (0, 1). Moreover, a differentiation of (4.19) discloses that

$$\mathrm{d}\theta = \frac{\sin\theta_1 \,\mathrm{d}u}{\sqrt{u(1 - u\sin^2\theta_1)}}. \tag{4.20}$$

Making use of Equations (4.16)–(4.20) we can rewrite Equation (4.15) in terms of our new variable u as

$$F_1(\nu) = L_1 \sin\theta_1 \left(\frac{\sin\pi h}{\pi h}\right) \sum_{l=0}^{\infty} \int_0^1 \frac{C^{(l)}\alpha_l^0}{\sqrt{u(1 - u\sin^2\theta_1)}} \times\left\{1 + 2\sum_{m=1}^{\infty} \frac{(-1)^m h^2}{h^2 - m^2}\, {}_2F_1\left(-\frac{m}{2}, \frac{m}{2}; \frac{1}{2}; u\sin^2\theta_1\right)\right\} \mathrm{d}u, \tag{4.21}$$

where, as before, $h \equiv 2\pi\nu$.

In order to evaluate the integrals on the r.h.s. of the foregoing equation remember that, by virtue of Equation (4.19), the argument c of the α_l^0's as given by Equation (3.58) can be rewritten as

$$c^2 = 1 - (1 - c_0^2)(1 - u); \tag{4.22}$$

and that, accordingly, the expansions (3.91)–(3.99) can be rewritten in ascending integral powers of $1 - u$. For $m = 0$, the r.h.s. of Equation (4.23) will, therefore, consist of integrals of the form

$$\int_0^1 (1 - u)^j [u(1 - u\sin^2\theta_1)]^{-1/2} \,\mathrm{d}u = B\left(\frac{1}{2}, j + 1\right) {}_2F_1\left(\frac{1}{2}, \frac{1}{2}; j + \frac{3}{2}; \sin^2\theta_1\right). \tag{4.23}$$

For $m > 0$, we can express the product

$$(1-u\sin^2\theta_1)^{-1/2}\,{}_2F_1\left(-\frac{m}{2},\frac{m}{2};\frac{1}{2};u\sin^2\theta_1\right)$$
$$={}_2F_1\left(\frac{1-m}{2},\frac{1+m}{2};\frac{1}{2};u\sin^2\theta_1\right) \tag{4.24}$$

in the form of a hypergeometric series which converges for $0\leqslant u\leqslant 1$ and $\theta_1<90°$; consequently,

$$\int_0^1(1-u)^j[u(1-u\sin^2\theta_1)]^{-1/2}\,{}_2F_1\left(-\frac{m}{2},\frac{m}{2};\frac{1}{2};u\sin^2\theta_1\right)\mathrm{d}u$$
$$=\int_0^1 u^{-1/2}(1-u)^j{}_2F_1\left(\frac{1-m}{2},\frac{1+m}{2};\frac{1}{2};u\sin^2\theta_1\right)\mathrm{d}u \tag{4.25}$$
$$=B\left(\frac{1}{2},j+1\right){}_2F_1\left(\frac{1-m}{2},\frac{1+m}{2};j+\frac{3}{2};\sin^2\theta_1\right).$$

An insertion of Equations (4.23) and (4.25) in (4.21) then yields

$$F_1\left(\frac{h}{2\pi}\right)=b^2L_1\sin\theta_1\left(\frac{\sin\pi h}{\pi h}\right)\sum_{l=0}^{\infty}\frac{C^{(l)}}{\nu}\sum_{n=0}^{\infty}\frac{\nu+2n+2}{(n+1)!}$$
$$\times(\nu+1)_n\left\{G_n(\nu+2,\nu+1,a)\right\}^2\sum_{i=0}^{n}\frac{(n+\nu+2)_i}{i!}\binom{n}{i}$$
$$\times\sum_{k=0}^{i}(-1)^{i+k+n}\binom{i}{k}(1-c_0^2)^{k+\nu+1}B\left(\frac{1}{2},k+\nu+2\right)$$
$$\times\left\{{}_2F_1\left(\frac{1}{2},\frac{1}{2},k+\nu+\frac{5}{2};\sin^2\theta_1\right)+\sum_{m=1}^{\infty}\frac{(-1)^m h^2}{h^2-m^2}\right.$$
$$\left.\times{}_2F_1\left(\frac{1-m}{2},\frac{1+m}{2},k+\nu+\frac{5}{2};\sin^2\theta_1\right)\right\} \tag{4.26}$$

for α_l^0 as given by Equation (3.91); and where, as before, $\nu=\frac{1}{2}(l+2)$ on the right-hand side.

If, moreover, Equation (3.94) is used for α_l^0 in place of Equation (3.91), the corresponding Fourier transform becomes

$$F_1\left(\frac{h}{2\pi}\right)=b^2L_1\sin\theta_1\left(\frac{\sin\pi h}{\pi h}\right)\sum_{l=0}^{\infty}\frac{C^{(l)}}{\nu}\sum_{n=0}^{\infty}\left(2n+\frac{1}{2}\right)$$
$$\times I_{2n}(0)F^{(4)}\left(-n+\frac{1}{2},n+1;\nu+1,2;a^2,b^2\right)$$
$$\times\sum_{i=0}^{n}(-1)^i\frac{(n+\frac{1}{2})_i}{(\frac{1}{2})_i}\binom{n}{i}(1-c_0^2)^iB\left(\frac{1}{2},i+1\right)$$
$$\times\left\{{}_2F_1\left(\frac{1}{2},\frac{1}{2};i+\frac{3}{2};\sin^2\theta_1\right)+2\sum_{m=1}^{\infty}\frac{(-1)^m h^2}{h^2-m^2}\right.$$
$$\left.\times{}_2F_1\left(\frac{1-m}{2},\frac{1+m}{2};i+\frac{3}{3};\sin^2\theta_1\right)\right\}. \tag{4.27}$$

If, lastly, Demircan's series (3.99) is inserted in Equation (4.21), we obtain an alternative expression for $F_1(\nu)$ of the form

$$F_1\left(\frac{h}{2\pi}\right) = b^2 L_1 \sin\theta_1 \left(\frac{\sin \pi h}{\pi h}\right) \sum_{l=0}^{\infty} \frac{C^{(l)}}{\nu} \sum_{n=1}^{\infty} \frac{2}{(n-1)!}$$

$$\times F^{(4)}(1-n, 1+n; 1+\nu, 2; a^2, b^2) \sum_{i=0}^{n} \frac{(n+i-1)!}{i!}$$

$$\times \binom{n}{i} \sum_{k=0}^{i} (-1)^{i+k} \binom{i}{k} (1-c_0^2)^k B\left(\frac{1}{2}, k+1\right)$$

$$\times \left\{ {}_2F_1\left(\frac{1}{2}, \frac{1}{2}; k+\frac{3}{2}; \sin^2\theta_1\right)\right.$$

$$\left. + 2 \sum_{m=1}^{\infty} \frac{(-1)^m h^2}{h^2 - m^2} {}_2F_1\left(\frac{1-m}{2}, \frac{1+m}{2}; k+\frac{3}{2}; \sin^2\theta_1\right)\right\}. \tag{4.28}$$

So far we have been concerned with an evaluation of the real part $F_1(\nu)$ of the Fourier transform (4.7). Let us next consider – as an alternative possibility – to set

$$f(-\theta) = 0; \tag{4.29}$$

which, for $l(\theta) = 1 - l$, would render our light curve a 'saw-tooth' curve as shown on Figure 1-5. In such a case, Equations (4.8) and (4.9) would reduce to

$$F_1(\nu) = \int_0^{\theta_1} f(\theta) \cos h\theta \, d\theta \tag{4.30}$$

and

$$F_2(\nu) = \int_0^{\theta_1} f(\theta) \sin h\theta \, d\theta, \tag{4.31}$$

i.e., the function $F_1(\nu)$ as defined now by Equation (4.30) would become one-half of that given previously by Equation (4.15); but – unlike for symmetrical light curves – the imaginary part $F_2(\nu)$ of Equation (4.7) will cease to be zero; and, in consequence, the Fourier series (4.11) will contain both the sine and cosine terms. The latter have already been evaluated; so that it remains to us only to evaluate the former.

In order to do so, we replace Equation (4.16) by an analogous expansion of $\sin h\theta$ of the form

$$\sin h\theta = 2 \frac{\sin \pi h}{\pi} \sum_{m=1}^{\infty} \frac{(-1)^m m}{h^2 - m^2} \sin m\theta, \tag{4.32}$$

where

$$\sin m\theta = m \sin\theta \, {}_2F_1\left(\frac{1-m}{2}, \frac{1+m}{2}; \frac{3}{2}; \sin^2\theta\right). \tag{4.33}$$

In consequence, an equation analogous to (4.22) for $F_2(\nu)$ will now be of the form

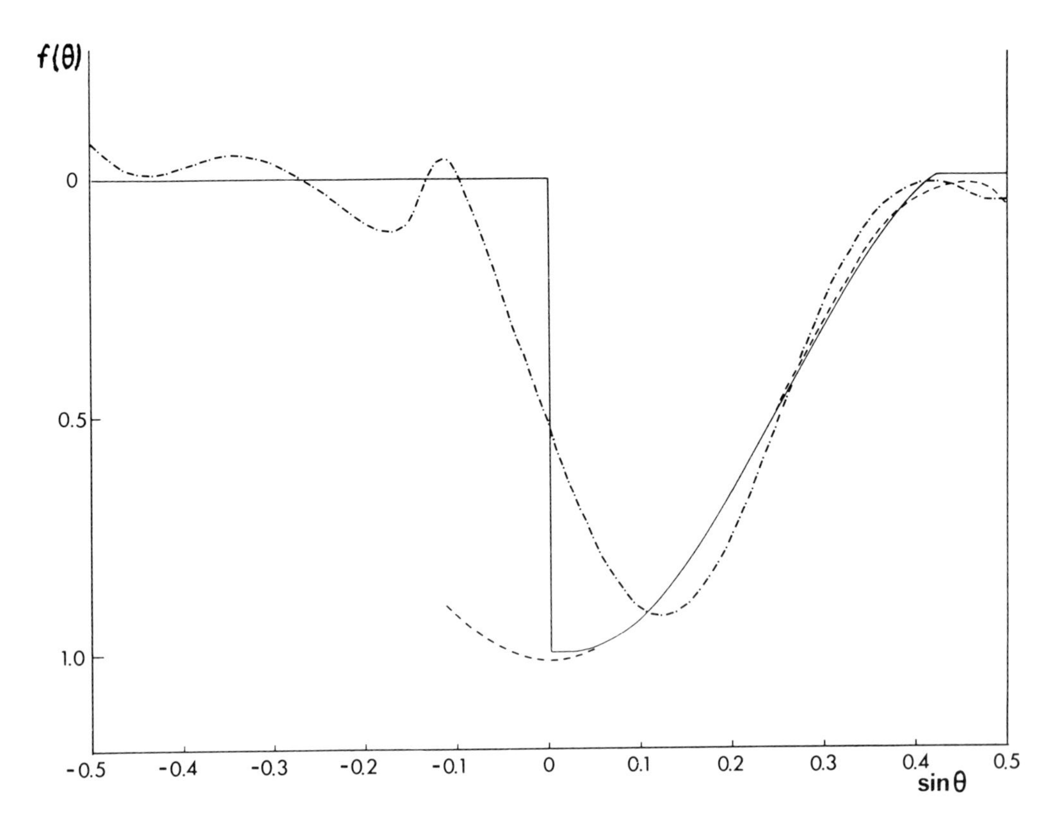

Fig. 1-5. A theoretical light curve of a stellar eclipse and its Fourier representation.

The solid curve represents the changes computed for a model discussed in Section 4 on the arbitrary assumption that $f(-\theta) = 0$. A Fourier representation of such a saw-tooth curve by five coefficients is shown as a broken curve of dots and dashes.

If, on the other hand, we regard the light curve as symmetrical with respect to the moment $\theta = 0$ of conjunction (i.e., set $f(-\theta) = f(\theta)$), a three-term Fourier approximation results in a fit represented by the dotted line barely distinguishable from the actual light curve on the scale of our diagram – except near both ends of the interval $(0, \theta_1)$.

$$F_2(\nu) = L_1 \sin^2\theta_1 \frac{\sin \pi h}{\pi} \sum_{l=0}^{\infty} \int_0^1 \frac{C^{(l)}\alpha_l^0}{\sqrt{1 - u \sin^2\theta_1}} \times \sum_{m=1}^{\infty} \frac{(-1)^m m^2}{h^2 - m^2} {}_2F_1\left(\frac{1-m}{2}, \frac{1+m}{2}; \frac{3}{2}; u \sin^2\theta_1\right) \mathrm{d}u, \tag{4.34}$$

where, analogously with Equation (4.24),

$$(1 - u\sin^2\theta_1)^{-1/2} {}_2F_1\left(\frac{1-m}{2}, \frac{1+m}{2}; \frac{3}{2}; u \sin^2\theta_1\right) = {}_2F_1\left(\frac{2-m}{2}, \frac{2+m}{2}; \frac{3}{2}; u \sin^2\theta_1\right). \tag{4.35}$$

A subsequent evaluation of $F_2(\nu)$ follows a course so similar to that of $F_1(\nu)$

that further details of it are unnecessary. Finite transforms of the type (4.30) and (4.31) have been used by Kitamura (1965) as a basis of his method of solution for the elements of eclipsing binary systems – a subject we shall take up in Chapters V and VI of this book – and tabulated (cf. Kitamura, 1967) by numerical quadratures in terms of the elements of the eclipse for the case of uniformly bright discs as well as of linear limb-darkening.

B. SYNTHESIS OF THE FOURIER TRANSFORM

Let us, however, return to the second part of the Fourier theorem, represented by Equation (4.10), which permits us to synthesize the continuous 'spectrum' of the frequencies ν constituting the Fourier transform $F(\nu)$ into our original function $f(\theta)$. Equation (4.10) discloses that, for symmetrical light curves $f(\theta)$ satisfying the condition (4.13),

$$\begin{aligned} f(\theta) \equiv L_1 \sum_{l=0}^{\infty} C^{(l)} \alpha_l^0 &= 2 \int_0^{\infty} F_1(\nu) \cos(2\pi\nu\theta)\, d\nu \\ &= \frac{1}{\pi} \int_0^{\infty} F_1(h) \cos h\theta \, dh, \end{aligned} \tag{4.36}$$

where the Fourier transform $F_1(\nu)$ is in our present case given by the alternative expressions (4.26)–(4.28). The only terms on the right-hand sides of these latter equations which depend on h are of the form $(h \sin \pi h)/(h^2 - m^2)$.

Since, however,

$$\int_0^{\infty} \frac{\sin \pi h \cos h\theta}{h^2 - m^2} h \, dh = \frac{1}{2} \pi (-1)^m \cos m\theta, \tag{4.37}$$

then by equating terms on both sides of Equations (4.36) factored by each one of the coefficients $C^{(l)}$ we find that

$$\begin{aligned} 2\pi\nu\alpha_l^0 = {} & b^2 \sin\theta_1 \sum_{n=0}^{\infty} \frac{\nu + 2n + 2}{(n+1)!} (\nu+1)_n \{G_n(\nu+2, \nu+1, a)\}^2 \\ & \times \sum_{i=0}^{n} \frac{(n+\nu+2)_i}{i!} \binom{n}{i} \sum_{k=0}^{i} (-1)^{n+i+k} \binom{i}{k} (1 - c_0^2)^{k+\nu+1} \\ & + B(\tfrac{1}{2}, k+\nu+2) \Big\{ {}_2F_1(\tfrac{1}{2}, \tfrac{1}{2}, k+\nu+\tfrac{5}{2}, \sin^2\theta_1) \\ & + 2 \sum_{m=1}^{\infty} {}_2F_1\left(\frac{1-m}{2}, \frac{1+m}{2}, k+\nu+\tfrac{5}{2}, \sin^2\theta_1\right) \cos m\theta \\ = {} & b^2 \sin\theta_1 \sum_{n=0}^{\infty} (2n+\tfrac{1}{2}) P_{2n}(0) F^{(4)} \\ & \times (-n+\tfrac{1}{2}, n+1; \nu+1, 2; a^2, b^2) \sum_{i=0}^{n} (-1)^i \frac{(n+\frac{1}{2})_i}{(\frac{1}{2})_i} \binom{n}{i} \\ & \times (1-c_0^2)^i B(\tfrac{1}{2}, i+1) \Big\{ {}_2F_1(\tfrac{1}{2}, \tfrac{1}{2}; i+\tfrac{3}{2}; \sin^2\theta_1) \end{aligned}$$

$$+2\sum_{m=1}^{\infty} {}_2F_1\left(\frac{1-m}{2}, \frac{1+m}{2}; i+\tfrac{3}{2}; \sin^2\theta_1\right)\cos m\theta\Bigg\}$$

$$= b^2 \sin\theta_1 \sum_{n=1}^{\infty} \frac{2}{(n-1)!} F^{(4)}(1-n, 1+n; 1+\nu, 2; a^2, b^2)$$

$$\times \sum_{i=0}^{\infty} \frac{(n+i-1)!}{i!}\binom{n}{i}\sum_{k=0}^{i}(-1)^{i+k}\binom{i}{k}(1-c_0^2)^k$$

$$\times B(\tfrac{1}{2}, k+1)\Bigg\{{}_2F_1(\tfrac{1}{2}, \tfrac{1}{2}; k+\tfrac{3}{2}; \sin^2\theta_1)$$

$$+2\sum_{m=1}^{\infty} {}_2F_1\left(\frac{1-m}{2}, \frac{1+m}{2}; k+\tfrac{3}{2}; \sin^2\theta_1\right)\cos m\theta\Bigg\}, \tag{4.38}$$

which represent nothing else but Fourier cosine series for α_l^0 in terms of *integral* multiples of the phase-angle θ; with coefficients given as explicit functions of the elements of the eclipse.

The reader may wish to compare the foregoing series (4.38)–(4.40) with Equation (3.42), representing another type of Fourier cosine series for a_l^0 in terms of integral multiples of an auxiliary angle ϕ, depending on θ by Equation (3.43) (see Figure 1-3). Lastly, a third type of a Fourier cosine series for α_l^0 in terms of integral multiples of the angle $\pi\theta/\theta_1$ can be obtained by an insertion of local values of the Fourier transforms $F_1(n/2\theta_1)$ from Equations (4.26)–(4.28) in the Fourier series (4.14). All these expansions differ in form, but all converge (at different rate in different domains) to the same answer. The asymptotic properties of finite sums of these Fourier series are generally superior to those of the expansions (3.91)–(3.99) of Hankel transforms investigated in Section I-3 – a fact which should render them more suitable for numerical machine work; but the reader had a full right to know the explicit forms of all alternatives.

However, one-branch light curves obtained by a synthesis of the Fourier transforms (4.30) and (4.31) are less satisfactory (see Figure 1-5); and, for this reason, we shall not go into this subject any further (for fuller details, cf. Kopal, 1977a). The real cause is the fact that the condition (4.29) underlying the transforms (4.30) and (4.31) renders the function $f(\theta)$ itself discontinuous at $\theta = 0$; and non-uniform convergence of the series (4.11) will give rise to oscillations (the Gibbs phenomenon!) in this region which can be damped out only by a summation of a great many terms.

Bibliographical Notes

I-**1**: The effects of linear limb-darkening (as represented by the 'cosine-law' approximation to a solution of the respective transfer problem) were taken account of in the analysis of the light curves of eclipsing variables first by A. Pannekoek (1902) and C. Rödiger (1902) at the turn of this century; followed 10 years later by H. N. Russell and H. Shapley (1912a, b).

Photometric effects of nonlinear limb-darkening – as represented by Equation (1.2) of this chapter and providing a closer approximation to the solution of the underlying transfer problem – were investigated first by Z. Kopal (1949), to whom all relevant results used in book are due.

I-**2**: The loss of light arising from the eclipses of spherical stars with uniformly bright apparent discs – as represented by Equation (2.14) – was known already to E. C. Pickering (1880), J. Harting

(1889) or E. Hartwig (1900). The first analytic investigations of the effects of limb-darkening (which required elliptic integrals for their formulation) were carried out by P. Harzer (1927) and, subsequently, by V. P. Tsesevich (1936) who provided also the most complete tabulations of such functions (cf. Tsesevich, 1939, 1940). A more limited set of such tables were also produced independently by K. Ferrari (1938, 1939); while a later work by J. E. Merrill (1950) was based on tables provided earlier by Tsesevich (1939, 1940).

For an exhaustive review of all these (and other tables published up to 1947 cf. Z. Kopal (1948).

I-3: A concept of the loss of light as 'aperture cross-correlation' discussed in this section, was introduced by Z. Kopal (1977b), and developed further by him (Kopal, 1977c) and his collaborators (O. Demircan, 1978, 1979) to the stage presented in the text.

The basic equations (3.32) or (3.34) expressing the fractional loss of light α_l^0 due to eclipses of arbitrarily-darkened stellar discs in terms of Hankel transforms were given first by Kopal (1977b), to whom also their evaluation in the form of expansion (3.91) or (3.94) and (3.98) is due (cf. Kopal, 1977c). Alternative forms for α_l^0 obtained by an expansion of the respective Hankel transforms in terms of different variables, as represented by Equation (3.99) (and others) are due to O. Demircan (1978d). For asymptotic expressions obtaining as the ratio of the radii $k \to 0$ cf. Z. Kopal (1977b).

I-4: The literature on the Fourier transformation (4.1)–(4.2) and techniques developed in its wake since 1807 when this theorem was first enunciated by Fourier, is truly enormous and cannot possibly be reviewed in this place. Of more recent sources, which emphasize applications and which the reader of the present section may peruse with profit, Chapter IV of C. Lanczos (1956), M. J. Lighthill (1958) or C. Lanczos (1966) should be of particular interest to astronomers.

A quantitative theoretical Fourier analysis of the light curves of eclipsing variables goes back to M. Kitamura (1965), who was first to evaluate (by quadratures) the Fourier coefficients a_h and b_h (in Kitamura's notations c_n and s_n) of the series (4.11), as defined by our Equations (4.30) and (4.31) under the assumption (4.29), and tabulated these extensively in terms of the elements of the eclipse (Kitamura, 1967). However, the Fourier series (4.11) constructed with their aid converge too slowly for practical use; and were later replaced by faster-converging cosine series by Z. Kopal (1976b).

The analytic expressions of finite cosine transforms $F_1(\nu)$ of the light curves in terms of a continuous frequency ν, as given by Equations (4.26)–(4.28) of this section, and their subsequent synthesis leading to Equations (4.38)–(4.40) are new.

CHAPTER II

LIGHT CHANGES OF CLOSE ECLIPSING SYSTEMS

In the preceding chapter of this book we gave an outline of the theory of light changes of eclipsing binary systems – applicable to any type of eclipses – of arbitrarily limb-darkened stars which are *spherical* in form, and would appear in projection on the celestial sphere as *circular* discs. With the completion of this task the question is bound to arise as to the extent to which the model of a binary system consisting of spherical stars can be regarded as a satisfactory representation of eclipsing systems actually encountered in the sky.

This would, in principle, be legitimate only in the absence of forces which could distort the shape of both components and thus cause them to depart from spherical form. Such forces are, however, bound to be ever present in the form of *axial rotation* of the components, and of their mutual *tidal action*. A theory of the extent to which these forces will force the stars of given structure to depart from spherical form has recently been summarized by the present writer in Chapter II of another volume (Kopal, 1978b), to which the reader is referred to fuller details that do not need to be repeated in this place.

The essential features of this problem rest on the fact that the *amplitude of photometric effects in close eclipsing systems, caused by both rotational as well as tidal distortion, grows rapidly with increasing proximity of their components.* In many known eclipsing systems the components are, to be sure, separated widely enough – and, consequently, their distortion is sufficiently small – that (within the limits of observational errors) they can be regarded as spheres; and for such systems a theory of the light changes as given in the preceding chapter should indeed remain adequate. However, if the components are closer to each other – and (apart from the far greater intrinsic interest in such binaries) observational selection strongly favours discovery of close systems – a situation arises requiring new approach. For a significant feature of 'close' eclipsing systems is the fact that observable light changes are no longer confined only to the times of the minima, but extend over the entire cycle. This is due to two reasons. First, since the components of close binaries possess, in general, the form of distorted ellipsoids with longest axes oriented constantly in the direction of the line joining their centres, their apparent cross-sections – and, therefore, the light – exposed to the observer should vary continuously in the course of a revolution ('ellipticity effect'). Secondly, it is also inevitable in close binaries that a fraction of the radiation of each component will be intercepted by its mate, to be absorbed and re-emitted (or scattered) in all directions – including that of the line of sight. The amount of light so 'reflected' by each component towards the observer will again vary with the phase ('reflection effect').

The changes of light arising from the ellipticity and reflection are independent of, and supplementary to, the light changes which arise if our binary system happens to be an eclipsing variable; and the magnitude of all increases with increasing proximity of the components. How to separate them from each other?

This question will be answered more fully in Chapter VI of this book. Before we come, however, to formulate it in more specific terms, we must first investigate the nature of the theoretical light curves of close eclipsing systems – between minima as well as within eclipses – which will serve as a basis for all subsequent developments. The aim of the present chapter will be to carry out such an investigation as a task preparatory for all subsequent work.

II-1. Light Changes of Distorted Stars Outside Eclipses

In order to investigate the light changes produced by axial rotation of distorted components of close binary systems in the course of their orbital revolution, let us adopt a rectangular frame of reference whose X-axis coincides with the line joining the centres of both components, and the Z-axis to be perpendicular to the orbital plane; its origin will be identified with the centre of gravity of the star undergoing eclipse (which, as before, will be referred to as the 'primary component'). Let, furthermore, the position of any point in this system be specified by the polar coordinates r, θ, ϕ; where ϕ denotes the angle, in the XY-plane, between a projected radius-vector and the X-axis; and θ is the polar distance.

The light $\mathfrak{L}$ of a star as seen by the observer as a great distance can then be generally expressed as

$$\mathfrak{L} = \int_S J \cos \gamma \, d\sigma, \tag{1.1}$$

where J denotes (as in our previous papers of this series) the intensity of light at any point of the visible surface; γ, the angle of foreshortening; and the surface element $d\sigma$ is given by the equation

$$\cos \beta \, d\sigma = r^2 \sin \theta \, d\theta \, d\phi, \tag{1.2}$$

where β stands for the angle between a radius-vector and surface normal. The range S of integration is to be extended over the whole visible hemisphere of the star between minima, and the visible portion of the apparent disc within eclipses.

Let, in what follows,

$$\begin{array}{lll} \lambda, & \mu, & \nu; \\ l, & m, & n; \\ l_0, & m_0, & n_0; \end{array}$$

denote the direction cosines of a radius-vector, surface normal, and the line of sight, respectively. In conformity with our definition of the system of XYZ-coordinates,

$$\lambda = \cos \phi \sin \theta, \quad \mu = \sin \phi \sin \theta, \quad \nu = \cos \theta; \tag{1.3}$$

while if ψ denotes the true anomaly (i.e., phase angle) of the secondary (eclipsing) component in the plane of its orbit measured from the moment of conjunction at which the eclipse takes place; and i, the inclination of the orbital plane to the celestial sphere,

$$l_0 = \cos\psi \sin i, \quad m_0 = \sin\psi \sin i, \quad n_0 = \cos i. \tag{1.4}$$

Let, moreover, the free surface of our star be regarded as one over which the total potential Ψ of all forces acting upon it remains constant. If so, the direction cosines l, m, n of a line normal to it at any point specified by the coordinates x, y, z will be given by the equations

$$\{l, m, n\} = -\frac{1}{g}\left\{\frac{\partial\Psi}{\partial x}, \frac{\partial\Psi}{\partial y}, \frac{\partial\Psi}{\partial z}\right\}, \tag{1.5}$$

where

$$g = \left\{\left(\frac{\partial\Psi}{\partial x}\right)^2 + \left(\frac{\partial\Psi}{\partial y}\right)^2 + \left(\frac{\partial\Psi}{\partial z}\right)^2\right\}^{1/2} \tag{1.6}$$

stands for the magnitude of the local gravity vector.

If the radius-vector r connecting any point of an equipotential surface

$$\Psi(r, \theta, \phi) = \text{constant} \tag{1.7}$$

with the origin of our coordinates is expressible as

$$r = a_1\left\{1 + \sum_{i,j} k_{i,j} T_j^i(\theta, \phi)\right\}, \tag{1.8}$$

where a_1 denotes the mean radius of such an equipotential; $T_j^i(0, \phi)$ the surface harmonics describing the symmetry of a distortion whose magnitude is specified by the coefficients $k_{i,j}$, it can be shown (cf. Kopal, 1959; Section IV-2) that

$$l = \lambda - \sum_{i,j} k_{i,j}\left\{\frac{\partial\lambda}{\partial\theta}\frac{\partial T_j^i}{\partial\theta} + \frac{1}{\sin^2\theta}\frac{\partial\lambda}{\partial\phi}\frac{\partial T_j^i}{\partial\phi}\right\}, \tag{1.9}$$

$$m = \mu - \sum_{i,j} k_{i,j}\left\{\frac{\partial\mu}{\partial\theta}\frac{\partial T_j^i}{\partial\theta} + \frac{1}{\sin^2\theta}\frac{\partial\mu}{\partial\phi}\frac{\partial T_j^i}{\partial\phi}\right\}, \tag{1.10}$$

$$n = \nu - \sum_{i,j} k_{i,j}\left\{\frac{\partial\nu}{\partial\theta}\frac{\partial T_j^i}{\partial\theta}\right\}, \tag{1.11}$$

in which we regard the $k_{i,j}$'s small enough for their squares or cross-products to be negligible.* Within this scheme of such an approximation, it is easy to show that

$$\cos\beta = \lambda l + \mu m + \nu n = 1 \tag{1.12}$$

and

$$\cos\gamma = l l_0 + m m_0 + n n_0, \tag{1.13}$$

where the direction cosines l_0, m_0, n_0 and l, m, n are given by Equations (1.4) and (1.9)–(1.11).

The surface brightness J can now be expressed – in conformity with Equation (1.2) of Chapter I – as

* For a removal of this restriction by retention of terms which are of second order in k_{ij} cf. Kopal and Kitamura (1968).

$$J = H(1 - u_1 - u_2 - \cdots - u_n + u_1 \cos \gamma + u_2 \cos^2 \gamma + \cdots + u_n \cos^n \gamma), \tag{1.14}$$

where $u_1, u_2, \ldots, u_n$ stand for the respective coefficients of limb-darkening; and H denotes the intensity of radiation emerging normally to the surface which, for distorted stars in radiative equilibrium should (cf. Kopal, 1978b; Section IV-4; pp. 191–197) vary in accordance with the equation

$$\frac{H - H_0}{H_0} = \frac{g - g_0}{g_0}, \tag{1.15}$$

where g and g_0 denote the local and mean surface gravity and H_0 the emergent intensity at $g = g_0$.

If, moreover, our star radiates like a black body, the total emission of which is proportional to the fourth power of the temperature T, it should follow from Equation (1.15) that

$$\frac{H}{H_0} = \frac{g}{g_0} = \left(\frac{T}{T_0}\right)^4, \tag{1.16}$$

where T denotes the local effective temperature and T_0, the mean effective temperature averaged over the entire surface of the distorted star. The surface brightness observed in the light of any particular wavelength λ then should (in accordance with Planck's law) vary as

$$\left(\frac{H_\lambda}{H_0}\right) = \frac{\exp(c_2/\lambda T_0) - 1}{\exp(c_2/\lambda T) - 1}, \tag{1.17}$$

where $c_2 = 1.438$ cm deg. If we substitute now in the preceding equation for T from Equation (1.16) and expand (1.17) in a Taylor series in ascending powers of $(g - g_0)/g_0$ in the neighbourhood of $T = T_0$, then to the first order in small quantities Equation (1.17) reduces to

$$\frac{H_\lambda}{H_0} = 1 - \tau\left(1 - \frac{g}{g_0}\right) + \cdots, \tag{1.18}$$

where the (linear) 'coefficient of gravity-darkening' is given (cf. Kopal, 1941) by

$$\tau = \left\{\frac{d \log H_\lambda}{d \log T^4}\right\}_{T_0} = \frac{c_2}{4\lambda T_0[1 - \exp(-c_2/\lambda T_0)]}. \tag{1.19}$$

For a generalization of this result to quadratic gravity-darkening, or to a dependence of g on T other than that given by Equation (2.16), cf. Kopal and Kitamura (1968).

On the other hand, from a theory of the distortion of stars of arbitrary structure and bounded by the equipotential $\Psi(a_1, \theta, \phi) = \text{constant}$, it is possible to show (cf., e.g., Kopal, 1978b; Chapter II-6) that

$$\frac{g - g_0}{g_0} = -\sum_{i,j} \{1 + \eta_j(a_i)\} k_{i,j} T_j^i(\theta, \phi), \tag{1.20}$$

where $\eta(a_1)$ represents the surface value of a function defined by well-known differential equation

$$a\frac{\mathrm{d}\eta_j}{\mathrm{d}a} + 6D(\eta_j + 1) + \eta_j(n_j - 1) = j(j+1), \tag{1.21}$$

subject to the initial condition $\eta_j(0) = j - 2$, where D stands for the ratio of the local density $\rho(a)$ to the mean density $\bar{\rho}(a)$ interior to an equipotential of mean radius a. For configurations of pronounced central condensation (when $D \ll 1$ throughout most part of the interior) it can be shown (cf. again Kopal, 1978b; Chapter II-5) that $\eta_j(a_1)$ tends asymptotically to the constant value of $j+1$; and in what follows we shall set $\eta_j(a_1) = j + 1$ which, for actual stars, should approximate the reality within errors generally not exceeding 1–2%.

In consequence, if we collect all results represented by Equations (1.8), (1.13), (1.14) and (1.18), we find that the integrand on the right-hand side of Equation (1.1) will consist of a series of terms of the form

$$r^2 H \cos^h \gamma = a_1^2 H_0 N^h \left\{1 - \sum_{i,j} [(j+2)\tau - 2] k_{i,j} T_j^i - h\left[\frac{l_0}{N} - \lambda\right] \sum_{i,j} k_{i,j} \frac{\partial T_j^i}{\partial \lambda} - h\left[\frac{m_0}{N} - \mu\right] \sum_{i,j} k_{i,j} \frac{\partial T_j^i}{\partial \mu} - h\left[\frac{n_0}{N} - \nu\right] \sum_{i,j} k_{i,j} \frac{\partial T_j^i}{\partial \nu}\right\}, \tag{1.22}$$

where $h = 1, 2, 3, \ldots,$ such that $h - 1 = n$ is the degree of the law of limb-darkening represented by Equation (1.14); and where we have abbreviated

$$N = \lambda l_0 + \mu m_0 + \nu n_0. \tag{1.23}$$

Moreover, if our surface harmonics T_j^i (as yet unspecified) could be regarded as homogeneous functions of λ, μ, ν of j-th degree, they would be bound (by Euler's theorem on homogeneous functions) to satisfy the relation

$$\lambda\frac{\partial T_j^i}{\partial \lambda} + \mu\frac{\partial T_j^i}{\partial \mu} + \nu\frac{\partial T_j^i}{\partial \nu} = jT_j^i, \tag{1.24}$$

by virtue of which Equation (1.22) can be further simplified to

$$r^2 H \cos^h \gamma = a_1^2 H_0 N^h \left\{1 - \sum_{i,j} \Omega_j^{(h)} k_{i,j} T_j^i - \frac{h}{N} \sum_{i,j} k_{i,j}\left[l_0 \frac{\partial T_j^i}{\partial \lambda} + m_0 \frac{\partial T_j^i}{\partial \mu} + n_0 \frac{\partial T_j^i}{\partial \nu}\right]\right\}, \tag{1.25}$$

where we have abbreviated

$$\Omega_j^{(h)} = (j+2)\tau - hj - 2. \tag{1.26}$$

If the distortion were ignored, and our star considered as spherical – with a distribution of brightness J on its apparent disc characterized by Equation (1.14) of n-th degree in $\cos\gamma$ – its light L_1 would be constant and given (cf. Equation (2.3) of Chapter I) by

$$L_1 = \pi \int_0^{a_1} J \sin^{-1}(r/a_1)\, \mathrm{d}r^2 = \pi a_1^2 H_0 \left\{1 - \sum_{l=0}^{n} \frac{l u_l}{l+2}\right\}. \tag{1.27}$$

In the presence of distortion, let us write

$$\mathfrak{L}_1 = L_1(1 + \delta\mathfrak{L}_1) \tag{1.28}$$

and, moreover, decompose $\delta\mathfrak{L}_1$ into a series of partial terms $\delta\mathfrak{L}^{(h)}$ denoting the individual contributions, to the resultant light variations $\delta\mathfrak{L}_1$, of photometric effects arising from the h-th term ($h = 1, 2, 3, \ldots, n+1$) of the series on the right-hand side of Equation (1.14). If so, it follows that

$$\delta\mathfrak{L}_1 = \sum_{h=1}^{n+1} C^{(h)} \delta\mathfrak{L}_1^{(h)}, \tag{1.29}$$

where (cf. Equations (2.8)–(2.9) of Chapter I)

$$\begin{aligned} C^{(h)} &= \frac{1 - \sum_{l=0}^{n} u_l}{1 - \sum_{l=0}^{n} \frac{l u_l}{l+2}} \quad \text{for } h = 1, \\ &= \frac{u_{h-1}}{1 - \sum_{l=0}^{n} \frac{l u_l}{l+2}} \quad \text{for } h > 1; \end{aligned} \tag{1.30}$$

and

$$\begin{aligned} \delta\mathfrak{L}_1^{(h)} = -\sum_{i,j} k_{i,j} \int_0^{\pi} \int_{\epsilon-\pi/2}^{\epsilon+\pi/2} N^h \left\{ \Omega_j^{(h)} T_j^i(\theta, \phi) + \frac{h}{N} \right. \\ \times \left. \left[l_0 \frac{\partial T_j^i}{\partial \lambda} + m_0 \frac{\partial T_j^i}{\partial \mu} + n_0 \frac{\partial T_j^i}{\partial \nu} \right] \right\} \sin\theta \, d\theta \, d\phi, \end{aligned} \tag{1.31}$$

where ϵ denotes the angle between the radius-vector R joining the centres of the two components and the line of sight, of direction cosines l_0, m_0, n_0 as given by Equation (1.4). Since (in conformity with the definition of our system of coordinates) the radius vector R coincides constantly with the X-axis of direction cosines $1, 0, 0$, it follows at once that

$$\cos\epsilon = l_0 = \cos\psi \sin i. \tag{1.32}$$

In order to be able to perform an integration of all terms on the right-hand side of Equation (1.31) between these limits, it is necessary to specify the form of the surface harmonics T_j^i in terms of their angular variables; and this depends on the nature of the forces producing the distortion. In the case of the *tidal* distortion symmetrical with respect to the X-axis,* these reduce to the zonal harmonics $P_j(\lambda)$; and the constants

$$k_{0,j} = \frac{m_2}{m_1}\left(\frac{a_1}{R}\right)^{j+1} \equiv w_1^{(j)}, \quad j = 2, 3, 4, \tag{1.33}$$

to the first order in small quantities for the primary component (with an

* The 'tidal lag' caused by dissipative phenomena (viscosity) operative in the system is (cf. Kopal, 1978b; Chapter III) usually much too small to be of consequence in this connection.

appropriate interchange of indices for its companion), where m_2/m_1 denotes the mass-ratio of the two stars; and R, their separation. In the case of *rotational* distortion – produced by a rotation of our star with a (constant) angular velocity about an axis perpendicular to the orbital plane, the sole first-order zonal harmonic $T_j^0(\lambda, \mu, \nu)$ of rotational origin turns out to be identical with $P_2(\nu)$; with a coefficient

$$k_{0,2} = -\frac{\omega_1^2 a_1^3}{3Gm_1} \equiv -\tfrac{1}{3} v_1^{(2)}, \tag{1.34}$$

where G denotes the constant of gravitation. If, moreover, the angular velocity ω_1 of axial rotation happens to coincide with the Keplerian angular velocity

$$\omega_\kappa^2 = G\frac{m_1 + m_2}{R^3} \tag{1.35}$$

of orbital revolution, Equation (1.34) may be rewritten as

$$k_{0,2} = -\tfrac{1}{3}\left(1 + \frac{m_2}{m_1}\right)\left(\frac{a_1}{R}\right)^3. \tag{1.36}$$

It may be noted that the coefficients $v_1^{(2)}$ and $w_1^{(j)}$ as introduced by the foregoing Equations (1.33) and (1.34) possess a simple geometrical meaning: namely $v^{(2)}$ defines the polar flattening of the respective configuration; while the $w^{(j)}$'s are proportional to the contribution to its equatorial ellipticity produced by the j-th partial tide.

For surface harmonics $T_j^i(\lambda, \mu, \nu)$ of the particular form $P_j(\lambda)$ or $P_2(\nu)$ it is more convenient to use Equation (1.22) in place of (1.25) to express the integrand $r^2 H \cos^h \gamma$ on the right-hand side of Equation (1.1). If, moreover, advantage is taken of an algebraic identity

$$\lambda P_j'(\lambda) = jP_j(\lambda) + P_{j-1}'(\lambda) \tag{1.37}$$

valid for zonal harmonics of arbitrary argument λ (or ν) – with primes denoting derivatives with respect to the argument in question – Equation (1.31) now assumes a more specific form

$$\begin{aligned} \delta\mathfrak{L}_1^{(h)} = {} & \tfrac{1}{3} v_1^{(2)} \int_0^{\pi} \int_{\epsilon-\pi/2}^{\epsilon+\pi/2} \{\Omega_2^{(h)} P_2(\nu) + h(n_0/N)P_2'(\nu) - hP_1'(\nu)\} N^h \sin\theta \, d\theta \, d\phi \\ & - \sum_{j=2}^{4} w_1^{(j)} \int_0^{\pi} \int_{\epsilon-\pi/2}^{\epsilon+\pi/2} \{\Omega_j^{(h)} P_j(\lambda) + h(l_0/N)P_j'(\lambda) - hP_{j-1}'(\lambda)\} \\ & \times N^h \sin\theta \, d\theta \, d\phi, \end{aligned} \tag{1.38}$$

In integrating the r.h.s. of the above equation we may note that *each disturbing term varying as $T_j^i(\theta, \phi)$ in the expansion* (1.8) *for the radius-vector will give rise to a photometric effect varying as the same harmonic $T_j^i(l_0, n_0)$ in the light curve*, factored by coefficients which depend on the extent of limb- and gravity-darkening (i.e., the values of h and $\Omega_j^{(h)}$). The outcome discloses that

$$\delta \mathfrak{L}_1^{(h)} = \frac{h(4+\beta_2)}{3(h+3)} v_1^{(2)} P_2(n_0) - \frac{h(4+\beta_2)}{h+3} w_1^{(2)} P_2(l_0) - \frac{(h-1)(h+1)}{(h+2)(h+4)} \times (10+\beta_3) w_2^{(3)} P_3(l_0) - \frac{h(h-2)}{(h+3)(h+5)} (18+\beta_4) w_1^{(4)} P_4(l_0) - \cdots, \tag{1.39}$$

where we have abbreviated

$$\beta_j = \{1+\eta_j(a_1)\}\tau \doteqdot (j+2)\tau. \tag{1.40}$$

Under these circumstances, an insertion of Equations (1.30) and (1.39) in (1.29) leads to the expression

$$\delta \mathfrak{L}_1 = X_2^{(n)}\{1+\tfrac{1}{4}\beta_2\}\{\tfrac{1}{3}v_1^{(2)}P_2(n_0) - w_1^{(2)}P_2(l_0)\} - X_3^{(n)}\{1+\tfrac{1}{10}\beta_3\} \times w_1^{(3)} P_3(l_0) - X_4^{(n)}\{1+\tfrac{1}{18}\beta_4\} w_1^{(4)} P_4(l_0) - \cdots, \tag{1.41}$$

where

$$X_j^{(n)} = 2(j+2)(j-1) \sum_{h=1}^{n+1} \frac{(h-j+2)(h-j+4)\ldots(h+j)}{(h+1)(h+2)\ldots(h+j+1)} C^{(h)}, \tag{1.42}$$

and the $C^{(h)}$'s continue to be given by Equations (1.30); with n denoting the degree of the adopted law of limb-darkening of the form (1.2) of Chapter I in powers of $\cos\gamma$. In particular, for uniformly bright discs ($n=0$),

$$X_2^{(1)} = 1, \quad X_3^{(1)} = 0, \quad X_4^{(1)} = -\tfrac{3}{4}; \tag{1.43}$$

for a linear law of limb-darkening ($n=1$)

$$X_2^{(2)} = \frac{15+u_1}{5(3-u_1)}, \quad X_3^{(2)} = \frac{5u_1}{2(3-u_1)}, \quad X_4^{(2)} = -\frac{9(1-u_1)}{4(3-u_1)}; \tag{1.44}$$

for quadratic limb-darkening ($n=2$),

$$X_2^{(3)} = \frac{2(15+u_1)}{5(6-2u_1-3u_2)}, \quad X_3^{(3)} = \frac{35u_1+48u_2}{7(6-2u_1-3u_2)}, \quad X_4^{(3)} = \frac{9(4u_1+7u_2-4)}{8(6-2u_1-3u_2)}; \tag{1.45}$$

etc.

A. DISCUSSION OF THE RESULTS

A glance at the light changes represented by Equation (1.41) discloses that – inasmuch as the coefficient $P_2(n_0)$ factoring the effects of rotational distortion is independent of the phase – the *polar flattening due to axial rotation will not cause any variation of light between eclipses*; all terms which will do so are due to the *tides*. Since, however, all terms of tidal origin depend on the phase only through the direction cosine $l_0 = \cos\psi \sin i$, the light changes produced by them will be *symmetrical* with respect to the moment of conjunctions (when $\psi = 0°$ or 180°).

The foregoing statements are, to be sure, strictly true only under certain dynamical restrictions. Thus if the orbit and equatorial plane of one (or both) components are not co-planar, the angle i of inclination of the orbital plane to the celestial sphere (and, therefore, the direction cosine n_0) will be subject to short-term as well as long-term periodic perturbations, due to precession and nutation (cf. Kopal, 1978b; Section V-2) which will affect also the observed variations of light. Similarly, if the two components revolve around the common centre of gravity in eccentric orbits so oriented that the longitude ω of the apsidal line is not 0° or 180°, the phase angle ψ in the direction cosine l_0 becomes identical with the *true* anomaly of the eclipsing component in its eccentric relative orbit; and, as a result, a symmetry of the light changes with respect to the conjunctions is no longer preserved (cf. Section IV-4 of this book). In general, however, both these complications are likely to be small; and in most cases, negligible; so that symmetry of the light changes should generally be fulfilled within the limits of observational errors – unless, of course, additional complications of physical nature (not incorporated in our present model) happen to be operative.

Let us consider next the nature of the effects, upon light changes, due to limb- and gravity-darkening of rotating distorted stars. In the case of a linear approximation ($n = 1$) to the actual law of limb-darkening, Equations (1.44) make it evident that no light variation arises from third-harmonic tidal distortion unless there is some limb-darkening; nor any from the fourth unless the darkening is incomplete. If gravity-darkening were absent (i.e., $\tau = 0$) Equation (1.39) would, for $u_1 = 0$, reduce to

$$\delta\mathfrak{L}^{(1)} = \tfrac{1}{3}v_1^{(2)}P_2(n_0) - w_1^{(2)}P_2(l_0) + \tfrac{3}{4}w_1^{(4)}P_4(l_0) + \cdots; \tag{1.46}$$

while for $u_1 = 1$

$$\delta\mathfrak{L}^{(2)} = \tfrac{8}{15}v_1^{(2)}P_2(n_0) - \tfrac{8}{5}w_1^{(2)}P_2(l_0) - \tfrac{5}{4}w_1^{(3)}P_3(l_0) + \cdots. \tag{1.47}$$

Equation (1.46) would express the changes of light due to the distorted geometry alone. A comparison of Equation (1.47) with (1.46) discloses, moreover, that limb-darkening tends to exaggerate the variation arising from second-harmonic distortion; gives rise to a significant term varying as third harmonic; and reduces that of the fourth harmonic. On the other hand, a full gravity-darkening ($\tau = 1$) of a centrally-condensed star would, for $u_1 = 0$, lead to

$$\delta\mathfrak{L}^{(1)} = \tfrac{2}{3}v_1^{(2)}P_2(n_0) - 2w_1^{(2)}P_2(l_0) + w_1^{(4)}P_4(l_0) + \cdots; \tag{1.48}$$

while, for $u_1 = 1$,

$$\delta\mathfrak{L}^{(2)} = \tfrac{16}{15}v_1^{(2)}P_2(n_0) - \tfrac{16}{5}w_1^{(2)}P_2(l_0) - \tfrac{15}{8}w_1^{(3)}P_3(l_0) + \cdots. \tag{1.49}$$

The reader may note that, in either case, the effects of full gravity-darkening upon changes of light invoked by the second, third, and fourth tidal-harmonic distortion of a centrally-condensed star is to multiply variations due to distorted geometry alone by the factors of 2, $\frac{3}{2}$ and $\frac{4}{3}$, respectively.

The effects of limb-darkening of higher orders on the variation of light between minima are altogether small: in total light, all terms on the r.h.s. of Equation (1.41) of degree higher than the first (corresponding to $h > 2$) are found to magnify the photometric effects of the second and third tidal harmonics by 1% and 4%, respectively; and to diminish the coefficients of the fourth harmonic by 2% (cf. Kopal, 1949). Therefore, an interpretation of even the most precise light curves between minima should not call for retention of more than the quadratic terms in $\cos \gamma$ on the r.h.s. of the adopted law of limb-darkening; while a failure to consider the quadratic term would vitiate the theoretical coefficients of harmonic light variation by more than a few per cent. This is, to be sure, true of the light changes exhibited between minima; while within eclipses (Section II-2) a very different situation will be encountered.

Throughout all the foregoing developments we have also confined our attention to the light changes due to distortion of the primary component of mass m_1 and radius a_1, which we placed at the origin of our coordinate system. The total light of a close binary between eclipses will, however, consist of a sum of the luminosities of both components; and the same is true of their light changes. Those of the secondary component are obviously governed by an equation of the same type as (1.41), in which indices referring to the primary and secondary components have been interchanged, and the phase angle ψ in l_0 shifted by 180°. This difference in phase will change the algebraic sign of the only odd harmonic (corresponding to $j = 3$) of first order occurring on the right-hand side of Equation (1.41), but not those of even harmonics ($j = 2$ and 4). As a result, no odd harmonics can appear in the combined light of a binary system consisting of identical components (for the effects of odd harmonics of one star would be neutralized by those due to its mate). On the other hand, the terms factored by harmonics of even orders will always tend to reinforce each other.

In forming the sum

$$\mathfrak{L} = \mathfrak{L}_1 + \mathfrak{L}_2 \tag{1.50}$$

and factoring out equal powers of $\cos \psi$ arising from both components we find that their combined light outside eclipses should vary as

$$\begin{aligned}
\mathfrak{L} = {} & 1 + \tfrac{1}{3}\sum_{i=1}^{2} L_i(X_2^{(k)})_i(1+\tau_i)v_i^{(2)}P_2(n_0) \\
& - \sum_{i=1}^{2}\sum_{j=2}^{4}(-1)^{(i+1)j}L_i(X_j^{(k)})_i\left\{1+\frac{\tau_i}{j-1}\right\}w_i^{(j)}P_j(l_0) \\
= {} & 1 + \tfrac{1}{3}\sum_{i=1}^{2} L_i(X_2^{(k)})_i(1+\tau_i)v_i^{(2)}P_2(\cos i) + \tfrac{1}{2}\sum_{i=1}^{2} L_i(X_2^{(k)})_i(1+\tau_i)w_i^{(2)} \\
& - \tfrac{3}{8}\sum_{i=1}^{2} L_i(X_4^{(k)})_i(1+\tfrac{1}{3}\tau_i)w_i^{(4)} \\
& + \tfrac{3}{2}\cos\psi\sin i\sum_{i=1}^{2}(-1)^{i+1}L_i(X_3^{(k)})_i(1+\tfrac{1}{2}\tau_i)w_i^{(e)} \\
& - \tfrac{3}{2}\cos^2\psi\sin^2 i\sum_{i=1}^{2} L_i\{(X_2^{(k)})_i(1+\tau_i)w_i^{(2)} - \tfrac{5}{2}(X_4^{(k)})(1+\tfrac{1}{3}\tau_i)w_i^{(4)}\} \\
& - \tfrac{5}{2}\cos^3\psi\sin^3 i\sum_{i=1}^{2}(-1)^{i+1}L_i(X_3^{(k)})_i(1+\tfrac{1}{2}\tau_i)w_i^{(3)}
\end{aligned}$$

$$-\tfrac{35}{8}\cos^4\psi\sin^4 i\sum_{i=1}^{2}L_i(X_4^{(k)})_i(1+\tfrac{1}{3}\tau_i)w_i^{(4)}+\cdots, \tag{1.51}$$

where, in accordance with Equations (1.33)–(1.36),

$$v_i^{(2)}=\frac{\omega_i^2 a_i^2}{Gm_i}=\left(1+\frac{m_{3-i}}{m_i}\right)\left(\frac{a_i}{R}\right)^3 \tag{1.52}$$

in case of synchronous rotation, and

$$w_i^{(j)}=\frac{m_{3-i}}{m_i}\left(\frac{a_i}{R}\right)^{j+1}. \tag{1.53}$$

The ratios a_i/R in these latter equations denote the fractional radii of the two components (i.e., radii of spheres having the same volume as the respective distorted configuration, and expressed in terms of their separation R taken as the unit of length). Lastly, the sum of the luminosities L_1 and L_2 of both components in their undistorted state has been adopted as our unit of light – so that $L_2 = 1 - L_1$.

Equation (1.51) constitutes the final outcome of this section, and will become the basis of our subsequent analysis in Chapter VI.

II-2. Light Changes of Distorted Systems Within Eclipses

In the preceding section we gave an account of the light changes of close binary systems which arise from the distortion of their components caused by rotation and tides. Provided only that the proximity of both components is sufficient to render distortion appreciable, such light changes are bound to arise irrespective of whether or not our binary happens also to be an eclipsing variable. Should, however, the inclination of the orbital plane to the line of sight be such as to cause the components to eclipse each other at the time of conjunctions, the changes of light them exhibited must likewise be affected by distortion of both components of such a couple: a distortion of the secondary (eclipsing) component must cause a corresponding deformation of its shadow cylinder cast in the direction of the line of sight; while a distortion of the component undergoing eclipse will affect not only a proportion of its apparent discs eclipsed by its mate, but also the distribution of brightness over the eclipsed part. The aim of the present section will be to describe such effects to the same order of accuracy to which we dealt with the light changes between minima in the preceding section.

In order to do so, let us change over from the 'astrocentric' system to coordinates XYZ, rotating in space with the radius-vector as introduced in the preceding section, to another rectangular system $X'Y'Z'$ rotating with respect to the observer around the line of sight; with origin at the centre of the primary star, but defined so that its Z'-axis coincides constantly with the line of sight, and its X'-axis is the direction of the projected centre of the secondary (eclipsing) component. Accordingly, the direction cosines of the Z'-axis in the unprimed system of coordinates will clearly be (l_0, m_0, n_0) as defined by Equation (1.4); while the direction cosines of the X-axis in the primed system will be $(l_0, 0, l_2)$, so that

$$l_2^2 = 1 - l_0^2 = \sin^2 \psi \sin^2 i + \cos^2 i = \delta^2, \tag{2.1}$$

where δ denotes the instantaneous projected distance of the centres of the two components in the $X'Y'$-plane, which is tangent to the celestial sphere at the origin of coordinates (cf. Figure 2-1; this figure is the same as Figure 1-2 on page 24 but is repeated below for the reader's convenience).

A transformation between the primed and unprimed system of orthogonal coordinates will evidently be governed by the scheme

	Z'	Y'	X'
X	l_0	0	l_2
Y	m_0	m_1	m_2
Z	n_0	n_1	n_2

where the values of the remaining direction cosines $m_{1,2}$ and $n_{1,2}$ can be determined in terms of l_0, m_0, n_0 from the orthogonality of both systems as

$$m_1 = -\frac{n_0}{\sqrt{1-l_0^2}}, \quad n_1 = \frac{m_0}{\sqrt{1-l_0^2}}; \tag{2.2}$$

and

$$m_2 = -\frac{l_0 m_0}{\sqrt{1-l_0^2}}, \quad n_2 = -\frac{l_0 n_0}{\sqrt{1-l_0^2}}; \tag{2.3}$$

respectively.

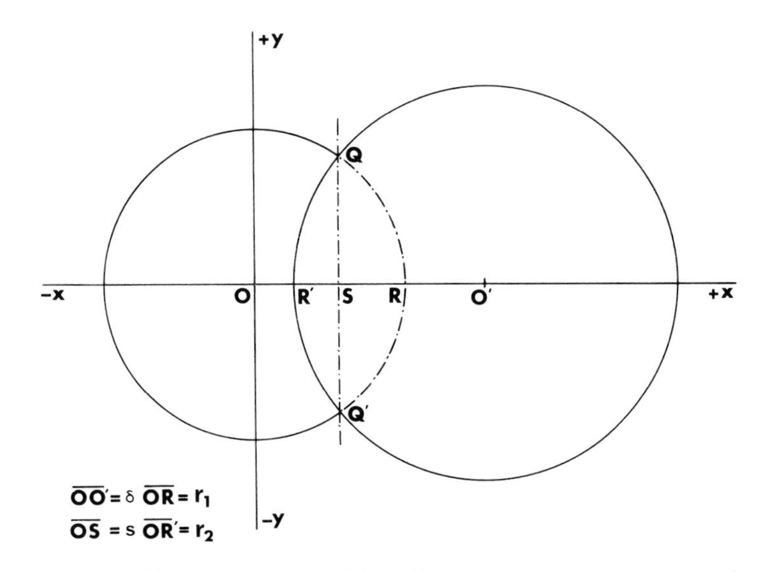

Fig. 2-1. Geometry of the eclipses of spherical stars.

A. THEORETICAL LIGHT CURVES

Let – as in the preceding parts of this book – $r_{1,2}$ denote the fractional radii of the respective components expressed in terms of the radius-vector R of the relative orbit of the two stars. The equation of the shadow cylinder cast by the secondary component in the direction of the line of sight in the primed coordinates then becomes

$$(\delta - X')^2 + Y'^2 = r_2^2(1 + \Delta\bar{r}_2)^2, \tag{2.4}$$

where $\Delta\bar{r}_2$ represents the deformation of the secondary's shadow cylinder in the $Z' = 0$ plane. If, furthermore, Δr_1 stands for the corresponding distortion of the primary component undergoing eclipse, the surface of the latter admits of the parametric representation of the form

$$X' = r_1(1 + \Delta r_1)L, \quad Y' = r_1(1 + \Delta r_1)M, \quad Z' = r_1(1 + \Delta r_1)N, \tag{2.5}$$

where L, M, N denote the direction cosines of an arbitrary radius in the primed system. These direction cosines satisfy the relation

$$L^2 + M^2 + N^2 = 1 \tag{2.6}$$

and can be regarded as rectangular coordinates, in the primed system, on the unit sphere.

If, however, we retain the radius-vector R of the relative orbit of the two stars as our unit of length, then

$$L = \frac{x}{r_1}, \quad M = \frac{y}{r_1}, \quad N = \frac{z}{r_1}. \tag{2.7}$$

Let us hereafter adopt x, y, z as our new coordinates in terms of which to express the direction cosines λ, μ, ν as defined by Equations (1.3). The two sets are related by equations of the form

$$\left.\begin{aligned} r_1\lambda &= l_0 z \qquad\quad + l_2 x, \\ r_1\mu &= m_0 z + m_1 y + m_2 x, \\ r_1\nu &= n_0 z + n_1 y + n_2 x. \end{aligned}\right\} \tag{2.8}$$

By virtue of these relations, the zonal harmonics $P_j(\lambda)$ or $P_2(\nu)$ encountered on the right-hand side of Equation (1.38) may now be rewritten as polynomials of the j-th degree in integral powers of x, y and z, with coefficients depending on the amount of distortion and the position of the components in their relative orbit. Since, moreover, the surface element

$$r_1 \sin\theta \, d\theta \, d\phi = \frac{dx \, dy}{z}, \tag{2.9}$$

the whole integrand in Equation (1.38) can be rewritten as an algebraic function of x, y and z; the general terms arising from the rotational and tidal distortion being of the form $x^m y z^n$ and $x^m z^n$, respectively, where $m \geqslant 0$ and $n \geqslant -1$.

In order to perform the integration on the right-hand side of Equation (1.38) over the fraction of the primary's disc visible at a particular phase of the eclipse, we must first specify the appropriate limits. Turning to the parametric Equations (2.5) we note that N, the cosine of the angle of foreshortening, would vanish at the limb of the primary star if the latter were spherical; and, for distorted bodies, becomes a small quantity whose squares we agreed to ignore. If so, however, then to this order of accuracy the intersection of the primary's surface with the xy-plane reduces to the circle

$$x^2 + y^2 = r_1^2, \tag{2.10}$$

a part of which represents the (dashed) arc QRQ' on Figure 2-1 limiting the eclipsed fraction of the primary's apparent disc.

In order to ascertain the form of the other arc $QR'Q'$ constituting the opposite limb of the eclipsed area, let us solve Equation (2.4) of the shadow cylinder with Equations (2.5) of the surface of the primary star. The equation of the arc $QR'Q'$ in the xy-plane is – to the first order in small quantities – then given by

$$(\delta - x)^2 + y^2 = r_2^2(1 + 2\Delta\bar{r}_2) - 2\{r_2^2 - \delta(\delta - x)\}\Delta r_1 + \cdots, \tag{2.11}$$

and the deviation Δy of the arc $QR'Q'$ from the circle

$$(\delta - x)^2 + y^2 = r_2^2 \tag{2.12}$$

assumes the form

$$\Delta y = \frac{r_2^2 \Delta\bar{r}_2}{\sqrt{r_2^2 - (\delta - x)^2}} - \frac{r_2^2 - \delta(\delta - x)}{\sqrt{r_2^2 - (\delta - x)^2}} \Delta r_1 + \cdots. \tag{2.13}$$

The first term on the right-hand side of this latter equation arises from the first-order distortion of the eclipsing component; and the second from that of the eclipsed star.

The fraction of the primary's light lost at any moment during eclipse can be decomposed in two parts: the 'circular' one – obtained by extending the limits of integration over the area bounded by the circles (2.11) and (2.12) – and the 'boundary corrections' arising from the distortion of the arcs QRQ' and $QR'Q'$. If, moreover,

$$s = \frac{r_1^2 - r_2^2 + \delta^2}{2\delta} \tag{2.14}$$

denotes the x-coordinate of the common chord of the circles (2.10) and (2.12) at any phase of the eclipse (cf. again Figure 2-1) the 'circular' part of the resulting fractional loss of light can be expressed in terms of the 'associated α-functions' of the form

$$\pi r_1^{m+n+2} \alpha_n^m = \left\{ \int_s^{r_1} \int_{-\sqrt{r_1^2 - x^2}}^{+\sqrt{r_1^2 - x^2}} + \int_{\delta - r_2}^{\delta} \int_{-\sqrt{r_2^2 - (\delta - x)^2}}^{+\sqrt{r_2^2 - (\delta - x)^2}} \right\} x^m z^n \, dx \, dy \tag{2.15}$$

if the eclipse is partial, and

$$\pi r_1^{m+n+2} \alpha_n^m = \int_{\delta - r_2}^{\delta + r_2} \int_{-\sqrt{r_2^2 - (\delta - x)^2}}^{\sqrt{r_2^2 - (\delta - x)^2}} x^m z^n \, dx \, dy \tag{2.16}$$

if it is annular. If $m = 0$, the foregoing Equations (2.15) and (2.16) reduce to Equations (2.10) and (2.13) of Chapter I; and represent a generalization of the latter for $m > 0$, necessary for a formulation of the light changes of distorted stars within eclipses.

The loss of light due to the distortion of the arcs QRQ' and $QR'Q'$ which limit the eclipsed area can, moreover, be obtained by a single integration of $2\Delta y$ with respect to x, within limits which range from $\delta - r_2$ to s as long as the eclipse remains partial, and from $\delta - r_2$ to $\delta + r_2$ when it becomes annular.

In order to be able to perform these latter integrations we must express the quantities Δr_1 as well as $\Delta \bar{r}_2$ on the right-hand side of Equation (2.13) as functions of x. Consistent with Equation (2.8) and its sequel,

$$\Delta r_{1,2} = \sum_{j=2}^{4} w_{1,2}^{(j)} P_j(\lambda) - \tfrac{1}{3} v_{1,2}^{(2)} P_2(\nu), \tag{2.17}$$

where the coefficients $v_{1,2}^{(2)}$ and $w_{1,2}^{(j)}$ continue to be given by Equations (1.52) and (1.53).

The arguments λ and ν of the zonal harmonics on the right-hand side of Equation (2.17) are given by Equations (2.8) in which appropriate values for y and z were inserted. Along the arc $QR'Q'$ at which the primary's surface is intersected by the shadow cylinder cast by the secondary component, from Equations (2.6) and (2.7) it follows that

$$z = \sqrt{r_1^2 - x^2 - y^2} = \sqrt{2\delta(s - x)} \tag{2.18}$$

if we insert for y from Equation (2.12); so that

$$\Delta r_1 = \sum_{j=2}^{4} w_1^{(j)} P_j(\lambda_1) - \tfrac{1}{3} v_1^{(2)} P_2(\nu_1), \tag{2.19}$$

where

$$r_1 \lambda_1 = l_0 \sqrt{2\delta(s - x)} \qquad + l_2 x, \tag{2.20}$$

$$r_1 \nu_1 = n_0 \sqrt{2\delta(s - x)} + n_1 \sqrt{r_2^2 - (\delta - x)^2} + n_2 x; \tag{2.21}$$

while along the curve at which the shadow cylinder is tangent to the secondary component (and, therefore, $z = 0$),

$$\Delta \bar{r}_2 = \sum_{j=2}^{4} w_2^{(j)} P_j(\bar{\lambda}_2) - \tfrac{1}{3} v_2^{(2)} P_2(\bar{\nu}_2), \tag{2.22}$$

where

$$r_2 \bar{\lambda}_2 = \qquad + l_2(\delta - x), \tag{2.23}$$

$$r_2 \bar{\nu}_2 = n_1 \sqrt{r_2^2 - (\delta - x)^2} + n_2(\delta - x). \tag{2.24}$$

The 'boundary corrections' due to the deformation of the arc $QR'Q'$ which arises from the distortion of the eclipsed component should, therefore, be expressible in terms of the functions $\mathfrak{J}_{\beta,\gamma}^{m}$ as defined by the equation

$$\pi r_1^{\beta+\gamma+m+3} \mathfrak{J}_{\beta,\gamma}^{m} = (r_2^2 - \delta^2) I_{0,\beta,\gamma}^{m} + \delta I_{0,\beta,\gamma}^{m+1}, \tag{2.25}$$

where β, γ and m are integers such that

$$\beta \geqslant -1, \quad \gamma \geqslant -1, \quad m \geqslant 0; \tag{2.26}$$

and

$$I^m_{0,\beta,\gamma} = \int_{\delta - r_2}^{c_2} [r_2^2 - (\delta - x)^2]^{\beta/2} [2\delta(s - x)]^{\gamma/2} x^m \, dx, \tag{2.27}$$

where $c_2 = s$ or $\delta + r_2$ depending on whether the eclipse is partial or annular.

Moreover, the remaining boundary corrections arising from $\Delta\bar{r}_2$ are all expressible in terms of an additional family of integrals of the form

$$\pi r_2^{\beta+\gamma+m+1} I^m_{\beta,\gamma} = \int_{\delta - r_2}^{c_2} [r_2^2 - (\delta - x)^2]^{\beta/2} [2\delta(s - x)]^{\gamma/2} (\delta - x)^m \, dx. \tag{2.28}$$

Note, in this connection, that while both three-index functions $\mathfrak{I}^m_{\beta,\gamma}$ and $I^m_{\beta,\gamma}$ as defined by the foregoing Equations (2.25) and (2.28) are nondimensional quantities, the $I^m_{0,\beta,\gamma}$'s are not.

In order to formulate explicit expressions for the loss of light during eclipses of distorted components of close binary systems, let us write – in conformity with Equation (1.29) – that

$$\Delta\mathfrak{L} = \sum_{h=1}^{n+1} C^{(h)} \{\alpha^0_{h-1} + \Delta\mathfrak{L}^{(h)}\}, \tag{2.29}$$

where the coefficients $C^{(h)}$ due to limb-darkening continue to be given by Equations (1.30), and the α^m_n's are associated α-functions as defined by Equations (2.15) or (2.16). The first part of the right-hand side of Equation (2.29) – consisting of $C^{(h)}\alpha^0_{h-1}$ – represents the loss of light during eclipses of spherical stars arbitrarily darkened at the limb; while the effects $\Delta\mathfrak{L}^{(h)}$ of distortion can be represented by

$$\Delta\mathfrak{L}^{(h)} = f^{(h)}_* + f^{(h)}_1 + f^{(h)}_2, \tag{2.30}$$

where $f^{(h)}_*$ will constitute the contributions of distortion and gravity-darkening over the circular portion of the eclipsed disc and expressible in terms of the associated α-functions; while $f^{(h)}_{1,2}$ represent the photometric contributions of the 'boundary corrections' along $QR'Q'$ arising from the distortion of the primary and secondary component.

In order to establish the explicit form of $f^{(h)}_*$ for limb-darkening of any degree, we rewrite the integrand on the right-hand side of Equation (1.38) in powers of x, y, z by means of Equations (2.8)–(2.9) and remember that, in the rotational terms, odd powers of y vanish on account of symmetry, while even powers can be expressed in terms of those of x and z by means of the relation $y^2 = r_1^2 - x^2 - z^2$ following from Equations (2.8)–(2.9). A term-by-term integration over circular limits then yields

$$f^{(h)}_* = \tfrac{1}{3}\{\tfrac{1}{2}\Omega^{(h)}_2[3(n_0^2 - n_1^2)\alpha^0_{h+1} + 3(n_2^2 - n_1^2)\alpha^2_{h-1} + 6n_0n_2\alpha^1_h$$

$$
\begin{aligned}
&+2P_2(n_1)\alpha^0_{h-1}] + h[2P_2(n_0)\alpha^0_{h-1} + 3n_0n_2\alpha^1_{h-2}]\}v^{(2)}_1 \\
&-\{\tfrac{1}{2}\Omega^{(h)}_2[3l_0^2\alpha^0_{h+1} + 6l_0l_2\alpha^1_h + 3l_2^2\alpha^2_{h-1} - \alpha^0_{h-1}] \\
&+ h[2P_2(l_0)\alpha^0_{h-1} + 3l_0l_2\alpha^1_{h-2}]\}w^{(2)}_1 - \{\tfrac{1}{2}\Omega^{(h)}_3[5l_0^3\alpha^0_{h+2} \\
&+ 15l_0^2l_2\alpha^1_{h+1} + 15l_0l_2^2\alpha^2_h + 5l_2^3\alpha^3_{h-1} - 3l_0\alpha^0_h - 3l_2\alpha^1_{h-2}] \\
&+ h[(l_0\alpha^0_h + 2l_2\alpha^1_{h-1})P'_3(l_0) - \tfrac{3}{2}l_0(\alpha^0_h - 5l_2^2\alpha^2_{h-2} + \alpha^0_{h-1})]\}w^{(3)}_1 \\
&-\{\tfrac{1}{8}\Omega^h_4[35l_0^4\alpha^0_{h+3} + 140l_0^3l_2\alpha^1_{h+2} + 210l_0^2l_2^2\alpha^2_{h+1} + 140l_0l_2^3\alpha^3_h \\
&+ 35l_2^4\alpha^4_{h-1} - 30l_0^2\alpha^0_{h+1} - 60l_0l_2\alpha^1_h - 30l_2^2\alpha^2_{h-1} + 3\alpha^0_{h-1}] \\
&+ \tfrac{1}{2}h[2l_0P'_4(l_0)\alpha^0_{h+1} + 15l_2^2(7l_0^2-1)\alpha^2_{h-1} - 2P'_3(l_0)\alpha^0_{h-1} \\
&+ 15l_0l_2[7l_0^2-2)\alpha^1_h + 35l_0l_2^3\alpha^3_{h-2} - 15l_0l_2\alpha^1_{h-2}]\}w^{(4)}_1 + \cdots,
\end{aligned}
\tag{2.31}
$$

where $\Omega^{(h)}_j$ continues to be given by Equation (1.26).

A similar integration of $2\Delta y$ as given by Equation (2.13) shows that

$$
\begin{aligned}
f^{(h)}_1 = {} & \tfrac{1}{3}\{3n_0^2\mathfrak{J}^0_{-1,h+1} + 3n_1^2\mathfrak{J}^0_{1,h-1} + 3n_2^2\mathfrak{J}^2_{-h-1} \\
&+ 6n_0n_2\mathfrak{J}^1_{-1,h} - \mathfrak{J}^0_{-1,h-1}\}v^{(2)}_1 \\
&-\{3l_0^2\mathfrak{J}^0_{-1,h+1} + 6l_0l_2\mathfrak{J}^1_{-1,h} + 3l_2^2\mathfrak{J}^2_{-1,h-1} - \mathfrak{J}^0_{-1,h-1}\}w^{(2)}_1 \\
&-\{5l_0^3\mathfrak{J}^0_{-1,h+2} + 15l_0^2l_2\mathfrak{J}^1_{-1,h+1} + 15l_0l_2^2\mathfrak{J}^2_{-1,h} \\
&+ 5l_2^3\mathfrak{J}^3_{-1,h-1} - 3l_0\mathfrak{J}^0_{-1,h} - 3l_2\mathfrak{J}^1_{-1,h-1}\}w^{(3)}_1 \\
&-\tfrac{1}{4}\{35l_0^4\mathfrak{J}^0_{-1,h+3} + 140l_0^3l_2\mathfrak{J}^1_{-1,h+2} + 210l_0^2l_2^2\mathfrak{J}^2_{-1,h+1} \\
&+ 140l_0l_2^3\mathfrak{J}^3_{-1,h} + 35l_2^4\mathfrak{J}^4_{-1,h-1} - 30l_0^2\mathfrak{J}^0_{-1,h+1} \\
&- 60l_0l_2\mathfrak{J}^1_{-1,h} - 30l_2^2\mathfrak{J}^2_{-1,h-1} + 3\mathfrak{J}^0_{-1,h-1}\}w^{(4)}_1 + \cdots,
\end{aligned}
\tag{2.32}
$$

and

$$
\begin{aligned}
(r_1/r_2)^{h+1}f^{(h)}_2 = {} & -\tfrac{1}{3}\{3n_1^2I^0_{1,h-1} + 3n_2^2I^2_{-1,h-1} - I^0_{-1,h-1}\}v^{(2)}_2 \\
&+\{3l_2^2I^2_{-1,h-1} - I^0_{-1,h-1}\}w^{(2)}_2 \\
&+\{5l_2^3I^3_{-1,h-1} - 3l_2I^1_{-1,h-1}\}w^{(3)}_2 \\
&+\tfrac{1}{4}\{35l_2^4I^4_{-1,h-1} - 30l_2^2I^2_{-1,h-1} + 3I^0_{-1,h-1}\}w^{(4)}_2 + \cdots.
\end{aligned}
\tag{2.33}
$$

The foregoing equations (2.30)–(2.33) specify the changes of light within minima exhibited by close eclipsing systems and arising from the first-order rotational as well as tidal distortion. In order to describe these in concise form we found it expedient to introduce three new families of auxiliary functions – namely, the associated alpha-functions α^m_n as defined by Equations (2.15)–(2.16), and the 'boundary integrals' $\mathfrak{J}^m_{\beta,\gamma}$ and $I^m_{\beta,\gamma}$ defined by Equations (2.25) and (2.28), respectively. Not all these functions are, to be sure, fully independent of each other. In particular, the $\mathfrak{J}^m_{\beta,\gamma}$-functions can be expressed as simple linear combinations of the α^m_n's for $n > 0$ – a fact which will enable us to combine Equations (2.15)–(2.16) and (2.25) largely into one (cf. Section VI-2). Before, however, we can justify such a procedure, diverse properties of these functions – introduced by way of a mathematical shorthand – must be appropriately investigated – a task which will constitute the subject of the next chapter.

Bibliographical Notes

II-**1**. The material of this section follows largely an earlier presentation of this subject in Chapter IV of the author's previous book on *Close Binary Systems* (Z. Kopal, 1959) which contains extensive references to earlier literature.

Up to the early 1930's, the form of the components in close binary systems was regarded as arbitrary and unrelated to other physical elements (size, mass-ratio) of the systems. The first investigator who considered this shape to be governed by the prevalent field of force in systems in hydrostatic equilibrium was S. Takeda (1934), who also considered first the effects of gravity-darkening on the light changes of close binaries between eclipses – a work developed further by Z. Kopal (1942).

II-**2**. The pioneer in the investigation of the light-changes due to mutual eclipses of distorted stars of shape appropriate for the prevalent field of force was again S. Takeda (1937), to whom the elements of the theory presented in this section are due. This theory was generalized and completed to its present form by Z. Kopal (1942). In particular, the term 'associated α-functions' to denote functions represented by Equations (2.15)–(2.16) goes back to this latter paper (Takeda called them 'circular integrals'); though the 'boundary correction' integrals of the form given by Equations (2.25) and (2.27) or (2.28) occur already under that name in Takeda (1937). The fundamental equations (2.29)–(2.33), on which much of our subsequent work will be based, made their first appearance as they stand in Chapter IV of the author's *Close Binary Systems* (1959); though their special form appropriate for the linear law of limb-darkening was already given in Z. Kopal (1942).

CHAPTER III

SPECIAL FUNCTIONS OF THE THEORY OF LIGHT CURVES

In the preceding Chapter II we completed the task of expressing the theoretical light changes of close eclipsing systems – due to any type of eclipse of one star by another – to the first order in small quantities in terms of 'circular' integrals of the form (2.15) or (2.16) of Chapter II defining the *associated α-functions* α_n^m of order n and index m; while the supplementary light changes arising from the limb distortion of the eclipsing and eclipsed components were found expressible in terms of additional 'boundary correction' integrals $\mathfrak{I}_{\beta,\gamma}^m$ and $I_{\beta,\gamma}^m$, as defined by Equations (2.25)–(2.28) of Chapter II, respectively. The aim of the present chapter will be to establish the explicit forms of all such functions in terms of the geometrical elements of the eclipse, and to investigate their properties (recursive, differential) a knowledge of which will prove helpful when we come to confront the problems awaiting us in the second part of this book.

III-1. Associated Alpha-Functions and Related Integrals

In order to embark on this task, we find it of advantage to rewrite Equations (2.15)–(2.16) of Chapter II defining the α-functions in the form

$$\pi r_1^{m+n+2}\alpha_n^m = \mathfrak{A}_n^m + \mathfrak{B}_n^m \tag{1.1}$$

where, for partial eclipses (occultations or transits),

$$\mathfrak{A}_n^m = \int_s^{r_1}\int_{-\sqrt{r_1^2-x^2}}^{\sqrt{r_1^2-x^2}} x^m z^n \,\mathrm{d}x\,\mathrm{d}y, \tag{1.2}$$

$$\mathfrak{B}_n^m = \int_{\delta-r_2}^{s}\int_{-\sqrt{r_2^2-(\delta-x)^2}}^{\sqrt{r_2^2-(\delta-x)^2}} x^m z^n \,\mathrm{d}x\,\mathrm{d}y. \tag{1.3}$$

Should the eclipse become annular, $\mathfrak{A}_n^m = 0$ and $\mathfrak{B}_n^m$ continues to be given by Equation (1.3) provided, however, that the quantity s in the upper limit of the first integral on the r.h.s. of Equation (1.3) has been replaced by $\delta + r_2$.

As to the expression (1.2) for $\mathfrak{A}_n^m$, integrating with respect to y we find at once that

$$\begin{aligned}\mathfrak{A}_n^m &= B\left(\frac{1}{2}, \frac{n+2}{2}\right)\int_s^{r_1} x^m (r_1^2 - x^2)^{(n+1)/2}\,\mathrm{d}x \\ &= B\left(\frac{1}{2}, \frac{n+2}{2}\right)\{D_{n+1}^m(r_1) - D_{n+1}^m(s)\},\end{aligned} \tag{1.4}$$

where B denotes a complete beta-function, and

$$D_{n+1}^{m}(\xi) = \int_{0}^{\xi} x^{m}(r_1^2 - x^2)^{(n+1)/2}\, dx \tag{1.5}$$

represents a binomial integral tractable by elementary means. In point of fact,

$$D_{n+1}^{m}(r_1) = \tfrac{1}{2}B\left(\frac{m+1}{2}, \frac{n+3}{2}\right) r_1^{m+n+2} \tag{1.6}$$

and

$$D_{n+1}^{m}(s) = \frac{1}{m+1}\left(\frac{s}{r_1}\right)^{m+1} {}_2F_1\left(-\frac{n+1}{2}, \frac{m+1}{2}; \frac{m+3}{2}; \frac{s^2}{r_1^2}\right) r_1^{m+n+2}; \tag{1.7}$$

where, for odd values of n, the hypergeometric series on the right-hand side reduces to a polynomial. In point of fact, for any value of n Equation (1.4) can be expressed as

$$\mathfrak{A}_n^m = \pi B\left(\frac{1}{2}, \frac{n+2}{2}\right) r_1^{m+n+2} J_{n+1,0}^{m}(\kappa_1) \tag{1.8}$$

in terms of the integrals $J_{\beta,\gamma}^m$ defined by Equation (3.31) later in this chapter, with the modulus

$$\kappa_1^2 = \frac{1}{2}\left(1 - \frac{s}{r_1}\right) = \frac{r_2^2 - (\delta - r_1)^2}{4\delta r_1}, \tag{1.9}$$

and would by itself constitute the entire loss of light if the eclipsing component would act as a straight occulting edge.

The evaluation of the function $\mathfrak{B}_n^m$ as defined by Equation (1.3) above proves to be somewhat more troublesome. Integrating Equation (1.3) with respect to y we find that, if $n \equiv 2\nu$ ($\nu = 0, 1, 2, \ldots$) is zero or an even integer,

$$\mathfrak{B}_{2\nu}^{m} = \frac{1}{\pi} B(\tfrac{1}{2}, 1+\nu) \sum_{j=0}^{r} B(\tfrac{1}{2}, \tfrac{1}{2} + \nu - j) I_{2j,1,2(\nu-1)}^{m}; \tag{1.10}$$

while if $n \equiv 2\nu - 1$ is odd,

$$\mathfrak{B}_{2\nu-1}^{m} = \frac{1}{\pi} B(\tfrac{1}{2}, \tfrac{1}{2} + \nu) \left\{ \sum_{j=1}^{\nu} B(\tfrac{1}{2}, j) I_{2(\nu-j),1,2j-1}^{m} + 2\Pi_{2\nu}^{m} \right\}, \tag{1.11}$$

where

$$I_{2\alpha,\beta,\gamma}^{m} = \int_{\delta-r_2}^{c_2} (r_1^2 - x^2)^{\alpha} [r_2^2 - (\delta - x)^2]^{\beta/2} [2\delta(s - x)]^{\gamma/2} x^m\, dx, \tag{1.12}$$

and

$$\Pi_{2\nu}^{m} = \int_{\delta-r_2}^{c_2} x^m (r_1^2 - x^2)^{\nu} \sin^{-1} \sqrt{\frac{r_2^2 - (\delta - x)^2}{r_1^2 - x^2}}\, dx, \tag{1.13}$$

where $c_2 = s$ or $\delta + r_2$ depending on whether the eclipse is partial or annular. The reader may note that, for $\alpha = 0$, the foregoing Equation (1.12) reduces to Equation (2.27) of Chapter II.

A. LITERAL EVALUATION

The methods of integration of Equation (2.12) depend in principle upon the values of the three subscripts α, β, γ. The first can be easily suppressed by putting

$$I^m_{2\alpha,\beta,\gamma} = \sum_{j=0}^{\alpha} (-1)^j \binom{\alpha}{j} r_1^{2(\alpha-j)} I^{m+2j}_{0,\beta,\gamma}. \tag{1.14}$$

Further, the nature of our problem is such that β can assume values of odd integers only.* Thus the character of Equation (1.12) depends on whether γ is odd or even. If it is zero or an even integer Equation (1.12) can be solved in terms of circular and algebraic functions by elementary methods.

If, however, γ is odd, Equation (1.12) becomes an elliptic integral. In order to solve it we change over to a new variable

$$t = x - h, \tag{1.15}$$

where h is a constant defined so as to render

$$[r_2^2 - (\delta - x)^2][s - x] = (t - e_1)(t - e_2)(t - e_3), \tag{1.16}$$

subject to conditions that

$$e_1 > e_2 > e_3 \quad \text{and} \quad e_1 + e_2 + e_3 = 0.$$

Evidently

$$h = \tfrac{1}{3}(2\delta + s), \tag{1.17}$$

and

$$e_1 = +\tfrac{1}{3}(\delta - s) + r_2, \quad e_2 = -\tfrac{2}{3}(\delta - s), \quad e_3 = +\tfrac{1}{3}(\delta - s) - r_2, \tag{1.18p}$$

if the eclipse is partial; and

$$e_1 = -\tfrac{2}{3}(\delta - s), \quad e_2 = +\tfrac{1}{3}(\delta - s) + r_2, \quad e_s = +\tfrac{1}{3}(\delta - s) - r_2, \tag{1.18a}$$

if it is annular. In either case we are therefore entitled to put

$$t = \wp(u), \tag{1.19}$$

where $\wp$ denotes the Weierstrass π-function of an argument u which replaces t as our independent variable.

The integrals on the right-hand side of Equation (1.14) in terms of this new variable become

$$I^m_{0,\beta,\gamma} = -2i^{\beta+\gamma}(2\delta)^{\gamma/2} \int_{\omega_2}^{\omega_1+\omega_2} \{[\wp(u) - e_1][\wp(u) - e_3]\}^{(\beta+1)/2}$$
$$\times \{\wp(u) - e_2\}^{(\gamma+1)/2} \{\wp(u) + h\}^m \, \mathrm{d}u, \tag{1.20}$$

* Terms with β zero or an even integer do not occur in the light curve on account of symmetry.

where $i \equiv \sqrt{-1}$ stands for the imaginary unit, with limits defined by

$$\wp(\omega_1) = e_1, \qquad \wp(\omega_2) = e_3. \tag{1.21}$$

If, as in Equations (1.10) or (1.11), $\beta = 1$, then Equation (1.20) can undergo further reduction. For, by definition, we have

$$2\sqrt{[\wp(u) - e_1][\wp(u) - e_2][\wp(u) - e_3]} = \wp'(u), \tag{1.22}$$

where accent denotes derivative with respect to u. Squaring Equation (1.22) and inserting in Equation (1.20) we obtain

$$I^m_{0,1,\gamma} = -\frac{i^{\gamma+1}}{2}(2\delta)^{\gamma/2} \int_{\omega_1}^{\omega_1+\omega_2} [\wp(u) - e_2]^{(\gamma-1)/2}[\wp(u) + h]^m[\wp'(u)]^2\, \mathrm{d}u. \tag{1.23}$$

But, if we abbreviate

$$g_2 = -4(e_1e_2 + e_1e_3 + e_2e_3), \quad g_3 = +4e_1e_2e_3, \tag{1.24}$$

it follows from Equation (1.22) that*

$$\{\wp'(u)\}^2 = 4\wp^3(u) - g_2\wp(u) - g_3 \tag{1.25}$$

and hence, γ being odd, the whole integrand of Equation (1.00) can be written out as a polynomial of the $\{\frac{1}{2}(\gamma - 1) + m + 3\}$th degree in powers of $\wp(u)$.

The last step in the evaluation of Equations (1.20) or (1.23) consists in reducing integrals

$$\int_{\omega_2}^{\omega_1+\omega_2} \{\wp(u)\}^j\, \mathrm{d}u, \quad j = 0, 1, 2, \ldots \tag{A}$$

to Legendre normal forms. This can proceed by expressing, by successive differentiation of Equation (1.25) the powers of $\wp(u)$ in terms of its derivatives. A general expression for $\wp^j(u)$ contains, in addition to $\mathrm{d}^{2(j-1)}\wp/\mathrm{d}u^{2(j-1)}$ and lower derivatives of even orders, also the first powers of $\wp(u)$ (for $j > 2$) and a constant. If we put

$$\int_{\omega_1}^{\omega_1+\omega_2} \mathrm{d}u = \omega_1 \quad \text{and} \quad \int_{\omega_2}^{\omega_1+\omega_2} \wp(u)\, \mathrm{d}u = -\eta_1$$

and remember that odd derivatives of $\wp(u)$ with arguments $\omega_1 + \omega_2$ or ω_2 vanish, we readily see† that

$$\int_{\omega_2}^{\omega_1+\omega_2} \{\wp(u)\}^2\, \mathrm{d}u = \tfrac{1}{12} g_2 \omega_1,$$

* This is the differential equation defining $\wp(u)$; cf. Whittaker and Watson, *Modern Analysis* (1920), Section 20.22.

† Cf. *op. cit.*, Section 20.52.

and for $j > 2$ all integrals of powers of $\wp(u)$ can be expressed as linear combinations of ω_1 and η_1, with coefficients involving powers and cross-products of the invariants g_2 and g_3. The functions ω_1 and η_1 rewritten in terms of Legendre normal integrals take finally the forms

$$\omega_1 = \frac{F\left(\frac{\pi}{2}, \kappa\right)}{\sqrt{e_1 - e_3}} \tag{1.26}$$

and

$$\eta_1 = \sqrt{e_1 - e_3} E\left(\frac{\pi}{2}, \kappa\right) - \frac{e_1}{\sqrt{e_1 - e_3}} F\left(\frac{\pi}{2}, \kappa\right), \tag{1.27}$$

where F and E denote the Legendre complete integrals of the first and second kind, with the modulus

$$\kappa^2 = \frac{e_2 - e_3}{e_1 - e_3}. \tag{1.28}$$

The reader should notice that the moduli appropriate for partial and annular eclipses are mutually reciprocal.

After having thus established the solution of Equation (1.12) in a finite number of terms for any value of the subscripts and of m, let us return to Equation (1.13). Integrating Π by parts we obtain

$$\Pi_{2\nu}^m = GD_{2\nu}^m(s) - \sqrt{\frac{\delta}{2}} \int_{c_2}^{c_2} \frac{XD_{2\nu}^m(x)\,\mathrm{d}x}{\sqrt{(x-c_1)(x-c_2)(x-c_3)}}, \tag{1.29}$$

where $G = \frac{1}{2}\pi$ or 0 depending on whether the eclipse is partial or annular,

$$X \equiv \frac{x^2 - 2sx + r_1^2}{r_1^2 - x^2}, \tag{1.30}$$

and

$$c_1 = \delta + r_2, \qquad c_2 = s, \qquad c_3 = \delta - r_2, \tag{1.31p}$$

if the eclipse is partial; and

$$c_1 = s, \qquad c_2 = \delta + r_2, \qquad c_3 = \delta - r_2, \tag{1.31a}$$

if it is annular. Substitute, as before,

$$x - \tfrac{1}{3} \sum_{j=1}^{3} c_j = \wp(u) \tag{1.32}$$

and expand

$$XD_{2\nu}^m(x) = \sum_{j=0}^{m+n+2} a_n^m(j)\{\wp(u)\}^j + b_n^m r_1^{m+n+2} \times \left\{\frac{r_1 - s}{\wp(u) + h - r_1} + (-1)^m \frac{r_1 + s}{\wp(u) + h + r_1}\right\}, \tag{1.33}$$

where the coefficients a_n^m are polynomials of the $(m + n + 2 - j)$th degree in r_1, δ, and s; and b_n^m is a positive fraction (numerical factor). Since, by definition,

$$dx = 2\sqrt{(x-c_1)(x-c_2)(x-c_3)}\,du \tag{1.34}$$

we see that the $\Pi_{2\nu}^m$, can be expressed in terms of integrals of powers of $\wp(u)$ which we have just solved, plus two integrals of the form

$$\int_{\omega_2}^{\omega_1+\omega_2} \frac{du}{\wp(u)+h\pm r_1}, \tag{B}$$

which are new and remain to be evaluated.

In order to do so we introduce new arguments $v_{1,2}$ defined by

$$-(h\pm r_1) = \wp(v_{1,2}). \tag{1.35}$$

Then, by means of a well-known theorem* we have

$$\int_{\omega_2}^{\omega_1+\omega_2} \frac{du}{\wp(u)-\wp(v_j)} = \frac{2}{\wp'(v_j)}\{w_1\zeta(v_i)-\eta_i v_j\}, \quad j=1,2, \tag{1.36}$$

where accent denotes derivative with respect to v_j and ζ is the Weierstrass zeta-function. As one can easily verify,

$$\wp'(v_{1,2}) = \mp 2i\sqrt{2\delta}(r_1\pm s) \tag{1.37p}$$

if the eclipse is partial, and

$$\wp'(v_{1,2}) = -2i\sqrt{2\delta}(r_1\pm s) \tag{1.37a}$$

if it is annular. In order to remove the imaginary unit we put

$$v_1 = iw_1, \qquad v_2 = iw_2+\omega_1. \tag{1.38}$$

Remembering that

$$\zeta(\omega_1)=\eta_1, \quad \wp(\omega_1)=e_1, \quad \wp'(\omega_1)=0, \tag{1.39}$$

the addition theorem for Weierstrass zeta-functions yields

$$\zeta(\omega_1+i\omega_j) = n_1+\zeta(iw_j)+\frac{i}{\sqrt{2\delta}}\{r_1-(\delta-r_2)\} \tag{1.40p}$$

if the eclipse is partial, and

$$\zeta(\omega_1+iw_j) = \eta_1+\zeta(iw_j)+i\sqrt{2\delta} \tag{1.40a}$$

if it is annular. If we further substitute

$$\zeta(iw) = -i\zeta^*(w), \tag{1.41}$$

where

$$\zeta^*(w;e_1,e_2,e_3) = \zeta(w;-e_1,-e_2,-e_3), \tag{1.42}$$

* Whittaker and Watson, *op. cit.*, Section 20.53.

we finally obtain that, for partial eclipses,

$$\sqrt{2\delta}(r_1+s)\int_{\omega_2}^{\omega_1+\omega_2}\frac{du}{\wp(u)-\wp(v_1)}=\omega_1\zeta^*(w_1)+\eta_1 w_1, \tag{1.43p}$$

and

$$\sqrt{2\delta}(r_1-s)\int_{\omega_2}^{\omega_1+\omega_2}\frac{du}{\wp(u)-\wp(v_2)}=-\omega_1\zeta^*(w_2)-\eta_1 w_2+\frac{\omega_1}{\sqrt{2\delta}}(r_1+r_2-\delta). \tag{1.43a}$$

If the eclipse is annular, Equation (1.43p) continues to hold good; but Equation (1.43a) is to be replaced by

$$\sqrt{2\delta}(r_1-s)\int_{\omega_2}^{\omega_1+\omega_2}\frac{du}{\wp(u)-\wp(v_2)}=\omega_1\zeta^*(w_2)+\eta_1 w_2-\omega_1\sqrt{2\delta}. \tag{1.44}$$

The functions $w_{1,2}$ and $\zeta^*(w_{1,2})$, expressed in terms of Legendre normal forms, become

$$w_{1,2}=\frac{F(\phi_{1,2},\kappa')}{\sqrt{e_1-e_3}} \tag{1.45}$$

and

$$\zeta^*(w_{1,2})=e_3 w_{1,2}+\sqrt{e_1-e_3}E(\phi_{1,2},\kappa')+\frac{1}{\sqrt{2\delta}}[r_1\pm(\delta-r_2)], \tag{1.46p}$$

$$\zeta^*(w_{1,2})=\dot{e}_3 w_{1,2}+\sqrt{e_1-e_3}E(\phi_{1,2},\kappa')+\sqrt{2\delta}, \tag{1.46a}$$

– as to whether the eclipse is partial (p) or annular (a) – where κ', the complementary modulus, is defined by

$$(\kappa')^2=\frac{e_1-e_2}{e_1-e_3}=1-\kappa^2, \tag{1.47}$$

and the amplitudes for partial and annular eclipses take the respective forms

$$\phi_1=\sin^{-1}\sqrt{\frac{2r_2}{r_1+r_2+\delta}},\quad \phi_2=\sin^{-1}\sqrt{\frac{2\delta}{r_1+r_2+\delta}}, \tag{1.48p}$$

and

$$\phi_1=\phi_2=\sin^{-1}\sqrt{\frac{r_1+r_2-\delta}{r_1+r_2+\delta}}. \tag{1.48a}$$

Let us put, for brevity's sake,

$$(r_1+s)\int_{\omega_2}^{\omega_1+\omega_2}\frac{du}{\wp(u)-\wp(v_1)}\mp(r_1-\delta)\int_{\omega_3}^{\omega_1+\omega_2}\frac{du}{\wp(u)-\wp(v_2)}\equiv\frac{1}{\sqrt{2\delta}}\mathfrak{E}_{1,2}. \tag{1.49}$$

By combination of the above formulae it follows that, for partial eclipses,

$$\mathfrak{E}_{1,2}=\left\{E\left(\frac{\pi}{2},\kappa\right)-F\left(\frac{\pi}{2},\kappa\right)\right\}\{F(\phi_1,\kappa')\pm F(\phi_2,\kappa')\}$$

$$+F\left(\frac{\pi}{2},\kappa\right)\{E(\phi_1,\kappa')\pm E(\phi_2,\kappa')+\kappa\cos\phi_1\sec\phi_2\}. \qquad (1.50)$$

If the upper sign is valid, this expression admits of a drastic simplification; for, by an obvious extension of a theorem due to Legendre,* the reader should have no difficulty to prove that

$$\mathfrak{E}_1=\frac{\pi}{2}+\sqrt{\frac{\delta}{r_2}}F\left(\frac{\pi}{2},\kappa\right). \qquad (1.51)$$

Hence, by subtraction of $\mathfrak{E}_1$ and $\mathfrak{E}_2$, the latter takes the form

$$\mathfrak{E}_2=\frac{\pi}{2}-2\left\{E\left(\frac{\pi}{2},\kappa\right)-F\left(\frac{\pi}{2},\kappa\right)\right\}F(\phi_2,\kappa')$$

$$-2\left\{E(\phi_2,\kappa')-\frac{1}{2}\sqrt{\frac{\delta}{r_2}}\right\}F\left(\frac{\pi}{2},\kappa\right), \qquad (1.52)$$

in which both kinds of incomplete integrals possess a common amplitude.

If, finally, the eclipse is annular we similarly obtain

$$\mathfrak{E}_1=\kappa\sqrt{\frac{\delta}{r_2}}F\left(\frac{\pi}{2},\kappa\right) \qquad (1.53)$$

and

$$\mathfrak{E}_2=2F(\phi_{1,2},\kappa')\left\{E\left(\frac{\pi}{2},\kappa\right)-F\left(\frac{\pi}{2},\kappa\right)\right\}+2F\left(\frac{\pi}{2},\kappa\right)$$

$$\times\left\{E(\phi_{1,2},\kappa')+\frac{\kappa}{2}\sqrt{\frac{\delta}{r_2}}\right\}. \qquad (1.54)$$

B. DISCUSSION OF THE RESULTS

This completes the evaluation of the 'circular integrals' α_n^m associated with the effects of *tidal* distortion. We found that the expressions for $\mathfrak{A}_n^m$ are all elementary (in fact, polynomial if n is odd); while those for $\mathfrak{B}_n^m$ are such only if n is zero or an even integer: if n were odd, the corresponding $\mathfrak{B}_n^m$'s can be evaluated exactly only in terms of elliptic integrals. Expressions of the form $I_{0,1,\gamma}^m$ where γ is a positive odd integer, or (if m is also odd) the $\Pi_{2\nu}^m$'s, can likewise be expressed in terms of Legendre complete integrals of the first and second kind. If, however, m is zero or even, the $\Pi_{2\nu}^m$'s are bound to involve also complete elliptic integrals of the third kind which belong to the 'circular' class and are, therefore, expressible in terms of incomplete integrals of the first and second kind with complementary moduli.

The photometric effects of second-harmonic *rotational* distortion introduce – in addition to terms already treated – new terms involving the first and second

* Cf. Whittaker and Watson (op. cit.), Section 22.735.

powers of y. Terms containing y vanish, however, on account of the symmetry of the eclipsed area with respect to the x-axis; while $y^2 = r_1^2 - x^2 - z^2$. Hence all photometric effects of the rotational as well as tidal distortion can be expressed in terms of the functions already treated in this section; with those of the particular type α_n^0 we got acquainted already in Chapter I.

For $m > 0$, by collecting the results established in this section we find that, if n is zero or an even integer (i.e., $n = 2\nu$; $\nu = 0, 1, 2, \ldots$) the corresponding associated α-functions assume the forms

$$\pi^2 r_1^{m+2(r+1)} \alpha_{2\nu}^m = B(\tfrac{1}{2}, \nu + 1)\{2G[D_{2\nu+1}^m(r_1) - D_{2\nu+2}^m(s)] + \sum_{j=0}^{\nu} B(\tfrac{1}{2}, \tfrac{1}{2} + \nu - j) I_{2j,1,2(\nu-j)}^m\}, \tag{1.55}$$

where (as before) $G \equiv \frac{1}{2}\pi$ or 0, depending on whether the eclipse is partial or annular; while if n is odd,

$$\pi^2 r_1^{m+2\nu+1} \alpha_{2\nu-1}^m = 2B(\tfrac{1}{2}, \tfrac{1}{2} + \nu) \times \left\{ G D_{2\nu}^m(r_1) + \tfrac{1}{2} \sum_{j=1}^{\nu} B(\tfrac{1}{2}, j) I_{2(\nu-j),1,2j-1}^m - \sqrt{\frac{\delta}{2}} \int_{c_3}^{e_2} \frac{X D_{2\nu}^m(x)\, dx}{\sqrt{(x-c_1)(x-c_2)(x-c_3)}} \right\}. \tag{1.56}$$

Functions of the type (1.55) can be evaluated in terms of inverse circular and algebraic functions; and their explicit forms for $n = 0, 2, 4$ and $m = 0(1)4$ such that $m + n \leqslant 5$ have been established by Kopal (1942).

In particular, Equation (2.14) of Chapter I for α_0^0 can be rewritten as

$$\alpha_0^0 = 2\{J_{1,0}^0(\phi_1) + (r_2/r_1)^2 J_{1,0}^0(\phi_2)\}, \tag{1.57}$$

where the $J_{\beta,\gamma}^0$-integrals of the arguments

$$\phi_{1,2} = \cos^{-1} \frac{\delta^2 + r_{1,2}^2 - r_{2,1}^2}{2\delta r_{1,2}} \tag{1.58}$$

will be fully discussed in Section III-3 of this chapter.

Functions of the type (1.56) for $n = -1, 1, 3, 5$ require elliptic integrals for their expression in a closed form; and those for $m = 0(1)4$ for $m + n \leqslant 5$ have likewise been so evaluated by the same author (Kopal, 1942). The explicit forms of those corresponding to odd values of n are too long to be reproduced here in full (for α_1^0, cf. Equations (2.16) or (2.19) of Chapter I); but those with even n's can be found in Appendix II.

As to the 'boundary corrections' $\mathfrak{I}_{\beta,\gamma}^m$ and $I_{\beta,\gamma}^m$ encountered in Section 2 of Chapter II, little remains to be added on their evaluation except to note that, in accordance with Equations (2.25) and (2.28) of Chapter II,

$$\pi r_1^{\beta+\gamma+m+3} \mathfrak{I}_{\beta,\gamma}^m = (r_2^2 - \delta^2) I_{0,\beta,\gamma}^m + \delta I_{0,\beta,\gamma}^{m+1} \tag{1.59}$$

and

$$\pi r_2^{\beta+\gamma+m+1} I_{\beta,\gamma}^m = \sum_{j=0}^{m} (-1)^j \binom{m}{j} \delta^{m-j} I_{0,\beta,\gamma}^j, \tag{1.60}$$

where the $I_{0,\beta,\gamma}^m$'s on the right-hand sides represent particular cases of integrals of the form (1.12) evaluated earlier in this section; and for their explicit forms cf. again Kopal (1942). Hence, all 'circular integrals' as well as 'boundary corrections' introduced in Section II-2 can be expressed in terms of functions already treated; and the mathematical solution of our problem in a closed form is thus complete.

An inspection of the equations defining the associated α-functions and the related boundary integrals in Chapter II disclose all these to be *homogeneous* functions of *three* variables (i.e., r_1, r_2 and δ or s) and can, therefore, be made to depend on *two* of their *ratios* which can be formed from them. The choice of such ratios is, in principle, arbitrary; and several variants will be used in this book (see, e.g., Equation (3.35) of Chapter I or Equation (0.6) of Chapter IV) to suit our convenience.

As to the particular values which the associated α-functions and related integrals assume at limiting phases of eclipses, at the moment of their first (or last) contacts (i.e., when $\delta = r_1 + r_2$), all α_n^m's, $\mathfrak{I}_{\beta,\gamma}^m$'s and $I_{\beta,\gamma}^m$'s vanish for every value of m, n or β, γ; and remain zero for $\delta > r_1 + r_2$. If the eclipse is an *occultation* (i.e., $r_1 < r_2$), both types of boundary integrals $\mathfrak{I}_{\beta,\gamma}^m$ and $I_{\beta,\gamma}^m$ vanish also at the moment of inner contact (for $\delta = r_1 - r_2$, which marks the commencement of *totality*) regardless of β, γ or m. However, at the beginning of totality the associated α-functions α_n^m vanish only if m is odd; while if $m \equiv 2\mu$ is even, it can be shown that

$$\alpha_{2\nu}^{2\mu} = \frac{\nu!\,\Gamma(\mu+\frac{1}{2})}{\sqrt{\pi}(\mu+\nu+1)!} \tag{1.61}$$

and

$$\alpha_{2\nu-1}^{2\mu} = \frac{\Gamma(\mu+\frac{1}{2})\Gamma(\nu+\frac{1}{2})}{\sqrt{\pi}\,\Gamma(\mu+\nu+\frac{3}{2})} \tag{1.62}$$

are constants depending on whether the index n is even (2ν) or odd ($2\nu - 1$).

If, on the other hand, the eclipse is a *transit* (i.e., $r_1 > r_2$) and the moment of the inner contact (at which $\delta = r_1 - r_2$) marks the beginning of the *annular* phase, the $\alpha_n^{2\mu}$'s at the moment of internal tangency remain functions of $k \equiv r_2/r_1$. For $\mu = 0$ it can be shown (cf. Kopal, 1975c) that, at the moment of internal tangency,

$$\alpha_n^0 = \frac{(n+2)!\,k^{(n+4)/2}}{\Gamma(2+\frac{1}{2}n)\Gamma(3+\frac{1}{2}n)}\,{}_2F_1(-\tfrac{1}{2}n, 2+\tfrac{1}{2}n; 3+\tfrac{1}{2}n; k). \tag{1.63}$$

For even n's the hypergeometric series on the right-hand side of the preceding equation will reduce to polynomials; and although for odd values of n the series on the r.h.s. of Equation (1.63) remain infinite, they can be expressed in terms of elementary functions (cf. Appendix I). However, for $m > 0$ no expressions of comparable simplicity have been discovered so far. As regards the values of the 'boundary integrals' $\mathfrak{J}^m_{\beta,\gamma}$ and $I^m_{\beta,\gamma}$ at the moment of internal tangency of a transit eclipse, they also vanish for $\delta = r_1 - r_2$ if $\gamma = 0$ and m is odd (cf. e.g., Kopal, 1947); but not otherwise.

The non-vanishing functions continue, moreover, to vary with δ throughout annular phase. Ultimately, at the moment of *central eclipse* ($\delta = 0$), the only functions whose coefficients in Equation (2.31) of Chapter II for the theoretical light curves do not vanish are those corresponding to $m = 0$, which for $\delta = 0$ can be shown to reduce to

$$\alpha_n^0 = \frac{2}{n+2}\{1-(1-k^2)^{(n+2)/2}\} = k^2\,{}_2F_1(-\tfrac{1}{2}n, 1; 2; k^2); \tag{1.64}$$

while

$$\pi\mathfrak{J}^0_{\beta,\gamma} = B(\tfrac{1}{2}, 1+\tfrac{1}{2}\beta)k^{\beta+3}(1-k^2)^{\gamma/2} \tag{1.65}$$

and

$$\pi I^0_{\beta,\gamma} = B(\tfrac{1}{2}, 1+\tfrac{1}{2}\beta)k^{-\gamma}(1-k^2)^{\gamma/2}, \tag{1.66}$$

where $k \leqslant 1$ continues to stand for the ratio of r_2/r_1.

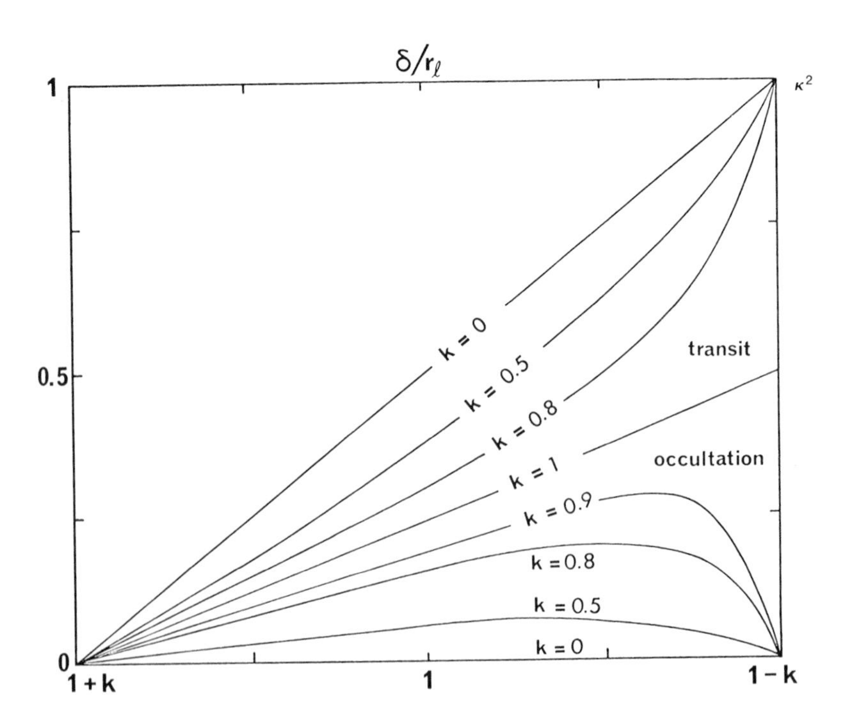

Fig. 3-1. Variation of the modulus κ^2 with δ during partial phases of the eclipse.

III-2. Recursion Relations

In Chapter II of this book we found it possible to express the light changes due to mutual eclipses of distorted components of close binary systems in terms of the associated α-functions and certain related integrals varying with the phase; and in the preceding section of this chapter we detailed the way in which all such functions can be evaluated in a closed form. Moreover, the explicit forms of all functions required by a theory of the light curves, complete to terms of first order in surficial distortion of both components, have been established already some time ago (cf. Kopal, 1942).

These have, however, proved to be generally so complicated as to be of limited use in practice: for computations by hand, many hours would be required to tabulate them for a single case as a function of the phase even with the aid of auxiliary tables specially constructed for this purpose (Kopal, 1947). Electronic computers of recent vintage would, to be sure, permit us today to reduce this time to minutes. However – and this is essential – *the evaluation of each single function from closed expressions established in the preceding section would call for a separate machine programme*; and the use of so many programmes would, in turn, tax the storage of the respective computer beyond the limits to which such computations may remain practicable.

In more specific terms, the number of special functions of the type α_n^m, $\mathfrak{I}_{\beta,\gamma}^m$ and $I_{\beta,\gamma}^m$ required for a description of the light curve correctly to quantities of first order in surficial distortion (i.e., including the effects of second, third, and fourth-harmonic tides) is equal to 24 of each family – i.e., a total of 72 functions – if the effects of limb-darkening are to be treated to a linear approximation; and if the adopted law of limb-darkening as given by Equation (1.2) of Chapter I were to be extended to terms of n-th order, the total number of special functions required for our purpose would increase $72n$-fold. It is clear that such requirements would make any attempt of meeting them by use of *literal* expressions investigated in the preceding section impracticable even with the aid of the fastest computing machines now in existence.

In order to circumvent this difficulty, and reduce the need of resorting to closed expressions in literal form to a minimum, we shall undertake next a search for the existence of *recursion formulae* connecting the associated α-functions for different m and n with each other, as well as with other types of special functions involved in our problem – formulae which should enable us to restrict functions requiring literal evaluation to a minimum, from which all others could be built up by purely algebraic steps. As will be shown in this section, such formulae indeed exist; and the basic set of functions prerequisite for the generation of all others reduces, in fact, to a *pair* of associated α-functions differing in degree n by one – such as α_0^0 and α_1^0, investigated already in Section I-2 – and aided by a family of integrals of the type $I_{\beta,\gamma}^1$ which are easiest to evaluate.

In point of fact, for the benefit of those investigators who may not have computers at their disposal suitable to evaluate these *ab initio* from the literal expressions (or their Fourier approximations established in Sections I-3 and I-4),

5-digit tables of α_0^0 as well as α_1^0 for every type of eclipse have been prepared by Tsesevich (1939, 1940) and others;* while those of $I_{\beta,\gamma}^0$ for $\beta = -1(2)3$ and $\gamma = -1(1)6$ by Kopal (1947).

A. RECURSION RELATIONS FOR THE α_n^m-FUNCTIONS

In order to establish the recursion formulae obtaining between the associated α-functions of different orders and indices, we find it convenient to re-define such functions in terms of plane-polar coordinates in the form

$$\pi r_1^{m+n+2}\alpha_n^m = 2\int_{\delta-r_2}^{r_1}\int_0^{\theta_0}(r_1^2 - r^2)^{n/2} r^{m+1}\cos^m\theta\,\mathrm{d}r\,\mathrm{d}\theta, \tag{2.1}$$

where

$$\cos\theta_0 = \frac{\delta^2 + r^2 - r_2^2}{2\delta r}. \tag{2.2}$$

This definition remains valid only during partial eclipses as long as $\delta \geqslant r_2$. When the converse becomes the case, and the eclipsing limb advances past the centre of the apparent disc of the star of fractional radius r_1, the limits of integration on the r.h.s. of Equation (2.1) must be adjusted so that

$$\pi r_1^{m+n+2}\alpha_n^m = \left\{\int_0^{r_1}\int_0^{\pi} - \int_{\delta-r_2}^{r_1}\int_0^{\pi-\theta_0}\right\} r^{m+1}(r_1^2 - r^2)^{n/2}\cos^m\theta\,\mathrm{d}\theta\,\mathrm{d}r; \tag{2.3}$$

moreover, during annular phases of transit eclipses (when $\delta < r_1 - r_2$) the upper limits r_1 and $\pi - \theta_0$ in the second pair of integrals on the right-hand side of the foregoing equation should be replaced by $\delta + r_2$ and π, respectively. Particular cases of the foregoing equations (2.1)–(2.3) for $m = 0$ were encountered already in Section I-2; here we merely generalized the latter for $m > 0$.

Integrating the right-hand sides of Equations (2.1) and (2.3) with respect to the angular variable θ we find that

$$\int_0^{\theta_0}\cos^m\theta\,\mathrm{d}\theta = \pi I_{-1,0}^m(\cos\theta_0), \tag{2.4}$$

where the $I_{-1,0}^m$'s are particular cases of the family of integrals defined by Equation (2.28) of Chapter II. Therefore, for $\delta \geqslant r_2$ Equation (2.1) can be rewritten as

$$r_1^{m+n+2}\alpha_n^m = 2\int_{\delta-r_2}^{r_1}(r_1^2 - r^2)^{n/2} I_{-1,0}^m(\cos\theta_0) r^{m+1}\,\mathrm{d}r \tag{2.5}$$

for any value of m or n. If $\delta < r_2$ – and the relevant form for is that given by

* For their bibliography cf. Kopal (1948b).

Equation (2.3) – the fact that, for $\theta_0 = \pi$

$$I_{-1,0}^{2\mu+1}(-1) = 0 \tag{2.6}$$

if $m \equiv 2\mu + 1$ is odd, and

$$I_{-1,0}^{2\mu}(-1) = \binom{\mu - \frac{1}{2}}{\mu} \tag{2.7}$$

if $m \equiv 2\mu > 0$ is even, Equation (2.5) can be expressed as

$$r_1^{2(\mu+1)+n+1}\alpha_n^{2\mu+1} = 2 \int_{r_2-\delta}^{r_1} (r_1^2 - r^2)^{n/2} I_{-1,0}^{2\mu+1}(\cos\theta_0) r^{2(\mu+1)}\,\mathrm{d}r \tag{2.8}$$

for odd values of m ($\mu = 0, 1, 2, \ldots$); while if m is even,

$$r_1^{2\mu+n+2}\alpha_n^{2\mu} = \frac{\Gamma(\mu+\frac{1}{2})\Gamma(1+\frac{1}{2}n)}{\sqrt{\pi}\,\Gamma(\mu+\frac{1}{2}n+2)} r_1^{2\mu+n+2} - 2 \int_{r_2-\delta}^{r_1} (r_1^2 - r^2)^{n/2} I_{-1,0}^{2\mu}(-\cos\theta_0) r^{2\mu+1}\,\mathrm{d}r. \tag{2.9}$$

The explicit form of the functions $I_{-1,0}^m(x)$ for any (non-negative) integral value of m are well known (cf. Equations (1)–(24) of Appendix IV), and their particular cases will be encountered in subsequent parts of this section. For our immediate needs it is sufficient to note that they satisfy the simple recursion formula

$$mI_{-1,0}^m(x) = (m-1)I_{-1,0}^{m-2}(x) + x^{m-1}I_{-1,0}^1(x), \tag{2.10}$$

of many others whose existence will be proved later on.

A combination of Equation (2.10) with Equations (2.1)–(2.3) for the associated α-functions discloses at once that the latter obey the recursion formula

$$(m+1)\alpha_n^{m+1} + m(\alpha_{n+2}^{m-1} - \alpha_n^{m-1}) = (2\delta/\pi r_1^{m+n+3}) I_{0,1,n}^m, \tag{2.11}$$

valid for $m = 0, 1, 2, \ldots$ and $n = -1, 0, 1, 2, \ldots$; where $I_{0,1,n}^m$ stands for the respective member of the family of integrals defined by Equation (2.27) of Chapter II. With suitable limits imposed on the latter, the foregoing recursion formula (2.11) holds good for any type of eclipse – be it occultation or transit; partial, total or annular.

Equation (2.11) does not represent the only recursion formula satisfied by the associated α-functions. Another can be constructed if we return to Equation (2.1) and integrate its right-hand side by parts, obtaining

$$r_1^{m+n+2}\alpha_n^m = \frac{2}{n+2} \int_{\delta-r_2}^{r_1} (r_1^2 - r^2)^{(n+4)/2} \frac{\mathrm{d}}{\mathrm{d}r}\{r^m I_{-1,0}^m(\cos\theta_0)\}\,\mathrm{d}r. \tag{2.12}$$

Replace now, in this equation, n by $n-2$ and subtract from Equation (2.12) as it stands; in doing so we find that

$$r_1^{m+n+2}\{n\alpha_{n-2}^m - (n+2)\alpha_n^m\} = 2\int_{\delta-r_2}^{r_1} (r_1^2 - r^2)^{n/2} \frac{d}{dr}\{r^m I_{-1,0}^m(\cos\theta_0)\} r^2\, dr. \tag{2.13}$$

If, furthermore, we make use of the fact that

$$\frac{d}{dr}\{r^m I_{-1,0}^m\} = mr^{m-1} I_{-1,0}^m + r^m \frac{dI_{-1,0}^m}{dr} \tag{2.14}$$

and insert this on the right-hand side of Equation (2.13), this latter equation assumes the form

$$r_1^{m+n+2}\{(m+n+2)\alpha_n^m - n\alpha_{n-2}^m\} = -2\int_{\delta-r_2}^{r_1} (r_1^2 - r^2)^{n/2} \frac{dI_{-1,0}^m}{dr} r^{m+2}\, dr. \tag{2.15}$$

By virtue of the fact that

$$\frac{dI_{-1,0}^m}{dr} = \cos^m\theta_0 \frac{d\theta_0}{dr}, \tag{2.16}$$

where the angle θ_0 continues to be defined by Equation (2.1), and by a resort to (1.59) Equation (2.15) can be rewritten as

$$(m+n+2)\alpha_n^m - n\alpha_{n-2}^m = 2\mathfrak{I}_{-1,n}^m, \tag{2.17}$$

valid for $m = 0, 1, 2, \ldots$ but $n = 1, 2, 3, \ldots$.

Equation (2.17) represents the second fundamental recursion relation for the associated α-functions of the same index and different orders, relating these with the $\mathfrak{I}_{-1\gamma}^m$'s. Like Equation (2.11), the foregoing formula (2.17) holds good for any type of eclipse; and extensive use of it will be made in the sequel. In particular, its use will permit us to rewrite the 'boundary corrections' $f_1^{(h)}$ as defined by Equation (2.31) of Chapter II in terms of the associated α-functions – a task we shall carry out in Chapter VI – care being only taken to note that, unlike Equation (2.11), Equation (2.17) does not hold good for $n = 0$ (and must be replaced by alternative appropriate relations).

Equations (2.11) and (2.17) represent two fundamental and mutually independent recursion formulae for associated α-functions, whose right-hand sides consist of integrals of the form $I_{0,1,n}^m$ or $\mathfrak{I}_{-1,n}^m$. The explicit forms of the latter are rather complicated, but their essential features in this connection are again recursion formulae which they obey. By virtue of the algebraic identity $2\delta(s-x) \equiv 2\delta s - 2\delta x$ it can be shown that

$$2\delta I_{0,\beta,\gamma}^m = 2\delta s I_{0,\beta,\gamma}^{m-1} - I_{0,\beta,\gamma+2}^{m-1} \tag{2.18}$$

and

$$2\delta r_1 \mathfrak{I}_{-1,n}^m = 2\delta s \mathfrak{I}_{-1,n}^{m-1} - r_1^2 \mathfrak{I}_{-1,n+2}^{m-1}. \tag{2.19}$$

The first, combined with Equation (2.11) leads to an 8-term recursion formula for the associated α-functions of the form

$$
\begin{aligned}
&2\delta r_1\{(m+2)[\alpha_{n+2}^{m+1}-\alpha_n^{m+1}]+(m+3)\alpha_n^{m+3}\}\\
&\quad-2\delta s\{(m+1)[\alpha_{n+2}^{m}-\alpha_n^{m}]+(m+2)\alpha_n^{m+2})\}\\
&\quad+r_1^2\{(m+1)[\alpha_{n+4}^{m}-\alpha_{n+2}^{m}]+(m+2)\alpha_{n+2}^{m+2}\}=0;
\end{aligned}
\tag{2.20}
$$

while a combination of Equation (2.17) with Equation (2.19) furnishes a 5-term recursion formula

$$
\begin{aligned}
&2\delta r_1\{(m+n+5)\alpha_{n+2}^{m+1}-(n+2)\alpha_n^{m+1}\}\\
&\quad-2\delta s\{(m+n+4)\alpha_{n+2}^{m}-(n+2)\alpha_n^{m}\}\\
&\quad+r_1^2\{(m+n+6)\alpha_{n+4}^{m}-(n+4)\alpha_{n+2}^{m}\}=0;
\end{aligned}
\tag{2.21}
$$

both valid for $m \geqslant 0$ and $n \geqslant -1$, which consist only of associated α-functions of different orders and indices, with coefficients representing algebraic functions of the elements of the eclipse.

How many associated α-functions must be known before the use of the recursion relations (2.11) and (2.17) can be invoked to generate any number of them for any phase of the eclipses? With those of zero index (α_n^0) we got acquainted already from Chapter I: although their literal evaluation for odd n's by the method of Section III-1 requires a resort to elliptic integrals of not too attractive a kind, the Fourier series established in Sections I-3 and I-4 for the α_n^0's, are valid regardless of whether n is even or odd, and can provide numerical approximations to local values of these functions with arbitrary accuracy.

The next family of functions easy to evaluate are associated α-functions of index one. In point of fact, the recursion formula (2.11) particularized for $m = 0$ discloses at once that

$$
r_1^{n+3}\alpha_n^1 = 2\delta r_2^{n+2} I_{1,n}^0 = \frac{2r_2^{n+3}}{n+2} I_{-1,n+2}^1,
\tag{2.22}
$$

where the $I_{\beta,\gamma}^m$'s represent integrals defined by Equation (2.28) of Chapter II. Such integrals are already familiar to us from preceding parts of this book; and their literal evaluation in a closed form by the methods of Section III-1 (in terms of elementary functions if n is an even integer, and of complete elliptic integrals of the first and second kind if n is odd) is more straightforward than that of any other special functions encountered in the theory of the light curves. They have been tabulated extensively (cf. Kopal, 1947); and obey many recursion formulae deduced in Section 3 of this chapter; of which Equation (3.51) combined with (2.22) leads to the four-term recursion relation

$$
\begin{aligned}
(n+10)\alpha_{n+6}^1 = {} & (n+2)(2\delta r_2/r_1^2)^3\mu(1-\mu^2)\alpha_n^1\\
&+[n+4-(3n+14)\mu^2](2\delta r_2/r_1^2)^2\alpha_{n+2}^1\\
&-(3n+22)\mu(2\delta r_2/r_1^2)\alpha_{n+4}^1, \quad n \geqslant 0,
\end{aligned}
\tag{2.23}
$$

in which we abbreviated $\mu \equiv (\delta - s)/r_2$. Equation (2.23) represents the only recursion formula for the α_n^m's which is homogeneous in functions of the *same* index m.

Moreover, the α_n^1's will likewise be found in the next section to be expressible as Hankel transforms of the form (3.76), approximable by Fourier series (3.77) which includes those given in Chapter I for α_n^0.

If we turn next to be the associated α-functions of index $m = 2$, the recursion formula (2.11) particularised for $m = 1$ furnishes the relation

$$2\alpha_n^2 = \alpha_n^0 - \alpha_{n+2}^0 + (s/r_1)\alpha_n^1 - (r_1/2\delta)\alpha_{n+2}^1, \tag{2.24}$$

which on insertion from Equation (2.22) can be rewritten as

$$2\alpha_n^2 = \alpha_n^0 - \alpha_{n+2}^0 + (r_2/r_1)^{n+1}\{(2\delta s/r_1^2)I_{1,0}^0 - I_{1,n+2}^0\}. \tag{2.25}$$

Moreover, for $m = 0$ the recursion formula (2.17) yields

$$(n+4)\alpha_{n+2}^0 - (n+2)\alpha_n^0 = 2\mathfrak{I}_{-1,n+2}^0 \tag{2.26}$$

where, by Equation (2.25) of Chapter II,

$$r_1^{n+4}\mathfrak{I}_{-1,n+2}^0 = r_2^{n+4}I_{-1,n+2}^0 - \delta r_2^{n+3}I_{-1,n+2}^1. \tag{2.27}$$

On the other hand, from Equation (2.22) and one of the Gauss recursion formulae for hypergeometric series representing $I_{-1,n}^0$ (cf. Kopal, 1978a; p. 443) it can be shown that

$$I_{-1,n+2}^0 = \frac{n+6}{4}\left(\frac{r_1}{\delta}\right)\left(\frac{r_1}{r_2}\right)^{n+4}\alpha_{n+2}^1 + \frac{n+2}{2}\left(\frac{\delta - s}{r_2}\right)\left(\frac{r_1}{r_2}\right)^{n+3}\alpha_n^1, \tag{2.28}$$

where

$$I_{-1,n+2}^1 = \frac{n+2}{2}\left(\frac{r_1}{r_2}\right)^{n+3}\alpha_n^1; \tag{2.29}$$

so that

$$4\mathfrak{I}_{-1,n+2}^0 = (n+6)(r_1/\delta)\alpha_{n+2}^1 - 2(n+4)(s/r_1)\alpha_n^1, \tag{2.30}$$

which combined with Equation (2.26) discloses that

$$(n+4)\alpha_{n+2}^0 = (n+2)\alpha_n^0 - (n+2)(s/r_1)\alpha_n^1 + \tfrac{1}{2}(n+6)(r_1/\delta)\alpha_{n+2}^1. \tag{2.31}$$

An elimination of α_{n+1}^0 between Equations (2.25) and (2.31) then leads to a relation of the form

$$(n+4)\alpha_n^2 = \alpha_n^0 + (n+3)(s/r_1)\alpha_n^1 - \tfrac{1}{2}(n+5)(r_1/\delta)\alpha_{n+2}^1; \tag{2.32}$$

which contains one term less than Equation (2.25).

Turning now to associated α-functions of the type α_n^3, a straightforward application of the recursion formula (2.20) for $m = 0$ discloses that

$$3\alpha_n^3 = \{2 + (s/r_2)^2\}\alpha_n^1 - \{2 + (s/\delta)\}\alpha_{n+2}^1 + (r_1/2\delta)^2\alpha_{n+4}^1; \tag{2.33}$$

while functions of the type α_n^4 expressed in terms of α_n^0 and α_n^1 follow from Equation (2.20) particularized for $m = 1$ and combined with Equations (2.32) and (2.33) as

$$\alpha_n^4 = \frac{3\alpha_n^0}{(n+4)(n+6)} + \frac{s}{4r_1}\left\{\frac{3(n^2+10n+20)}{(n+4)(n+6)} + \frac{s^2}{r_1^2}\right\}\alpha_n^1$$
$$- \frac{3r_1}{8\delta}\left\{\frac{n+6}{n+4} + \frac{n+5}{n+6}\left(\frac{2\delta s}{r_1^2}\right) + \frac{s^2}{r_1^2}\right\}\alpha_{n+2}^1$$

$$+\frac{3r_1}{8\delta}\left\{\frac{n+7}{n+6}+\frac{s}{2\delta}\right\}\alpha^1_{n+4}-\frac{1}{4}\left(\frac{r_1}{2\delta}\right)^2\alpha^1_{n+6}. \tag{2.34}$$

This completes a survey of algebraic recursion formulae for generating all associated α-functions required to describe the light curves of close eclipsing systems within minima correctly to quantities of first order in terms of the distortion of their components, and for any degree of the law of limb-darkening of their apparent discs (be n even or odd). Literal forms of some (corresponding to even values of n) will be given in Appendix II; and different expansions approximating their numerical values with arbitrary precision have been constructed in Sections 3 and 4 of Chapter I of this book.

B. RECURSION RELATIONS FOR THE BOUNDARY INTEGRALS

As to the 'boundary integrals' of the form $\Im^m_{-1,n}$ and $I^m_{-1,n}$, the recursion formula (2.11) of the present section permits us to express all $\Im^m_{-1,n}$'s as a simple linear combination of the α^m_n's provided that $n>0$. Should $n=0$, the recursion formula (2.19) can be particularized to yield

$$\begin{aligned}2\delta r_1\Im^m_{-1,0}&=2\delta s\Im^{m-1}_{-1,0}-r_1^2\Im^{m-1}_{-1,2}\\&=2\delta s\Im^{m-1}_{-1,0}+r_1^2\{\alpha^{m-1}_0-\tfrac{1}{2}(m+3)\alpha^{m-1}_2\}\end{aligned} \tag{2.34}$$

by Equation (2.17); which starting from

$$\begin{aligned}\Im^0_{-1,0}&=\left(\frac{r_2}{r_1}\right)^2 I^0_{-1,0}-\frac{\delta r_2}{r_1^2}I^1_{-1,0}\\&=\alpha^0_0-\frac{1}{\pi}\cos^{-1}\frac{s}{r_1}\end{aligned} \tag{2.35}$$

and

$$\Im^1_{-1,0}=\tfrac{3}{2}\alpha^1_0-\frac{1}{\pi}\sqrt{1-\frac{s^2}{r_1^2}} \tag{2.36}$$

permits us to generate successively all requisite $\Im^m_{-1,0}$'s for $m>1$ with the aid of the explicit forms of $I^m_{-1,0}$ listed in Appendix IV.

If we turn to the integrals of the type $I^m_{-1,n}$, a successive application of the algebraic identity

$$2\delta r_2 I^m_{-1,n}=2\delta(\delta-s)I^{m-1}_{-1,n}+r_\nu^2 I^{m-1}_{-1,n+2} \tag{2.37}$$

to reduce all $I^m_{-1,n}$'s to appropriate combinations of the $I^1_{-1,n}$'s which, in turn, are expressible (by Equation (2.22)) in terms of the associated α-functions of the form α^1_n. Doing so we eventually find that

$$I^m_{-1,n}=\sum_{j=1}^m\frac{n+2j-2}{2^j}\binom{m-1}{j-1}\left(\frac{r_1}{r_2}\right)^{n+j}\left(\frac{\delta-s}{r_2}\right)^{m-j}\left(\frac{r_1}{\delta}\right)^{j-1}\alpha^1_{n+2j-4}, \tag{2.38}$$

valid for $m>0$ and $n>0$. Should $n=0$, all expressions of the form $I^m_{-1,0}$ are elementary (see Appendix IV); while if $m=0$, Equation (2.28) yields

$$I^0_{-1,n} = \frac{n}{2}\left(\frac{\delta - s}{r_2}\right)\left(\frac{r_1}{r_2}\right)^{n+1} \alpha^1_{n-2} + \frac{n+4}{4}\left(\frac{r_1}{\delta}\right)\left(\frac{r_1}{r_2}\right)^{n+2} \alpha^1_n, \tag{2.39}$$

regardless of whether $n > 0$ is even or odd. If both m and n are zero,

$$I^0_{-1,0} = \frac{1}{\pi} \cos^{-1} \frac{\delta - s}{r_2}, \tag{2.40}$$

as can be verified by a direct resort to Equation (2.28) of Chapter II.

Lastly, Equations (2.32)–(2.33) of Chapter II for $f^{(h)}_{1,2}$ contain still certain rotational terms – of the form $\Im^0_{1,h-1}$ and $I^0_{1,h-1}$ – which have not been explicitly evaluated so far. The latter follows, to be sure, immediately from Equation (2.22) as

$$2\delta(r_2/r_1)^{h+1} I^0_{1,h-1} = r_1 \alpha^1_{h-1}, \tag{2.41}$$

but an evaluation of $\Im^0_{1,h-1}$ is more involved. In order to do so, let us return to Equation (1.59) defining $\Im^m_{\beta,\gamma}$, particularize it for $\beta = 1$, $\gamma = h - 1$; and insert for the $I^m_{0,1,h-1}$'s on its right-hand side from Equation (2.11) in terms of the requisite α-functions; the outcome discloses that

$$\begin{aligned} 2\delta r_1 \Im^m_{1,h-1} &= (m+2)\delta r_1 \alpha^{m+2}_{h-1} + (m+1)(r_2^2 - \delta^2)\alpha^{m+1}_{h-1} + (m+1)\delta r_1 \\ &\quad \times (\alpha^m_{h+1} - \alpha^m_{h-1}) + m(r_2^2 - \delta^2)(\alpha^{m-1}_{h-1} - \alpha^{m-1}_{h-1}) \end{aligned} \tag{2.42}$$

for any value of $m \geqslant 0$ – an equation which can be compared with Equation (2.17) for $\Im^m_{-1,n}$. Particularize now Equation (2.42) for $m = 0$ and take advantage of the recursion formula (2.24) for the α^2_n's; we find that

$$4\delta r_2 \Im^0_{1,h-1} = (r_1^2 + r_2^2 - \delta^2)\alpha^1_{h-1} - r_1^2 \alpha^1_{h+1}, \tag{2.43}$$

a relation of which use will be made later (cf. Chapter VI) in a modified form.

III-3. Differential Properties

In the preceding sections of this chapter we investigated the algebraic form and recursion properties of different families of special functions encountered in the theory of the light changes of close eclipsing systems. What are the *differential* properties of such functions, and which differential equations do they satisfy if we consider them as functions of the phase?

A. DIFFERENTIAL PROPERTIES OF α-FUNCTIONS

In order to investigate this subject, let us depart from the fact known to us from preceding sections: namely, that all functions of interest of the associated alpha-functions α^m_n as well as the boundary integrals $\Im^m_{\beta,\gamma}$ and $I^m_{\beta,\gamma}$ – are *homogeneous* of (zero degree) in the parameters r_1, r_2 and δ. Accordingly, Euler's theorem on homogeneous functions permits us to anticipate at once that

$$r_1 \frac{\partial \alpha^m_n}{\partial r_1} + r_2 \frac{\partial \alpha^m_n}{\partial r_2} + \delta \frac{\partial \alpha^m_n}{\partial \delta} = 0; \tag{3.1}$$

and, similarly,

$$\left\{r_1 \frac{\partial}{\partial r_1} + r_2 \frac{\partial}{\partial r_2} + \delta \frac{\partial}{\partial \delta}\right\} (\mathfrak{I}^m_{\beta,\gamma}, I^m_{\beta,\gamma}) = 0 \tag{3.2}$$

for any values of m, n or β, γ for which these functions have been defined.

Equations (3.1) or (3.2) represent, to be sure, merely algebraic relations obtaining between the first partial derivatives of the respective functions. In order to detail these in a more specific form, let us return to Equation (2.1) for α^m_n and differentiate both sides of it with respect to r_1: in doing so we find that

$$\begin{aligned} r_1^{m+n+1}&\left\{(m+n+2)\alpha^m_n + r_1 \frac{\partial \alpha^m_n}{\partial r_1}\right\} \\ &= 2nr_1 \int_{\delta - r_2}^{r_1} (r_1^2 - r^2)^{(n-2)/2} I^m_{-1,0}(\cos \theta_0) r^{m+1}\, dr \\ &= nr_2^{m+n+1} \alpha^m_{n-2}, \end{aligned} \tag{3.3}$$

yielding

$$(m+n+2)\alpha^m_n - n\alpha^m_{n-2} = -r_1 \frac{\partial \alpha^m_n}{\partial r_2} = 2\mathfrak{I}^m_{-1,n} \tag{3.4}$$

by Equation (2.17).

Differentiating Equation (2.5) next with respect to r_2, we find that

$$r_1^{m+n+2} \frac{\partial \alpha^m_n}{\partial r_2} = 2 \int_{\delta - r_2}^{r_1} (r_1^2 - r^2)^{n/2} \frac{\partial I^m_{-1,0}}{\partial r_2} r^{m+1}\, dr \tag{3.5}$$

where, by Equations (2.3) and (2.4),

$$\begin{aligned} \pi \frac{\partial I^m_{-1,0}}{\partial r_2} &= \cos^m \theta_0 \frac{\partial \theta_0}{\partial r_2} \\ &= \left\{\frac{\delta^2 + r^2 - r_\nu^2}{2\delta r}\right\}^m \frac{2r_2}{\sqrt{(2\delta r)^2 - (\delta^2 + r^2 - r_2^2)^2}}. \end{aligned} \tag{3.6}$$

Therefore, on introducing a variable x defined by

$$\delta^2 + r^2 - r_2^2 = 2\delta x, \tag{3.7}$$

so that $r_1^2 - r^2 = 2\delta(s - x)$ and $\delta^2 + r_2^2 - r^2 = 2\delta(\delta - x)$, we can rewrite (3.5) as

$$\pi r_1^{m+n+2} \frac{\partial \alpha^m_n}{\partial r_2} = 2r_2 \int_{\delta - r_2}^{s} [r_2^2 - (\delta - x)^2]^{-1/2} [2\delta(s - x)]^{n/2} x^m\, dx = 2r_2 I^m_{0,-1,n} \tag{3.8}$$

in accordance with Equation (2.27) of Chapter II. Lastly, by invoking the use of Euler's theorem (3.1) and inserting from (3.4) and (3.8) we find that

$$\begin{aligned} \pi r_1^{m+n+2} \delta \frac{\partial \alpha^m_n}{\partial \delta} &= 2\{\mathfrak{I}^m_{-1,n} - r_1^2 I^m_{0,-1,n}\} \\ &= 2\delta\{I^{m+1}_{0,-1,n} - \delta I^m_{0,-1,n}\}. \end{aligned} \tag{3.9}$$

Therefore, for all values of $m \geqslant 0$ and $n \geqslant -1$ the partial derivatives of $\alpha_n^m(r_1, r_2, \delta)$ with respect to its arguments can be expressed in a closed form in terms of the $I_{0,-1,n}^m$-integrals as

$$\pi r_1^{m+n+2} \frac{\partial \alpha_n^m}{\partial r_1} = -2\{(r_2^2 - \delta^2) I_{0,-1,n}^m + \delta I_{0,-1,n}^{m+1}\}, \tag{3.10}$$

$$\pi r_1^{m+n+2} \frac{\partial \alpha_n^m}{\partial r_2} = 2 r_2 I_{0,-1,n}^m, \tag{3.11}$$

$$\pi r_1^{m+n+2} \frac{\partial \alpha_n^m}{\partial \delta} = 2\{I_{0,-1,n}^{m+1} - \delta I_{0,-1,n}^m\}. \tag{3.12}$$

A linear combination of these equations discloses, moreover, the existence of a differential recursion formula

$$r_1 \frac{\partial \alpha_n^{m+1}}{\partial r_2} = \delta \frac{\partial \alpha_n^m}{\partial r_2} + r_2 \frac{\partial \alpha_n^m}{\partial \delta}. \tag{3.13}$$

If, in particular, the index $m = 0$, the foregoing equations (3.10)–(3.12) reduce further to

$$r_1 \frac{\partial \alpha_n^0}{\partial r_1} = \frac{2}{r_2} \left(\frac{r_2}{r_1}\right)^{n+2} \{\delta I_{-1,n}^1 - r_2 I_{-1,n}^0\}, \tag{3.14}$$

$$r_2 \frac{\partial \alpha_n^0}{\partial r_2} = 2 \left(\frac{r_2}{r_1}\right)^{n+2} I_{-1,n}^0, \tag{3.15}$$

$$\delta \frac{\partial \alpha_n^0}{\partial \delta} = -2 \frac{\delta}{r_2} \left(\frac{r_2}{r_1}\right)^{n+2} I_{-1,n}^1; \tag{3.16}$$

where the functions $I_{-1,n}^0$ and $I_{-1,n}^1$ on the right-hand sides of Equations (3.14)–(3.16) represent particular cases of integrals defined by Equation (2.28) of Chapter II, corresponding to $\beta = -1$, $\gamma = n$, and $m = 0, 1$. If, moreover, we normalize their limits to the interval (0, 1), we recognize in them integral representation of certain hypergeometric series of the type ${}_2F_1$. In point of fact, it can be shown that, for $m = 0$ or 1 only, but any value of n,

$$2\pi I_{-1,n}^m = \left(\frac{\delta}{r_2}\right)^{n/2} B\left(\frac{1}{2}, \frac{n+2}{2}\right) (2\kappa)^{n+1} {}_2F_1\left(\tfrac{1}{2} - m, \tfrac{1}{2} + m; \frac{n+3}{2}; \kappa^2\right) \tag{3.17}$$

if the eclipse is partial, and

$$I_{-1,n}^m = \left(\frac{n}{4\kappa^2}\right)^m \left(\frac{\delta}{r_2}\right)^{n/2} (2\kappa)^n {}_2F_1\left(m - \frac{n}{2}, m + \tfrac{1}{2}; 2m + 1; \frac{1}{\kappa^2}\right) \tag{3.18}$$

if it is annular; where the modulus

$$\kappa^2 = \tfrac{1}{2}\left(1 - \frac{\delta - s}{r_2}\right) = \frac{r_1^2 - (\delta - r_2)^2}{4\delta r_2} \tag{3.19}$$

is identical with that previously given by Equation (2.17) of Chapter I, or Equation (1.28) of this chapter. If an eclipse happens to be an occultation (i.e., $r_1 < r_2$), then the value of the modulus κ^2 is constrained by the inequality

$0 \leq \kappa^2 \leq \frac{1}{2}$; while for transit eclipses ($r_1 > r_2$) it can be anywhere within $0 < \kappa^2 < 1$ (see Fig. 3-1 on page 82) during partial eclipses, while during annular phase of transit eclipses $\kappa^2 > 1$.

Accordingly, during *partial* phases of the eclipses,

$$\pi r_2 \frac{\partial \alpha_n^0}{\partial r_2} = \left(\frac{r_2}{r_1}\right)^{n+2} \left(\frac{\delta}{r_2}\right)^{n/2} B\left(\frac{1}{2}, \frac{n+2}{2}\right) (2\kappa)^{n+1} {}_2F_1\left(\frac{1}{2}, \frac{1}{2}; \frac{n+3}{2}; \kappa^2\right) \quad (3.20)$$

and

$$\pi r_1 \frac{\partial \alpha_n^0}{\partial \delta} = -\left(\frac{r_2}{r_1}\right)^{n+1} \left(\frac{\delta}{r_2}\right)^{n/2} B\left(\frac{1}{2}, \frac{n+2}{2}\right) (2\kappa)^{n+1} {}_2F_1\left(-\frac{1}{2}, \frac{3}{2}; \frac{n+3}{2}; \kappa^2\right); \quad (3.21)$$

while if the eclipse becomes *annular*,

$$r_2 \frac{\partial \alpha_n^0}{\partial r_2} = 2\left(\frac{r_2}{r_1}\right)^{n+2} \left(\frac{\delta}{r_2}\right)^{n/2} (2\kappa)^n {}_2F_1\left(-\frac{n}{2}, \frac{1}{2}; 1; \frac{1}{\kappa^2}\right) \quad (3.22)$$

and

$$r_1 \frac{\partial \alpha_n^0}{\partial \delta} = -n\left(\frac{r_2}{r_1}\right)^{n+1} \left(\frac{\delta}{r_2}\right)^{n/2} (2\kappa)^{n-2} {}_2F_1\left(\frac{2-n}{2}, \frac{3}{2}; 3; \frac{1}{\kappa^2}\right). \quad (3.23)$$

The reader may note that the hypergeometric series on the right-hand sides of Equations (3.20) and (3.21) – though *not* in Equations (3.22)–(3.23) – represent spherical harmonics of fractional orders. In fact, Equations (3.20) and (3.21) can be rewritten alternatively as

$$\delta \frac{\partial \alpha_n^0}{\partial r_2} = \frac{\Gamma(1+\frac{1}{2}n)}{\sqrt{2\pi}} \left(\frac{2\delta r_2}{r_1^2}\right)^{(n+2)/2} (1-\mu^2)^{(n+1)/4} P_{-1/2}^{-(n+1)/2}(\mu) \quad (3.24)$$

and

$$\delta \frac{\partial \alpha_n^0}{\partial \delta} = -\frac{\Gamma(1+\frac{1}{2}n)}{\sqrt{2\pi}} \left(\frac{2\delta r_2}{r_1^2}\right)^{(n+2)/2} (1-\mu^2)^{(n+1)/4} P_{1/2}^{-(n+1)/2}(\mu), \quad (3.25)$$

where the $P_{\pm 1/2}^{-(n+1)/2}(\mu)$'s satisfy the Legendre differential equation

$$\frac{d}{d\mu}\left\{(1-\mu^2)\frac{d}{d\mu}\right\} P_{\pm 1/2}^{-(n+1)/2} + \frac{1}{4}\left\{1 \pm 2 - \frac{(n+1)^2}{1-\mu^2}\right\} P_{\pm 1/2}^{-(n+1)/2} = 0 \quad (3.26)$$

and

$$\mu \equiv \frac{\delta - s}{r_2} = \frac{r_2^2 - r_1^2 + \delta^2}{2\delta r_2} = 1 - 2\kappa^2. \quad (3.27)$$

Next, differentiating Equation (3.21) further with respect to δ we find that

$$\pi r_1 \frac{\partial^2 \alpha_n^0}{\partial \delta^2} = -\left(\frac{r_2}{r_1}\right)^{n+1} B\left(\frac{1}{2}, \frac{n+2}{2}\right) \frac{\partial}{\partial \delta}\left\{\left(\frac{\delta}{r_2}\right)^{n/2} (2\kappa)^{n+1} \times {}_2F_1\left(-\frac{1}{2}, \frac{3}{2}; \frac{n+3}{2}; \kappa^2\right)\right\}. \quad (3.28)$$

Since, by a well-known hypergeometric identity,

$$\frac{\partial}{\partial \kappa^2}\left\{\kappa^{n-1}\,{}_2F_1\left(-\frac{1}{2},\frac{3}{2},\frac{n+3}{2};\kappa^2\right)\right\}=\frac{n+1}{2}\,\kappa^{n-1}\,{}_2F_1\left(-\frac{1}{2},\frac{3}{2};\frac{n+1}{2};\kappa^2\right), \tag{3.29}$$

it can be shown (cf. Kopal, 1975c) that, for $n>0$, the functions satisfy a linear differential recursion formula of the form

$$\frac{\partial^2\alpha_n^0}{\partial\delta^2}-\frac{n}{2\delta}\frac{\partial\alpha_n^0}{\partial\delta}+\frac{n}{2}\frac{\delta^2+r_1^2-r_2^2}{\delta r_1^2}\frac{\partial\alpha_{n-2}^0}{\partial\delta}=0 \tag{3.30}$$

of second order, valid for *any* type of eclipse (be it partial or annular).

From all the foregoing developments it has transpired that the '*boundary correction' integrals of the type* $\mathfrak{I}_{\beta,\gamma}^m$ *and* $I_{\beta,\gamma}^m$ *can also be regarded as partial derivatives of the associated-functions with respect to the parameters* r_1, r_2 *and* δ. In particular, the $I_{\beta,\gamma}^m$-integrals have turned out play so ubiquitous a role in the entire theory of the light curves of close eclipsing systems that their particular properties deserve further study.

B. DIFFERENTIAL AND RECURSION PROPERTIES OF THE BOUNDARY CORRECTIONS: THE J-INTEGRALS

In order to investigate such properties, let us set

$$I_{\beta,\gamma}^m=(\delta/r_2)^{\gamma/2}J_{\beta,\gamma}^m, \tag{3.31}$$

and reduce the integrals $J_{\beta,\gamma}^m$ so defined to a more tractable form. If the eclipse is *partial* (i.e., $c_2=s$ in Equation (2.28) of Chapter II), this can be done by introducing a new variable u defined by

$$\delta-x=r_2(1-2\kappa^2u), \tag{3.32}$$

where κ^2 continues to be given by Equation (3.19). If so, an introduction of Equation (3.32) in Equation (2.28) of Chapter II discloses that

$$2\pi J_{\beta,\gamma}^m=(2\kappa)^{\beta+\gamma+2}\int_0^1 u^{\beta/2}(1-\kappa^2u)^{\beta/2}(1-u)^{\gamma/2}(1-2\kappa^2u)^m\,\mathrm{d}u, \tag{3.33}$$

i.e., that $J_{\beta,\gamma}^m$ turns out to depend on the geometry of eclipses only through the modulus κ. The aim of the substitution (3.31) has, in fact, been to decompose the functions $I_{\beta,\gamma}^m$ of two variables into a product of two parts, each depending on the single variable δ/r_2 or κ.

For $m=0$ or 1, the explicit forms of the respective J-integrals is already known to us from Equations (3.17) and (3.18). If, however, $m>1$, the integral on the r.h.s. of Equation (3.33) can again be recognized as an integral representation of one of Appell's generalized hypergeometric series (see Appendix III). In fact, if we abbreviate

$$a=\tfrac{1}{2}(\beta+2)\quad\text{and}\quad\beta=\tfrac{1}{2}(\beta+\gamma+4), \tag{3.34}$$

Equation (3.33) can be rewritten as

$$2\pi J^m_{\beta,\gamma} = (2\kappa)^{2(b-1)} B(a, b-a) F^{(1)}(a; 1-a, -m; b; \kappa^2, 2\kappa^2), \tag{3.35}$$

where $F^{(1)}$ denotes the first one of Appell's generalized hypergeometric functions (cf. Appell and Kampé de Fériet, 1926) in two variables $x \equiv \kappa^2$ and $y \equiv 2\kappa^2$.

The function $F^{(1)}$ is known to satisfy, in general, a certain system of two simultaneous partial differential equations of second order. If, however, as in the present case the ratio x/y is constant, this system can be reduced to a single ordinary differential equation which was obtained by Burchnall (1942). From his results we deduce that if t denotes the operator

$$t \equiv \kappa^2 \frac{\mathrm{d}}{\mathrm{d}\kappa^2}, \tag{3.36}$$

the function $F^{(1)}$ on the r.h.s. of Equation (3.35) satisfies an ordinary differential equation

$$\begin{aligned} &t(t+b-1)(t+b-2)F^{(1)} - \kappa^2(3t+1-a-2m)(t+b-1)(t+a)F^{(1)} \\ &\quad + 2\kappa^4(t+1-a-m)(t+a)(t+a+1)F^{(1)} = 0 \end{aligned} \tag{3.37}$$

of third order. Moreover, a substitution

$$J^m_{\beta,\gamma} = \kappa^{2(b-1)} F^{(1)} \tag{3.38}$$

in Equation (3.37) leads to

$$\begin{aligned} &t(t-1)(t+1-b)J^m_{\beta,\gamma} - \kappa^2(3t+4-a-3b-2m)t(t+a+1-b)J^m_{\beta,\gamma} \\ &\quad + 2\kappa^4(t+2-a-b-m)(t+a+1-b)(t+a+2-b)J^m_{\beta,\gamma} = 0, \end{aligned} \tag{3.39}$$

disclosing that the function $J^m_{\beta,\gamma}$ – and, therefore, the derivative $\partial\alpha^0_n/\partial\delta$ by Equation (3.16) – satisfies a linear differential equation of *third* order – and, therefore, α^0_n itself will satisfy such an equation of *fourth* order (for fuller details cf. Alkan, 1978).

A solution of Equation (3.39) for $J^m_{\beta,\gamma}$ can be sought as a series in ascending powers of κ^2. Since the operator t occurs as a factor of the first as well as the second term on the l.h.s. of Equation (3.34), the unit difference between the roots of the respective indicial equation will not give rise to any logarithmic singularity at the origin. If we construct such a particular solution by standard methods, we find it to assume the form

$$2\pi J^m_{\beta,\gamma} = 4^{b-1} B(a, b-a) \sum_{j=0}^{\infty} \frac{a_j(1-a)_j}{j!(b)_j} {}_2F_1(-j, -m; a-j; 2)\kappa^{2(b+j-1)}, \tag{3.40}$$

where $(a)_j \equiv a(a+1)(a+2)\ldots(a+j-1)$, $(a)_0 = 1$ denotes the customary Pochhammer symbols, and ${}_2F_1(-j, -m, a-j, 2)$ stands for a Jacobi polynomial $G_j(-m-j, a-j, 2)$ of degree j. The foregoing series (3.40) constitutes the most general representation of the integral $J^m_{\beta,\gamma}$ valid for all (not necessarily integral) values of β, γ or m-provided only that their combination is such as to make the series on the r.h.s. of Equation (3.40) convergent. Since, in our problem m is restricted to assume the values of zero or a positive integer, while $\beta + \gamma > -2$, the absolute and uniform convergence of the expansion on the r.h.s. of Equation (3.40) is assured.

It may also be noted that, for three particular values of the parameters β, γ and m, the differential equation (3.37)) governing the function $F^{(1)}$ reduces to one of second order. This will happen first if $m = 0$; for the expansion on the r.h.s. of Equation (3.40) reduces then to an ordinary hypergeometric series; and, as a result,

$$2\pi J^0_{\beta,\gamma} = 4^{b-1}B(a, b-a)\kappa^{2(b-1)}{}_2F_1(a, 1-a; b; \kappa^2), \tag{3.41}$$

in conformity with Equation (3.17).

The second reducible case arises when $\gamma = 0$ and, therefore, $a = b - 1$. If so, Equation (3.27) can be reduced (by removal of a factor) to

$$\{t - \kappa^2(3t + 2 - b - 2m) + 2\kappa^4(t + 2 - b - m)\}(t + b - 1)F^{(1)} = 0, \tag{3.42}$$

which is also one of second order (cf. Heun, 1889). A third reducible case arising when $2(\beta + m + 2) + \gamma = 0$ (Chaundy, 1943) has no relevance to our applications, and is being mentioned only for the sake of completeness.

All equations given so far hold good only for partial eclipses. Should the latter become *annular*, the substitution required to reduce $J^m_{\beta,\gamma}$ to tractable form becomes

$$\delta - x = r_2(1 - 2v) \tag{3.43}$$

in place of Equation (3.32) used formerly; and

$$\pi J^m_{\beta,\gamma} = 2^{\beta+\gamma+1}\kappa^\gamma \int_0^1 v^{\beta/2}(1-v)^{\beta/2}(1-\kappa^{-2}v)^{\gamma/2}(1-2v)^m \, dv; \tag{3.44}$$

and the integral on the r.h.s. represents once more Appell's generalized hypergeometric series of the form

$$2\pi J^m_{\beta,\gamma} = 4^{b-1}B(a, a)\kappa^{2(b-a-1)}F^{(1)}(a; -m, a-b+1; 2a; 2, \kappa^{-2}). \tag{3.45}$$

The differential equation governing this function is again of the form (3.39), and can be obtained from it by an appropriate permutation of parameters. Its solution is, in turn, expressible as

$$2\pi J^m_{\beta,\gamma} = 4^{b-1}B(a, a) \sum_{j=0}^{\infty} \frac{(a)_j(a-b+1)_j}{j!(2a)_j} {}_2F_1(-m, a+j; 2a+j; 2)\kappa^{2(b-a-j-1)}. \tag{3.46}$$

If $m = 0$, this series reduces to

$$2\pi J^0_{\beta,\gamma} = 4^{b-1}B(a, a)\kappa^{2(b-a-1)}{}_2F_1(a, a-b+1, 2a; \kappa^{-2}); \tag{3.47}$$

while if, in addition, $\gamma = 0$ (i.e., $a = b - 1$),

$$2\pi J^0_{\beta,0} = 4^a B(a, a). \tag{3.48}$$

This latter expression no longer involves κ and, therefore, does not vary during annular phase.

As proved to be the case with associated α-functions, not all J-integrals need to be literally evaluated; for the task of establishing their numerical values for particular geometric configurations can be greatly facilitated by numerous *recursion formulae* relating the $J^m_{\beta,\gamma}$'s for different values of β, γ and m in algebraic manner. Some of these are trivial (representing simple identities); while others (less obvious) are based on more complicated recursion properties of generalized hypergeometric functions (not all of which are yet adequately explored).

Of such *three-term* recursion formulae, two are trivial: namely,

$$J^m_{\beta+2,\gamma} = J^m_{\beta,\gamma} - J^{m+2}_{\beta,\gamma} \tag{3.49}$$

and

$$J^m_{\beta,\gamma+2} = 2\{J^{m+1}_{\beta,\gamma} - \mu J^m_{\beta,\gamma}\}, \tag{3.50}$$

where μ continues to be given by Equation (3.27); but the third,

$$\gamma J^m_{\beta+2,\gamma-2} = (\beta + m + 2)J^{m+1}_{\beta,\gamma} - mJ^{m-1}_{\beta,\gamma} \tag{3.51}$$

is not; and was discovered only recently by Lanzano (1976b); though its particular form for $m = 0$ was known already to Kopal (1959; p. 215).*

Four-term recursion formulae are indeed many; and have been deduced mainly by Lanzano (1976a, b, c) from recursion formulae satisfied by Appell's generalized hypergeometric series. Thus (cf. Lanzano, 1976a)

$$\begin{aligned}(2\beta + \gamma + 2m + 4)J^m_{\beta,\gamma+2} + 4(\beta + \gamma + m + 2)\mu J^m_{\beta,\gamma}\\ = 4\gamma(1 - \mu^2)J^m_{\beta,\gamma-2} + 4mJ^{m-1}_{\beta,\gamma},\end{aligned} \tag{3.52}$$

while

$$\begin{aligned}(2\beta + \gamma + 2m + 4)J^m_{\beta,\gamma+2} + 2(4\beta + 3\gamma + 4m + 6)\mu J^m_{\beta,\gamma}\\ = 8(\gamma - 2)\mu(1 - \mu^2)J^m_{\beta,\gamma-4} - 4[2(\beta + \gamma) - (2\beta + 3\gamma + 2m)(1 - \mu^2)]J^m_{\beta,\gamma}\end{aligned} \tag{3.53}$$

valid for any type of eclipse.

Moreover, a combination of the foregoing Equations (3.52) and (3.53) with the three-term recursion formulae (3.49)–(3.51) leads to other similar relations of interest. Thus an elimination of $J^m_{\beta,\gamma+2}$ between Equations (3.50) and (3.52) leads to a relation of the form

$$(2\beta + \gamma + 4m + 4)J^{m+1}_{\beta,\gamma} + \gamma\mu J^m_{\beta,\gamma} = 2\gamma(1 - \mu^2)J^m_{\beta,\gamma-2} + 2mJ^{m-1}_{\beta,\gamma}; \tag{3.54}$$

while an elimination of $J^m_{\beta,\gamma}$ between the same equations yields

$$4(\beta + \gamma + m + 2)J^{m+1}_{\beta,\gamma} - \gamma J^m_{\beta,\gamma+2} = 4\gamma(1 - \mu^2)J^m_{\beta,\gamma-2} + 4mJ^{m-1}_{\beta,\gamma}. \tag{3.55}$$

Another important four-term recursion formula

$$4\{J^m_{\beta+2,\gamma-2} + \mu J^m_{\beta,\gamma}\} = 4(1 - \mu^2)J^m_{\beta,\gamma-2} - J^m_{\beta,\gamma+2} \tag{3.56}$$

* Cf. also Equation (2.28), deduced by Kopal (1978a), which in the present notations is equivalent to $(\gamma + 4)J^1_{-1,\gamma+2} = 2(\gamma + 2)(J^0_{-1,\gamma} - \mu J^1_{-1,\gamma})$.

has been discovered by Lanzano (1976b), which allows for a simultaneous variation of β and γ for constant m.

Furthermore, an elimination of $J^m_{\beta,\gamma+2}$ between Equations (3.50) and (3.56) yields

$$2J^m_{\beta+1,\gamma-2} + \mu J^m_{\beta,\gamma} = 2(1-\mu^2)J^m_{\beta,\gamma-2} - J^{m+1}_{\beta,\gamma}; \tag{3.57}$$

while an elimination of $J^m_{\beta,\gamma+2}$ between Equations (3.52) and (3.56) discloses that

$$\begin{aligned}(2\beta+\gamma+2m+4)J^m_{\beta+2,\gamma-2} + (\beta+m+2)\mu J^m_{\beta,\gamma}\\ = 2(\beta+\gamma+m)(1-\mu^2)J^m_{\beta,\gamma-2} - mJ^{m-1}_{\beta,\gamma}.\end{aligned} \tag{3.58}$$

If, in (3.58), we set $m=0$ and replace γ by $\gamma+2$, we obtain the second of Equations (5-55) on p. 216 of Kopal (1959). Similarly, eliminating $J^m_{\beta,\gamma-2}$ between Equations (3.49) and (3.56) we obtain

$$2\gamma J^m_{\beta+2,\gamma-2} + 2mJ^{m-1}_{\beta,\gamma} = (\beta+m+2)\{J^m_{\beta,\gamma+2} + 2\mu J^m_{\beta,\gamma}\}; \tag{3.59}$$

which for $m=0$ reduces to the first of Equations (5-54) of Kopal (*op. cit.*). Lastly, an elimination of $J^m_{\beta,\gamma-2}$ between Equations (3.54) and (3.57) leads to a three-term recursion formula (3.51) noted before.

Rewriting again Equation (3.50) for $\gamma-2$ and eliminating $J^m_{\beta,\gamma-2}$ between Equations (3.50) and (3.57) we find that

$$2\mu J^m_{\beta+2,\gamma-2} + \mu J^{m+1}_{\beta,\gamma} = 2(1-\mu^2)J^{m+1}_{\beta,\gamma-2} - J^m_{\beta,\gamma}. \tag{3.60}$$

Lanzano (1976b) also proved that

$$\beta J^m_{\beta-2,\gamma+2} + (2\beta+\gamma+2m+2)\mu J^m_{\beta,\gamma} = 2\gamma(1-\mu^2)J^m_{\beta,\gamma-2} + 2mJ^{m-1}_{\beta,\gamma}; \tag{3.61}$$

which for $m=0$ reduces again to one of Equations (5-55) on p. 216 of Kopal (1959). Moreover, an elimination of $J^m_{\beta,\gamma-2}$ between Equations (3.54) and (3.61) leads to another three-term recursion formula

$$\beta J^m_{\beta-2,\gamma+2} = (2\beta+\gamma+2m+4)J^{m+1}_{\beta,\gamma} - 2(\beta+m+1)\mu J^m_{\beta,\gamma}; \tag{3.62}$$

while an elimination of $J^m_{\beta,\gamma-2}$ between Equations (3.52) and (3.61) yields

$$2\beta J^m_{\beta-2,\gamma+2} = 2(\gamma+2)\mu J^m_{\beta,\gamma} + (2\beta+\gamma+2m+4)J^m_{\beta,\gamma+1}, \tag{3.63}$$

which for $m=0$ reduces to the second of Equations (5-54) on p. 216 of Kopal (1959). Lastly, an elimination of $J^m_{\beta,\gamma-2}$ between Equations (3.58) and (3.61) leads to a four-term recursion formula

$$\begin{aligned}(2\beta+\gamma+2m+4)\{\gamma J^m_{\beta+2,\gamma-2} + mJ^{m-1}_{\beta,\gamma}\}\\ = (\beta+m+2)\{\beta J^m_{\beta-2,\gamma+2} + 2(\beta+m+1)\mu J^m_{\beta,\gamma}\}.\end{aligned} \tag{3.64}$$

In addition, the $J^m_{\beta,\gamma}$-integrals satisfy a great many *five-term* recursion formulae (some of which can be found in Lanzano (1976b, c); but in view of their lesser practical value we shall not reproduce them in this place.

C. DIFFERENTIAL PROPERTIES OF HANKEL TRANSFORMS

In conclusion of this chapter, let us turn to the differential properties of the associated α-functions re-written in terms of the *Hankel transforms* introduced in Section 3 of Chapter I. The reader may recall that, in that section, we found it possible to express the associated α-functions of zero order exactly as

$$\alpha_n^0 = 2^\nu \Gamma(\nu) \int_0^\infty (kx)^{-\nu} J_\nu(kx) J_2(x) J_0(hx)\,\mathrm{d}x, \tag{3.65}$$

where

$$\nu = \frac{n+2}{2}; \tag{3.66}$$

while h, k and x continue to be defined by Equations (3.33) and (3.35) of Chapter I.

Let us differentiate now the foregoing Equation (3.65) with respect to the parameters r_1, r_2 and δ (involved in h and k) behind the integral sign. Since each one of these parameters occurs in the argument of only one of the Bessel functions constituting the integrand on the right-hand side of Equation (3.66), the use of the well-known recursion formulae

$$\frac{\mathrm{d}}{\mathrm{d}x}\left\{\frac{J_\nu(x)}{x^\nu}\right\} = -\frac{J_{\nu+1}(x)}{x^\nu} \tag{3.67}$$

and

$$\frac{2\nu}{x} J_\nu(x) = J_{\nu-1}(x) + J_{\nu+1}(x), \tag{3.68}$$

satisfied by Bessel functions of any order (integral or fractional), a partial differentiation with respect to $r_{1,2}$ and δ discloses that

$$r_1 \frac{\partial \alpha_n^0}{\partial r_1} = -2^\nu \Gamma(\nu) k^{1-\nu} \int_0^\infty x^{1-\nu} J_{\nu+1}(kx) J_1(x) J_0(hx)\,\mathrm{d}x, \tag{3.69}$$

$$r_2 \frac{\partial \alpha_n^0}{\partial r_2} = 2^\nu \Gamma(\nu) k^{-\nu} \int_0^\infty x^{1-\nu} J_\nu(kx) J_0(x) J_0(hx)\,\mathrm{d}x, \tag{3.70}$$

$$\delta \frac{\partial \alpha_n^0}{\partial \delta} = -2^\nu \Gamma(\nu) h k^{-\nu} \int_0^\infty x^{1-\nu} J_\nu(kx) J_1(hx) J_1(x)\,\mathrm{d}x. \tag{3.71}$$

If we compare now the foregoing equations with Equations (3.10)–(3.12) specifying the same derivatives in terms of the $I^m_{0,-1,n}$ integrals, it follows at once that, *for* $0 \leqslant m < 2$,

$$\begin{aligned} I^m_{-1,n} &= 2^{\nu-1}\Gamma(\nu) k^\nu \int_0^\infty x^{1-\nu} J_\nu(kx) J_m(x) J_m(hx)\,\mathrm{d}x \\ &= (2/b^2)^{\nu-1} a^\nu \Gamma(\nu) \int_0^\infty y^{1-\nu} J_\nu(ay) J_m(by) J_m(cy)\,\mathrm{d}y, \end{aligned} \tag{3.72}$$

where a, b, c and y are defined by Equations (3.54) and (3.56)–(3.58) of Chapter I.

Since, moreover, by Bateman's expansion (3.81) of that chapter,

$$J_\nu(ay)J_m(by) = (2/y)a^\nu b^{-m} \sum_{j=0}^{\infty} (-1)^j(m+\nu+2j+1)\frac{\Gamma(m+\nu+j+1)\Gamma(\nu+j+1)}{j!\Gamma(m+j+1)\{\Gamma(\nu+1)\}^2} \times \{G_{m+j}(\nu+1-m, \nu+1, a)\}^2 J_{m+\nu+2j+1}(y) \tag{3.73}$$

while, by Equation (3.86) of Chapter I,

$$\int_0^\infty y^{-\nu}J_m(cy)J_{m+\nu+2j+1}(y)\,dy = \frac{\Gamma(m+j+1)c^m(1-c^2)^\nu}{2^\nu m!\Gamma(\nu+j+1)} \times G_j(m+\nu+1, m+1; c^2), \tag{3.74}$$

it follows that

$$I^m_{-1,n} = \left(\frac{a}{b}\right)^{2\nu}\left(\frac{c}{b}\right)^m \frac{b^2(1-c^2)^\nu}{m!\nu\Gamma(\nu+1)} \sum_{j=0}^{\infty} \frac{(-1)^j}{j!}(m+\nu+2j+1)\Gamma(m+\nu+j+1) \times \{G_{m+j}(\nu+1-m, \nu+1, a)\}^2 G_j(m+\nu+1, m+1; c^2), \tag{3.75}$$

valid for $m = 0, 1$ and any value of n.

The foregoing equation (3.75) represents a similar Fourier approximation to the $I^m_{-1,n}$-integrals as Equation (3.91) of Chapter I did for the α^0_n's. Moreover, by a comparison of Equations (2.22) and (3.72) it follows at once that, for the indices $m = 0$ as well as 1,

$$\begin{aligned} \alpha^m_n &= 2^\nu\Gamma(\nu)\int_0^\infty (kx)^{-\nu}J_{m+\nu}(kx)J_1(x)J_m(hx)\,dy \\ &= 2^\nu\Gamma(\nu)b\int_0^\infty (ay)^{-\nu}J_{m+\nu}(ay)J_1(by)J_m(cy)\,dy, \end{aligned} \tag{3.76}$$

which can likewise be represented (by the same method) by an expansion of the form

$$\alpha^m_n = \frac{(ac)^m(1-c^2)^{\nu+1}}{m!(\nu)_{m+1}} \sum_{j=0}^{\infty} \frac{(-1)^j}{(j+1)!}(m+\nu+2j+2)(\nu+j+2) \times (j+1)_m(m+\nu-1)_j\{G_{j+1}(m+\nu, m+\nu+1, a)\}^2 \times G_j(m+\nu+2, m+1; c^2), \tag{3.77}$$

which for $m = 0$ reduces indeed to Equation (3.91) of Chapter I. Equation (3.76) or its Fourier expansion (3.77) is, however, valid only for $m = 0$ and 1; and *not* for an arbitrary value of m. (For $m > 1$, cf. Appendix V.)

It should, moreover, be kept in mind that *Fourier approximations of the form*

(3.75) *or* (3.77) *are integrable, but not differentiable.* Therefore, if we would differentiate the series on the r.h.s. of Equation (3.77) for α_n^0 with respect to the parameters r_1, r_2 or δ, the results would *not* agree with Equation (3.75) above – in spite of the fact that Equations (3.14)–(3.16) as well as (3.72) are exact; because the Fourier series (3.75) or (3.77) representing (3.72) or (3.76) are not differentiable.

Furthermore, by a comparison of Equations (3.4) and (3.69) of this chapter it follows at once that the 'boundary integrals' of the form $\mathfrak{I}^0_{-1,n}$ are likewise expressible as Hankel transforms

$$\mathfrak{I}^0_{-1,n} = 2^{\nu-1}\Gamma(\nu)\int_0^\infty (kx)^{1-\nu}J_{\nu+1}(kx)J_1(x)J_0(hx)\,\mathrm{d}x, \tag{3.78}$$

as the reader can verify independently by a combination of the recursion formula (2.17) for the associated α-functions with their Hankel transform expressions (3.65) and the recursion formulae (3.67)–(3.68) valid for the Bessel functions; in fact, the existence of the recursion formula (2.17) for the α-functions is (in the Hankel-transform formalism) a direct consequence of Equation (3.68).

The value of the integral on the r.h.s. of Equation (3.78) can again be approximated by Fourier expansions similar to Equations (3.75) or (3.77) valid for the α_n^m's or $I^m_{-1,n}$'s: in point of fact, by an appeal to Bateman's expansion (3.73) and integral formula (3.74) we similarly establish that

$$\mathfrak{I}^0_{-1,n} = \frac{a^2(1-c^2)^\nu}{\nu(\nu+1)\Gamma(\nu+2)}\sum_{j=0}^\infty \frac{(-1)^j}{j!}(\nu+2j+3)\Gamma(\nu+j+3) \times\{G_{j+1}(\nu+1,\nu+2;a)\}^2 G_{j+1}(\nu+1,1,c^2). \tag{3.79}$$

Moreover, it has recently been shown by Demircan (1977) that functions of the type α_n^0 and $\mathfrak{I}^0_{-1,n}$ are related by the rapidly-converging expansion of the form

$$\alpha_n^0 = \frac{\alpha_0^0}{\nu} + \sum_{j=1}^\infty \frac{\nu\{\Gamma(\nu)\}^2 H_{2j}}{\Gamma(\nu+1-j)\Gamma(\nu+1+j)}, \tag{3.80}$$

where the H_{2j}'s stand for the integrals

$$H_{2j} \equiv \int_0^\infty J_{2j}(kx)J_1(x)J_0(hx)\,\mathrm{d}x. \tag{3.81}$$

For $j=0$, a comparison of Equation (3.81) with (3.65) discloses at once that $H_0 \equiv \frac{1}{2}\alpha_0^0$. For $j=1$, a similar comparison of Equation (3.81) with (3.78) shows that

$$H_2 \equiv \mathfrak{I}^0_{-1,0}; \tag{3.82}$$

and a repeated use of recursion formulae (3.67)–(3.68) valid for the Bessel functions on the r.h.s. of Equation (3.81) reveals that

$$H_4 \equiv 3\mathfrak{I}^0_{-1,2} - \mathfrak{I}^0_{-1,0}, \tag{3.83}$$

$$H_6 \equiv 10\mathfrak{I}^0_{-1,4} - 8\mathfrak{I}^0_{-1,2} + \mathfrak{I}^0_{-1,0}, \tag{3.84}$$

$$H_8 \equiv 35\mathfrak{I}^0_{-1,6} - 45\mathfrak{I}^0_{-1,4} + 15\mathfrak{I}^0_{-1,2} - \mathfrak{I}^0_{-1,0}, \tag{3.85}$$

etc. Since, in general, $\mathfrak{I}^0_{-1,n+2} \ll \mathfrak{I}^0_{-1,n}$, a convergence of the expansion on the r.h.s. is very rapid: indeed, Demircan (*op. cit.*) demonstrated that a retention of only the first four terms in the summation on the r.h.s. of Equation (3.80) yields values of α_1^0 correct to six decimal places.

Lastly, it can be shown that the associated α-functions of zero index satisfy also a *linear* homogeneous differential equation in h, with *constant* coefficients, of *infinite* order (though this order can be reduced to a finite one for any desired degree of approximation. In order to demonstrate this, differentiate Equation (3.71) repeatedly with respect to h. If we abbreviate

$$h \frac{\mathrm{d}}{\mathrm{d}h} \equiv t, \tag{3.86}$$

a repeated application of this operator on both sides of Equation (3.71) discloses that

$$t(t-2)(t-4)\ldots(t-2n)\alpha_l^0$$
$$= (-1)^{n+1} 2^\nu \Gamma(\nu) \int_0^\infty (kx)^{-\nu} J_\nu(kx) J_1(x) J_{n+1}(hx)(hx)^{n+1}\,\mathrm{d}x. \tag{3.87}$$

Since however,

$$J_n(hx) = \sum_{j=0}^{\infty} \frac{(-1)^j (hx)^{n+2j}}{2^{n+2j}(j!)\Gamma(n+j+1)}, \tag{3.88}$$

it follows that

$$\sum_{n=0}^{\infty} \frac{J_n(hx)}{2^n(n!)}(hx)^n = 1; \tag{3.89}$$

so that if we form the sum of an infinite number of equations of the type (3.87), each divided by the factor $2^n(n!)$, we obtain

$$\left\{1 + \sum_{n=1}^{\infty} (-1)^n \frac{t(t-2)(t-4)\ldots(t-2n+2)}{2^n n!}\right\} \alpha_l^0$$
$$= 2^\nu \Gamma(\nu) \int_0^\infty \frac{J_\nu(kx)}{(kx)^\nu} J_1(x) \sum_{n=0}^{\infty} \frac{J_n(hx)}{2^n n!}(hx)^n\,\mathrm{d}x$$
$$= 2^\nu \Gamma(\nu) \int_0^\infty (kx)^{-\nu} J_\nu(kx) J_1(x)\,\mathrm{d}x, \tag{3.90}$$

which represents a linear differential equation for α_l^0 as a function of h, with constant coefficients and constant absolute term on the right-hand side.

The integral on the right-hand side of the foregoing equation can, moreover, be easily evaluated in a closed form; for if $k \equiv r_2/r_2 \leq 1$ (corresponding to an occultation eclipse),

$$2^\nu \Gamma(\nu) \int_0^\infty (kx)^{-\nu} J_\nu(kx) J_1(x)\, dx = \frac{1}{\nu} = \frac{2}{l+2} \tag{3.91}$$

is equal to the maximum value of α_l^0 attained during totality; while if the eclipse is a transit (i.e., $k \geq 1$),

$$2^\nu \Gamma(\nu) \int_0^\infty (kx)^{-\nu} J_\nu(kx) J_1(x)\, dx = k^{-2} {}_2F_1(1, 1-\nu; 2; k^{-2})$$

$$= \frac{1 - \nu(1 - k^{-2})^\nu}{\nu} \tag{3.92}$$

signifies the loss of light at the moment of central eclipse (when $\delta = 0$). If, moreover, we set

$$\alpha_l^0(h, k) = \begin{cases} \dfrac{2}{l+2} - f_l(h, k), & \text{(occultation)} \\[2ex] \dfrac{2}{l+2}\{1 - (1 - k^{-2})^\nu\} - f_l(h, k), & \text{(transit)} \end{cases} \tag{3.93}$$

it follows that the differential Equation (3.90) rewritten in terms of f_l will assume the form

$$\left\{1 + \sum_{n=1}^{\infty} (-1)^n \frac{t(t-2)(t-4)\ldots(t-2n+2)}{2^n n!}\right\} f_l = 0, \tag{3.94}$$

which (cf. Demircan, 1978d) is tantamount to a requirement that

$$(t-2)(t-4)\ldots(t-2n) f_l = 0. \tag{3.95}$$

Equations (3.94) or (3.95) are linear and homogeneous in f_l; moreover, they possess the remarkable property that the coefficients of the derivatives $t^n f_l$ on their left-hand sides are pure numbers!

Like Equation (3.90), Equations (3.94) or (3.95) are exact only for $n \to \infty$; however, for finite values of n, they continue to approximate the exact solutions more closely the larger the value of n. The ratio k of the radii of the two components enters the solutions of Equations (3.94) or (3.95) only through the boundary conditions required to specify their desired particular solutions.

Bibliographical Notes

III-**1**. The material of this section has been taken very largely (with some subsequent additions) from Z. Kopal (1942). It was in this latter paper that the associated α-functions of any order and index (as well as the requisite 'boundary corrections' integrals) were evaluated explicitly in terms of the elements of the eclipse in a closed form.

III-**2**. The first recursion relations satisfied by the associated α-functions, and the related boundary integrals of the form $\mathfrak{I}^m_{\beta,\gamma}$ or $I^m_{\beta,\gamma}$ were reported in Z. Kopal (1947). A more complete account of them can be found in Section IV.5 of Kopal's *Close Binary Systems* (1959); in particular, the fundamental recursion formulae (2.11) and (2.17), or (2.20) and (2.21) appeared in that source for the first time. A large part of the material presented in this section is, however, new and has not previously been published.

III-**3**. For the first differential properties of the α^m_n's, $\mathfrak{I}^m_{\beta,\gamma}$'s or $I^m_{\beta,\gamma}$'s, cf. Chapter VIII of Z. Kopal (1946), augmented by those reported since by the same author in Kopal (1947) as well as Section IV.5 of his *Close Binary Systems* (Kopal, 1959). It is interesting that the validity of Euler's condition (3.1) in the particular case of $m = 0$ and $n = 0,1$ was noted already by A. B. Wyse (1939). The general differential properties of the α^m_n's represented by Equations (3.10)–(3.12) go back to Kopal (1959), and so does the difference-differential equation (3.13); but that represented by Equation (3.30) was not discovered till much later (Kopal, 1975c).

The relation (3.29) between the $J^m_{\beta,\gamma}$-integrals and Appell's generalized hypergeometric functions of the form $F^{(1)}$ goes back to Section IV.5 of Z. Kopal (1959); and so does the differential equation (3.39) satisfied by them; while the explicit form of a fourth-order equation of α^0_n has recently been worked out by a combination of Equations (3.16), (3.31) and (3.39) by H. Alkan (1978).

As regards the recursion relations for the $J^m_{\beta,\gamma}$-integrals, three-term recursion formulae given in this section have mostly been given already in Kopal (1959; Section IV.5). Four-term recursion formulae given in the text have all been discovered more recently by P. Lanzano (1976a, b, c). The same author noted also the existence of (some) five term recursion formulae satisfied by these integrals; but our knowledge of them is so far incomplete.

Lastly, the differential and recursion formulae for the α^0_n, $\mathfrak{I}^m_{-1,n}$ and $I^m_{-1,n}$ functions (with $m = 0$ or 1) based on their formulation in terms of the Hankel transforms, as given in the text, are all due to Z. Kopal (1977b, c) – with the exception of Equation (3.80) which is due to O. Demircan (1977).

PART II

DECODING OF THE LIGHT CHANGES

In the first part of this book (Chapters I–III) an outline has been given of the theory of light changes of close eclipsing systems – between minima as well as within eclipses – which should permit us to evaluate the instantaneous light of the system, at any moment of its orbital cycle, in terms of its geometrical elements (i.e., fractional radii and orbital inclination) as well as of the physical parameters (fractional luminosities, mass-ratio, coefficients of limb- or gravity-darkening). These elements are, however, not known to us *a priori*; or indicated only indirectly from other (spectroscopic) evidence; the principal avenue of approach to their determination is indeed the observed light changes. Therefore, in the second half of this volume, commencing with Chapter IV, we wish now to address ourselves to a problem *inverse* to that treated in the first: namely, to an extraction of the elements of an eclipsing system from an analysis of its observed light changes.

The tasks confronting us in this connection can be aptly compared with the *decoding* of the messages which reach us from such systems on the wings of their variable light; and for such tasks theoretical light curves discussed in Chapters I–III furnish a physically acceptable code. The code now in our hands is, however, still far from being completely known: its principal limitation rests on the fact that its foundations are based wholly on the *equilibrium* model of the respective systems; and pays as yet no respect to dynamical phenomena which may complicate the observed photometric features.

Even within the framework of restrictions on which our model is based, the number of the elements required for its specification is considerable – consisting as it does of the *orbital elements*

semi-major axis of the relative orbit	a
orbital eccentricity	e
longitude of periastron	ω
longitude of ascending node	Ω
orbital period	P
epoch of principal conjunction	t_0;

the *elements of the eclipse*

fractional radius of the component undergoing eclipse	r_1
fractional radius of the eclipsing star	r_2
inclination of their orbit to the celestial sphere	i
fractional luminosities of the components	$L_{1,2}$;

and the elements characterizing the *physical properties* of the two stars: namely, their

surface brightness $J_{1,2}$
coefficients of limb-darkening $u_{1,2}$
coefficients of gravity darkening $\tau_{1,2}$
mass-ratio of the system m_1/m_2

in addition to the absolute values of the masses, radii and luminosities of the two components.

In order to determine the latter, the photometric evidence alone may not suffice: additional spectroscopic evidence may be required to specify absolute masses and dimensions of the two stars;* while in order to determine their absolute luminosities also a knowledge of the parallax (and, for more distant stars, of the absorption of light in interstellar space along the line of sight) becomes prerequisite. Of the elements specifying the position of the orbital plane in space, the longitude Ω of the ascending node remains generally inaccessible to determination on the basis of photometric or spectroscopic data; and only two known eclipsing binaries – ϵ Aurigae and VV Cephei – exhibit orbits which are large enough to be measured astrometrically (for ϵ Aur, cf. Strand, 1959; for VV Cep, Fredricks, 1960; or van de Kamp, 1978). As regards the inclination i of the orbital plane to the celestial sphere, the phenomena of eclipses will enable us to determine its absolute value, but *not* its algebraic sign. The orbital eccentricity e and the longitude of periastron ω can, in principle, be established independently from either spectroscopic or photometric data alone; though, in general, their photometric determination by the methods of Section IV-5 is superior in accuracy to values deduced from the asymmetry of the radial-velocity curves.

For a determination of the absolute properties of eclipsing binary systems – such as masses and dimensions of their constituent components – spectroscopic determinations of the radial velocity changes are necessary, but *not* sufficient; the clue to them rests in a knowledge of the angle of inclination i which can be determined from the photometric observations of such systems if – and only if – they happen to be eclipsing variables. It is true that the *ratios* of the masses of the two components as well as of their radii can be determined either from spectroscopic or (for close binary systems) photometric observations alone without a knowledge of the angle of inclination i; but for a determination of the individual values of masses or dimensions of the constituent star a knowledge of i constitutes a necessary prerequisite. Lastly, the orbital period P – a determination of which furnishes (through Kepler's third law of periodic times) a relation between the total mass and absolute dimensions of the system – can, in general, be deduced from extended series of observations to a degree of accuracy far surpassing that of any other element by methods which are sufficiently well known to warrant repetition in this place (cf., e.g., Hagen, 1921); and the time t_0 of the minimum of light can likewise be determined directly from the observations, and independently of all other elements.

* An exception arises for systems containing X-ray sources (or pulsars) as one of their components. Periodic oscillations of the arrival time of their flicker, reflecting the absolute motion of the respective component along the line of sight (and due to the 'light equation' in the binary orbit; cf. Kopal, 1978b; pp. 304–309) contain exactly the same information on their orbital motion as would a spectroscopic orbit of the same star.

CHAPTER IV

ANALYSIS OF THE LIGHT CHANGES IN THE TIME DOMAIN

After the preliminary remarks contained in the foregoing paragraphs, let us turn now to the actual task of an analysis of the observations of instantaneous brightness of an eclipsing variable in the course of time, with the aim of deducing from them the elements of the system enumerated in the introduction. Photometric observations of such systems performed by different methods (visual, photographic, photoelectric) furnish, in general, a series of discrete values of instantaneous brightness of the system prevalent at different time or phase of the orbital cycle. If both components are spherical, the light of the system should remain constant between eclipses,* and will hereafter be adopted as a unit in terms of which the instantaneous luminosity l observed at a time t, or phase angle

$$\theta = (2\pi/P)(t - t_0) \tag{0.1}$$

can be expressed. If, moreover, the orbital plane of our binary is inclined by an angle of $90° \pm i$ to the line of sight and R stands for the separation of both components hereafter regarded as constant,† the apparent separation δ of the centres of the two stars projected on the celestial sphere will be given by the equation

$$\delta^2 = \sin^2\theta \sin^2 i + \cos^2 i = 1 - \cos^2\theta \sin^2 i. \tag{0.2}$$

On the other hand, let l denote the fractional light of the system, expressed in terms of its maximum light between minima taken as a unit; and λ, its value attained at the time of maximum eclipse (when $\theta = 0$); while $L_{1,2}$ stand for the fractional luminosities of the two components, defined so that

$$L_1 + L_2 = 1. \tag{0.3}$$

Let, moreover, the subscript 1 refer, as before, to the component undergoing eclipse; and subscript 2 to the eclipsing star – regardless of which one of these is the larger (or more luminous) of the two. If, lastly, the fractional loss of light of the star undergoing eclipse is denoted by α, the fractional light l of the system can be expressed as

$$l = (1 - \alpha)L_1 + L_2 = 1 - \alpha L_1 \tag{0.4}$$

by (0.3).

The fractional loss of light α arising from eclipses and occurring in Equation (0.4) has already been studied extensively in the first part of this book. As long as the apparent disc of the star undergoing eclipse can be regarded as circular, we found it possible to express α in the form

* Apart from the reflection effect, which we disregard at this stage.

† The effects arising from a variation of R arising from orbital eccentricity will be taken up in Section IV-5.

$$\alpha = \sum_{l=0}^{\infty} C^{(l)} \alpha_l^0, \tag{0.5}$$

where the coefficients $C^{(l)}$, as defined by Equations (2.8)–(2.9) of Chapter I, depend on the extent of limb-darkening of the eclipses star, and the α_l^0's are the associated α-functions of order l and index 0.

The latter were found (cf. Section I-3) to depend on the circumstances of the eclipse through two parameters which, for the sake of normalization, can be taken as

$$k = \frac{r_a}{r_b} \quad \text{and} \quad p = \frac{\delta - r_b}{r_a}, \tag{0.6}$$

where r_a stands for the fractional radius of the *smaller* component (the eclipse of which we have referred to as an *occultation*); and r_b, that of the *larger* star (eclipsed in the course of a *transit*). If so, then the ratio k of the radii as defined by (0.6) has been constrained to satisfy the inequality

$$0 \leqslant k \leqslant 1; \tag{0.7}$$

while the 'geometrical depth' p of the eclipse was similarly made to vary within

$$1 \geqslant p \geqslant -1 \tag{0.8}$$

between the moment of the first contact (when $\delta_1 = r_a + r_b$) of the eclipse and that of the internal tangency ($\delta_2 = r_b - r_a$) of both discs.

If, therefore, $\alpha \equiv \alpha(k, p)$, this relation can, in principle, be inverted to yield $p \equiv p(k, \alpha)$ the geometrical depth of the eclipse as a function of the fractional loss of light α and the ratio of the radii k for any degree of darkening of the eclipsed star. This inversion cannot be performed analytically; but adequate tables of $p(k, \alpha)$ for several intermediate degrees of (linear) darkening at the limb have been constructed by Tsesevich (1939, 1940).

Let, moreover,

$$p_0 = \frac{\delta_0 - r_b}{r_a} = \frac{\cos i - r_b}{r_a} \tag{0.9}$$

denote the *maximum* geometrical depth of the respective eclipse (occurring at the moment of conjunction when $\theta = 0$), and $\alpha(k, p_0) \equiv \alpha_0$ be the maximum fractional loss of light. By Equation (0.4),

$$\alpha = \frac{1 - l}{L_1}, \tag{0.10}$$

which at the time of maximum eclipse ($l = \lambda$) becomes

$$\alpha_0 = \frac{1 - \lambda}{L_1}; \tag{0.11}$$

so that, on elimination of L_1 between (0.10) and (0.11),

$$\alpha = \frac{1 - l}{1 - \lambda} \alpha_0 \equiv n \alpha_0. \tag{0.12}$$

If so, however, the apparent separation δ of the centres of the two stars can be expressed in terms of observable quantities in two different ways: either, from the time t of observations by Equations (0.1) and (0.2); or from the fractional light l of the system by use of the identity

$$\delta \equiv r_b\{1 + kp(k, \alpha)\}, \tag{0.13}$$

where α is given by Equation (0.12). Squaring Equation (1.13) and equating with that given by Equation (0.2) we arrive at a relation between the fractional light l observed at a time t (or – which is equivalent – between α and θ) of the form

$$\sin^2 \theta \sin^2 i + \cos^2 i = r_b^2\{1 + kp(k, \alpha)\}^2, \tag{0.14}$$

containing three unknown constants (i.e., r_b, $r_a \equiv kr_b$, and i) to be determined by an appropriate analysis.

Equations (0.12) and (0.14) constitute the basic relations, fundamental to all methods proposed for the solution of our problem in the time domain so far. They are not unique, nor necessary to our end as they stand. In point of fact, Equation (0.14) constitutes a mere identity $\delta^2(\theta) \equiv \delta^2(\alpha)$; and could be replaced by any more general identity of the form $f\{\delta(\theta)\} \equiv f\{\delta(\alpha)\}$, where $f(\delta)$ may stand for a more general than merely quadratic function of δ (as long as it remains bounded between $\cos i < \delta < r_a + r_b$). However, no approach alternative to the use of Equation (0.14) has been considered – let alone elaborated – so far; and, therefore, in what follows we shall adopt Equations (0.12) and (0.14) as a basis for all subsequent analysis.

Moreover, the fact that these equations contain three unknowns discloses that a minimum of *three* observations of $l(\theta)$ during partial phases of the eclipse are necessary to render a solution for them determinate. Thus – as in other problems of celestial mechanics – three observations are found to be theoretically sufficient for specification of the geometrical elements of an eclipsing system. However, the individual photometric measurements of $l(\theta)$ are so much less accurate than the positional measurements in astrometry (the former are seldom subject to errors of less than one per cent; while the errors of the latter can be made no more than one part in several millions of the quadrant) that it is well-nigh impossible to do so in practice. The difficulties characteristic of eclipsing orbit work, based on the best observations that can be made so far, should be comparable to those that would confront the computer of asteroidal or cometary orbits if the positional measurements at his disposal were inaccurate within half a degree! In such a situation a relatively low accuracy of the underlying data would have to be offset by large numbers of observations; and these would require statistical rather than individual treatment: the weight of the solution would have to be distributed over a wide stretch of the data rather than being based on a few discrete points.

And – as we shall see – this is what we must do in an analysis of the light changes of eclipsing binary systems as well. The aim of the present chapter will be to outline the ways in which this can be done in the time-domain; while a transfer of the problem to the frequency-domain will be taken up in the next chapter.

IV-1. Direct methods: Historical

A history of efforts to solve the light curves of eclipsing variables directly for the geometry of their eclipses goes back to Pickering's work on Algol (Pickering, 1880) almost one hundred years ago; and after some isolated efforts by Harting (1889), Myers (1898), Hartwig (1900) or Roberts (1903) which gradually extended the scope of the problem the first sustained effort to systemize the field was undertaken by Russell (1912a, b) who in collaboration with Harlow Shapley (Russell and Shapley, 1912a, b) developed direct methods associated with their names which will be briefly outlined below.

The method proposed by Russell in 1912 to solve the light curves of eclipsing binary systems directly for their elements can be summarized as follows. Let θ_j, p_j $(j = 1, 2, 3)$ denote the phase angles and geometrical depths at three arbitrary moments within partial phases of the eclipse. Equations of the form (0.14) particularized for these phases contain r_a, r_b and i as the only unknowns. Now – in contrast with his predecessors – Russell conceived the idea of eliminating the fractional radius r_b and the orbital inclination i between them in order to isolate the ratio $k = r_a/r_b$. Equations (0.14) are homogeneous in the unknown constants $\sin^2 i$, $\cos^2 i$ and r_b; therefore, their consistency for $j = 1, 2, 3$ requires the vanishing of the determinant

$$\begin{vmatrix} \sin^2\theta_1 & (1+kp_1)^2 & 1 \\ \sin^2\theta_2 & (1+kp_2)^2 & 1 \\ \sin^2\theta_3 & (1+kp_3)^2 & 1 \end{vmatrix} = 0, \tag{1.1}$$

where $p_j \equiv p(k, \alpha_j)$, and which represents an equation containing k as the only unknown; once k has been obtained, the other elements follow without difficulty.

How to solve this equation? If the eclipse is *total*, the maximum fractional loss of light

$$\alpha_0 \equiv \alpha(k, -1) = 1; \tag{1.2}$$

and, consequently, for $p > -1$ the values of

$$\alpha \equiv n = \frac{1-l}{1-\lambda} \tag{1.3}$$

follow directly from the observations. In such a case, Russell proposed to assign to two points $\theta_{1,2}$, $p(k, \alpha_{1,2})$ a specific location on the light curve, corresponding to $\alpha_1 = 0.6$ and $\alpha_2 = 0.9$, while retaining the third as arbitrary (the subscript of which may, therefore, be dropped). By developing the determinant on the left-hand side of Equation (1.1) we find that the equation itself can be rewritten in the form

$$\sin^2\theta = A + B\psi(k, p), \tag{1.4}$$

where

$$A \equiv \sin^2\theta_1, \quad B \equiv \sin^2\theta_1 - \sin^2\theta_2 \tag{1.5}$$

are constants, and

$$\psi(k;\alpha,\alpha_2,\alpha_3)=\frac{2(p-p_1)+k(p^2-p_1^2)}{2(p_1-p_2)+k(p_1^2-p_2^2)} \tag{1.6}$$

is a function of k and α in addition to the adopted values of $\alpha_{1,2}$. A knowledge of any third point of coordinates θ, α on the light curve then renders Equation (1.4) one for k alone; and once it has been solved for this ratio (with the aid of suitable tables of the function $\psi(k,\alpha)$ prepared originally by Russell in 1912 (and, more recently, by Russell and Merrill (1952) for specific degrees of (linear) limb-darkening of the eclipses star), the remaining geometrical elements of the eclipse can be obtained from the equations

$$\cot^2 i=\frac{A}{\Phi_1(k)} \tag{1.7}$$

and

$$(r_b \csc i)^2=\frac{B}{(1+kp_1)^2\Phi_1(k)}, \tag{1.8}$$

where the constants A, B continue to be given by Equations (1.5), and

$$\Phi_1(k)=1-\left(\frac{1+kp_2}{1+kp_1}\right)^2 \tag{1.9}$$

stands for an auxiliary function of k likewise available in tabular form.

The foregoing equations (1.1)–(1.9) represent (within the scheme of assumptions on which the whole procedure is based) a direct solution of our problem if the light minimum order investigation is due to a total or annular eclipse; and examples of such light curves are shown (see Fig. 4-1 on page 140). Should the eclipse become *partial* (as disclosed by a continuous variation of light throughout both minima; for typical examples see Figs. 1-1 or 5.7), our problem becomes complicated by the fact that – although the fractional loss α of light during eclipse continues to be given by Equation (0.12), the maximum obscuration $\alpha_0\equiv\alpha(k,p_0)$ is no longer unity as was the case for total eclipses, but becomes a function of the elements we seek to determine.

In order to obtain a direct solution in such a case, let us depart again from Equation (0.14) which, at the moment of maximum eclipse (when $\theta=0$ and $\alpha=a_0$) reduces to

$$\cos^2 i=r_b^2(1+kp)^2,\quad p\equiv p(k,\alpha_0)>-1. \tag{1.10}$$

Subtracting (1.10) from (0.14) we find that

$$\sin^2\theta=(r_ar_b\csc^2 i)\{2(p-p_0)+k(p^2-p_0^2)\}. \tag{1.11}$$

In order to eliminate the unknown multiplicative constant $r_ar_b\csc^2 i$ on the r.h.s. of Equation (1.11), Russell proposed to divide the latter by an analogous equation pertaining to a specific phase of the eclipse – and he chose for this purpose a point corresponding to $p_1\equiv p(k,\frac{1}{2})$. If so, it follows immediately that

$$\frac{\sin^2\theta(n)}{\sin^2\theta(\frac{1}{2})}=\frac{2(p-p_0)+k(p^2-p_0^2)}{2(p_1-p_0)+k(p_1^2-p_0^2)}\equiv\chi(k,\alpha_0;n), \tag{1.12}$$

where the function $\psi(k, \alpha_0; n)$ thus introduced will play, for partial eclipses, a role similar to that played by the function $\psi(k, \alpha)$ if the eclipse is total. In point of fact, these two functions are related by the equation

$$\psi(k, \alpha_0; n) \equiv \frac{\psi(k, n\alpha_0) - \psi(k, \alpha_0)}{\psi(k, \frac{1}{2}\alpha_0) - \psi(k, \alpha_0)}. \tag{1.13}$$

The new function $\psi(k, \alpha_0; n)$ depends on three parameters (i.e., $n_1, n_2; n$) of which one (n) can belong to any particular phase. Suppose that the values of $\sin^2\theta$ have been determined from the light curve at points corresponding to the values of n_1, n_2 and $\frac{1}{2}$. If so, a solution of the relations

$$\psi\{k, \alpha(k, p_0); n_1\} = \frac{\sin^2\theta(n_1)}{\sin^2\theta(\frac{1}{2})} \equiv c_1 \tag{1.14}$$

and

$$\psi\{k, \alpha(k, p_0); n_2\} = \frac{\sin^2\theta(n_2)}{\sin^2\theta(\frac{1}{2})} \equiv c_2 \tag{1.15}$$

should be sufficient to specify the values of k and α_0 (or k and p_0). This is, unfortunately, true only in theory, but not in practice; for the $\psi(k, \alpha_0; n)$-functions corresponding to different values of n have long been known to *simulate* functional dependence to such an extent as to render a determination of the elements k and α_0 of partially-eclipsing systems from a pair of equations of the form (1.14)–(1.15) pertaining to one and the same minimum effectively impossible. To do so would require an empirical knowledge of the constants $c_{1,2}$ on the r.h.s. of Equations (1.14) and (1.15) much more accurate than that possible from present-day observations (especially if $\alpha_0 \ll 1$); and in order to render our problem solvable in practice, another independent relation between k and α_0 must be sought.

Such a relation can indeed be obtained by a simultaneous appeal to *both* alternate minima of light and their observed properties in the following manner. Let $\lambda_{a,b}$ denote the fractional light of the system at the moment of maximum occultation and transit eclipses alternating in each system; and, moreover, $\alpha_0' \equiv \alpha(a, c_0)$, $\alpha_0'' \equiv \alpha(b, c_0)$ be the fractional losses of light of the components of luminosities L_a, L_b, respectively, where the arguments

$$a = \frac{r_1}{r_1 + r_2}, \quad b = \frac{r_2}{r_1 + r_2}, \quad c_0 = \frac{\delta_0}{r_1 + r_2}, \tag{1.16}$$

of the α-functions are identical with those defined by Equations (3.56)–(3.58) of Chapter I.

If so, then for an *occultation* eclipse ($r_a \equiv r_1 < r_2$),

$$\lambda_a = 1 - \alpha_0' L_a; \tag{1.17}$$

while half a revolution later, for a *transit* ($r_b \equiv r_1 > r_2$)

$$\lambda_b = 1 - \alpha_0'' L_b. \tag{1.18}$$

An elimination of L_a and L_b between Equations (1.17)–(1.18) and (0.3) asserting that $L_a + L_b = 1$ discloses that, for an occultation eclipse,

$$\alpha_0' = 1 - \lambda_a + \frac{1 - \lambda_b}{k^2 Y(a, c_0)}; \tag{1.19}$$

while for a transit,

$$\alpha_0'' = 1 - \lambda_b + (1 - \lambda_a) k^2 Y(a, c_0), \tag{1.20}$$

where the value of the ratio of the radii $k = r_1/r_2$ may be anywhere between 0 (corresponding to the eclipse by a straight occulting edge) and ∞ (point transit); and where we have abbreviated

$$\frac{\alpha(b, c_0)}{\alpha(a, c_0)} \equiv \frac{\alpha(1 - a, c_0)}{\alpha(a, c_0)} \equiv k^2 Y(a, c_0). \tag{1.21}$$

The function $Y(a, c_0)$ so defined varies but slowly with the parameters a, c_0 as well as with the coefficients of limb-darkening of the two stars; and can likewise be suitably tabulated.

Equations (1.19) and (1.20) represent the second desired relations between the parameters a (or $b = 1 - a$) and c_0, based on the depths of the two light minima; and whichever of them should be adjoined to Equations (1.14) or (1.15) to obtain a solution for the elements of the eclipse depends on the nature of the observed minima: should the deeper minimum (under analysis) be due to an occultation eclipse, Equation (1.19) should be used; while if it happens to be a transit, Equation (1.20) should be adopted. If, moreover, the occultation eclipse happens to be total (i.e., if $\alpha_0' = 1$), Equation (1.19) can be solved to yield

$$k^2 = \frac{1 - \lambda_b}{\lambda_a Y(a, c_0)} \tag{1.22}$$

the 'depth' $k \leqslant 1$ of the respective system; while for an annular transit a similar equation holds good provided that the fractional luminosity λ_b refers to the light intensity of the system at the commencement of the annular phase.

In general, only one of the alternatives $\lambda_1 = \lambda_a$ or λ_b will lead to a real solution for the elements of the eclipse; and the alternative which does so should be accepted as true. Moreover, once the elements a and c_0 (or k and p_0) have thus been obtained, the remaining geometrical elements may be obtained by a straightforward use of Equation (0.14) for $\theta = 0$, and the solution of our problem would appear to be complete.

So much for the solution obtained on the basis of assumptions set forth at the commencement of this chapter, which in a mathematical sense borders on triviality. If we pause, however, to reconsider the premises on which such direct solution is based, our view of the situation is bound to become considerably less optimistic. In point of fact, a mathematician might inquire what right we have to believe that we obtained any solution at all, if certain parts of it (such as the location of the pivotal points corresponding to $\alpha = 0.5$, 0.6, or 0.9) had to be anticipated in advance, with no possibility of subsequent improvement. This is indeed a serious objection to the Russell–Shapley (or any other) method aiming at a direct algebraic solution for the photometric elements of eclipsing binary systems. For the positions of particular points on the light curves (at which the loss of light may attain certain round values) can never be determined from the observations with a higher precision than those of other points; and any errors in their assigned locations would obviously go through all subsequent algebra – vitiating the outcome to an extent to which direct methods of solution offer no clue.

In order to lessen the errors arising from this source, Russell and Shapley proposed to establish the location of the pivotal (or indeed any other) point on the light curve, not directly from the observations, but from a smooth curve drawn by free hand to follow the course of the individual observations or normal points. A graphical interpolation of this nature may have constituted a shrewd move, justificable in the early days of astronomical photometry when the relative accuracy of the individual normals was between 1 and 10% of the measured light intensity. With the gradually increasing precision of photometric observations the time was, however, bound to come when an empirical interpolation and smoothing of the observed data, as represented by the drawing of free-hand curves, was bound to become a liability. For it is a fact that a substituion of such curves for the actual observations may amount to an unwarranted, and possible risky, interference with the basic data, which may vitiate all results based upon them in a systematic manner. Moreover, no matter what the scale of such curves, or the skill with which they could be drawn, their exclusive use would never permit us to learn the *uncertainty* with which the elements of eclipsing binary systems are defined by the available observational data.

As in all other branches of astronomy or of physical sciences in general, a determination of the uncertainty of any results based on photometric observations of finite accuracy should represent an integral part of each solution; for without it we should not know the extent to which such photometric elements can be trusted. The methods of Russell and Shapley, developed with the aim of directness and convenience rather than completeness of rigour, turned their back on this all-important requirement – and, instead, placed undue emphasis on mere graphical representation of the observed light changes. Any method relying on the use of fixed points read off a free-hand curve, and aiming primarily at a graphical representation of the observed data, assigns arbitrary weights to different parts of the light curve (in fact, infinite weights to the selected pivotal points) and, as a result, cannot enable us to ascertain any errors of the solution, or its stability. A mere inspection of the deviations of normal points from free-hand curves can ordinarily disclose very little about the way in which observational errors can affect individual elements. In point of fact, the less determinate the solution, the easier it is to obtain a seemingly good graphical representation of the observations by a right combination of wrong elements – as only too many investigators in the past have learned the hard way to their sorrow!

The present writer pointed out many years ago (Kopal, 1941) that some of the drawbacks inherent in the need to adopt fixed locations for selected pivotal points can be removed by a reversal of the order in which the unknown elements of the eclipse are to be extracted from the observed data. The fundamental strategy of the Russell–Shapley methods has been to single out the ratio k of the radii for determination first, and subsequent elements later. It was this circumstance which led the 1912 investigators to manipulate the fundamental identity (0.14) to the rather complicated forms (1.6) or (1.12) in which k remained the single unknown. Suppose, however, that we forego this strategy and restrict our means to the use of Equation (0.14) as it stands; by abbreviating

$$\sin^2\theta \equiv x \quad \text{and} \quad \{1 + kp(k, \alpha)\}^2 \equiv y(k, \alpha) \tag{1.23}$$

we can rewrite Equation (0.14) in the form

$$x = (r_b \csc i)^2 y - \cot^2 i, \tag{1.24}$$

which in the xy-coordinates represents a straight line.

The quantities x and y as defined by Equation (1.23) are, however, not independent, but constrained by the geometry of the eclipses. In order to establish the ratio of the radii k we may seek – by trial and error, if not otherwise – such value of it which will 'rectify' the relation (1.24) and represent it as a straight line fitting best the observed points (l, θ); the intercepts of this line, in the xy-coordinates, should then specify the values of $r_b^2 \csc^2 i$ and $\cot^2 i$. The principal advantage of such a strategy is the ability to make use of the observed photometric data as they stand, without having to give undue weight to any part of them, or having to resort to the use of any smooth curves drawn by free hand at any stage of our analysis. But once we are willing to do this, we can do more: namely, to formulate our approach to the solution of the problem in such a way as to determine the most probable values of all elements of the eclipse, and the uncertainty within which these are defined by the observations, not merely graphically, but analytically, along the lines developed in the next section.

IV-2. Iterative Methods: Total and Annular Eclipses

Suppose that photometric observations of the fractional light l of the system at a phase θ indicate the minimum under investigation to be due to a total (or annular) eclipse; and that our task is to extract from its shape the geometrical elements $r_{1,2}$ and i which are responsible for it by direct use of the observed data as they stand.

The fundamental equation which relates the geometrical elements with the observed quantities continues to be of the form (0.14), which at the moments of the first or second contact (corresponding to the geometrical depth $p = \pm 1$) reduces to

$$\sin^2\theta_{1,2} \sin^2 i + \cos^2 i = r_b^2(1 \pm k)^2; \tag{2.1}$$

the angles $\theta_{1,2}$ denoting the phases of the first and second contact. Subtracting Equation (2.1) from (0.14) and solving for $\sin^2\theta$ we find that

$$(p^2 - 1)C_1 + 2(p \mp 1)C_2 + C_3 = \sin^2\theta, \tag{2.2}$$

where

$$C_1 = r_a^2 \csc^2 i, \tag{2.3}$$

$$C_2 = r_a r_b \csc^2 i, \tag{2.4}$$

$$C_3 \equiv \sin^2 \theta_{1,2} = (r_a \pm r_b)^2 \csc^2 i - \cot^2 i, \tag{2.5}$$

are unknown constants to be determined.

Of the quantities involved in Equation (2.2), two are supplied directly by the observations (i.e., θ and α in $p(k, \alpha)$ and three (r_a, r_b, i) are unknown. Therefore, at least three equations of the form (2.2) are required to render our problem determinate, and more if the values of $l(\theta)$ are affected by observational errors. *As many equations of the form* (2.2) *can obviously be set up as there are observations* (or normal points), *characterized by a known pair of l and θ, and regarded as equations of condition whose least-squares solution should yield the most probable values of C_1, C_2 and C_3.*

An essential feature of our problem is, moreover, the fact that – on account of an inherent non-linearity (in fact, transcendental nature) of our problem – the unknown geometrical elements r_a, r_b and i are contained not only in C_1, C_2 and C_3, but also (through p) in their variable coefficients on the left-hand side of Equation (2.2). Therefore, the most probable values of the unknowns cannot result from a single solution of a set of equations of condition of the form (2.2), but can be obtained only by *iteration.*

In more specific terms, inasmuch as the geometrical element involved in p is k, it is desirable to commence the iterative processes with an *adopted* value (say, K) of this parameter, evaluate the corresponding geometrical depths $p(K, \alpha)$, form the coefficients of $C_{1,2}$ in Equation (2.2), and solve for the C_j's by the method of least squares. If the adopted value K were correct, the ratio

$$\frac{C_1}{C_2} = \frac{r_a}{r_b} \equiv k \tag{2.6}$$

resulting from our solution should have verified it. If, however, this is not the case and the foregoing ratio (2.6) differs significantly from K, the solution is to be repeated – each time on the basis of the ratio C_1/C_2 for k resulting from the previous approximation – until the assumed and resulting values of the ratio of the radii agree within the limits of observational errors.

With which value of K should we commence the iterations? If both minima have been observed and empirical values of the fractional intensities $\lambda_{a,b}$ are available, recourse may be had at this stage to Equation (1.22) furnishing the 'depth' k in their terms if we set, to a first approximation, the numerical value of the auxiliary function $Y(a, c_0) = 1$. Equation (1.22) yields, to be sure, *two* values of k – depending on whether $\lambda_1 = \lambda_a$ or λ_b; and unless a proper identification is possible on the basis of extraneous (spectroscopic) evidence, further criteria to discriminate between the two alternatives must be invoked. Fortunately, in many cases such criteria are readily on hand. Thus if the orbital plane is parallel with the line of sight – an expectation which for total (annular) eclipses is unlikely to be far off – a simple geometry shows that if $\theta_{1,2}$ denote (as before) the phase angles of the first and second contacts of a total or annular eclipse,

$$\frac{\sin \theta_2}{\sin \theta_1} = \frac{1-k}{1+k}, \tag{2.7}$$

from which it follows that

$$k = \frac{\tan \frac{1}{2}(\theta_1 - \theta_2)}{\tan \frac{1}{2}(\theta_1 + \theta_2)}; \tag{2.8}$$

and provided that θ_1 is not too large,

$$k = \frac{\theta_1 - \theta_2}{\theta_1 + \theta_2} \tag{2.9}$$

will give a somewhat larger, but still sufficiently close value of k. For $i < 90°$, the foregoing Equations (2.8) or (2.9) yield the *lower limit* of a possible value of k which is consistent with the observed durations of partial phases of the eclipses. If one of the 'depth' k's deduced from Equation (2.8) is significantly smaller than this limit, it is inadmissible and the identification of the type of eclipse (total or annular) on which it was based is disproved.

This method to approximate k uniquely is quite effective if the depths of both minima are very unequal. If, in particular, one minimum is very much deeper than the other, the odds are very much in favour of its being due to total eclipse. If, on the other hand, the depths of both minima are comparable – and, therefore, of necessity rather shallow,* the two values of k consistent with Equation (1.22) may not vary much and Equations (2.8) or (2.9) may not help us too much to discriminate between them. In the absence of an extraneous (spectroscopic) evidence to the contrary, it is legitimate to embark on our analysis on the basis of an assumption that the deeper minimum is due to a total eclipse. Such an assumption always leads to a geometrically possible solution; while that of an annular eclipse will do so only if $1 - \lambda_1 < k^2 Y(a, c_0)$.

The whole situation can be summarized by noting that if the depths of *both* minima can be extracted from the given light curve with sufficient accuracy, Equation (1.22) should provide a relatively good value (or a pair of values) of the ratio of the radii k; whereas if only *one* minimum is available for analysis, the value of k can be estimated (less accurately) by means of Equations (2.8) or (2.9). Let us denote this provisional value by K. Once it is adopted, the values of $p(K, \alpha)$ corresponding to it can be found for each observed point of the light curve from Tsesevich's tables for the respective types of eclipse, the coefficients $p^2 - 1$ and $2(p \mp 1)$ of the quantities C_1 and C_2 in Equation (2.2) can be ascertained; and Equations (2.2) *regarded as equations of condition whose least-squares solution should yield the most probable values of the unknown constants* C_1, C_2, C_3, as defined by Equations (2.3)–(2.5) and corresponding to the adopted value of K. If, moreover, the ratio (2.6) differs significantly from K, the solution will have to be repeated – each time with the ratio C_1/C_2 resulting from the previous approximation taken as a basis of the next cycle – until the assumed and resulting values of this ratio agree within the limits of observational errors.

* Equation (1.22) makes it evident that (since $k \leq 1$) in no eclipsing system can the fractional loss of light in *both* minima exceed 0.5 in intensity units, or 0.75 of a magnitude.

A. FORMATION AND WEIGHTING OF THE EQUATIONS OF CONDITION

Before we actually embark on an iterative solution of this nature, due care must be exercised in forming the normal equations of our problem (the number of which is equal to the number of our unknowns) from the equations of condition based on the individual observations (whose number is much larger). The *weights* with which equations of this latter type should enter our solution are, in general, likely to be very different for two reasons. First, the observations of the instantaneous brightness of the system may be of unequal quality (or the 'normal points' on which our equations are based may not consist of an equal number of individual observations). Whether or not this is so must be made clear by the observer himself; and his information should guide us in ascribing to each observation (or normal point) an empirical weight to which it may be entitled.

The second reason which is bound to render the weights of our equations of condition unequal has, however, nothing to do with the quality of the underlying observations – but is intrinsic to our approach to the solution. Each observation on which our equations of condition are based is specified by a pair of the measured quantities l and θ. The accuracy of the measurement of θ (i.e., of the time) is, in general, so much greater than that of the instantaneous brightness l that practically the entire uncertainty of our solution will depend on the errors of the measured values of l. This implies, in turn, that it is the coefficients p^2-1 and $2(p \pm 1)$ of the unknown quantities $C_{1,2,3}$ rather than the absolute terms $\sin^2\theta$ of the equations of condition of the form (2.2) which are principally affected by observational errors. An error Δl committed in measuring the instantaneous brightness of the system is, however, essentially equivalent to an error of

$$\frac{\mathrm{d}\sin^2\theta}{\mathrm{d}l}\,\Delta l$$

on the right-hand side of Equation (2.2); and if so, the weight $\sqrt{w}$ of the respective equation will be given by

$$\sqrt{w} = \frac{\mathrm{d}l}{\mathrm{d}\sin^2\theta}\,\frac{\epsilon}{|\Delta l|}, \tag{2.10}$$

where ϵ denotes the error of a light measurement of unit weight. Now, provided that there is no difference in empirical weight between the individual normals and that the errors of measurement are constant on the intensity scale, we may put $\epsilon = |\Delta l|$ – in which case

$$\sqrt{w} = \frac{\mathrm{d}l}{\mathrm{d}\sin^2\theta}; \tag{2.11a}$$

whereas if the errors of observation are constant on the logarithmic (magnitude) scale, we may put $\epsilon = |\Delta l|/l$, obtaining

$$\sqrt{w} = \frac{1}{l}\,\frac{\mathrm{d}l}{\mathrm{d}\sin^2\theta}. \tag{2.11b}$$

The derivative $\mathrm{d}l/\mathrm{d}\sin^2\theta$ in the preceding equations characterizes the slope of the light curve plotted in the $(l - \sin^2 \theta)$ coordinates, and its value could be read off a smooth curve drawn freely to follow the course of observed normal points. It can, however, also be evaluated directly – without recourse to any free-hand curve – by a differentiation of the fundamental equation (2.2). Since

$$\frac{\mathrm{d}\sin^2\theta}{\mathrm{d}l} = \frac{\mathrm{d}\sin^2\theta}{\mathrm{d}\alpha}\frac{\mathrm{d}\alpha}{\mathrm{d}l} = -\frac{1}{1-\lambda}\frac{\mathrm{d}\sin^2\theta}{\mathrm{d}\alpha} \tag{2.12}$$

by Equation (1.3); and by differentiating Equation (2.2) we find that

$$\frac{\mathrm{d}\sin^2\theta}{\mathrm{d}\alpha} = 2C_2(1+kp)\frac{\partial p}{\partial \alpha}, \tag{2.13}$$

it follows that, if the errors of observation are constant on the *intensity* scale,

$$\sqrt{w} = -\frac{1-\lambda}{2C_2(1+kp)(\partial p/\partial\alpha)}; \tag{2.14a}$$

while if they are constant on the *magnitude* scale,

$$\sqrt{w} = -\frac{1-\lambda}{2lC_2(1+kp)(\partial p/\partial\alpha)}. \tag{2.14b}$$

It should be noted that the quantity C_2 in the denominator of the preceding expressions is not known at the stage when the individual equations of conditions are weighted; but since it is constant we can eliminate it by multiplying both sides of each equation by it. The term $1-\lambda$ in the numerator, representing the fractional loss of light at the bottom of the minimum under investigation, is likewise a constant, and therefore immaterial as long as only *one* minimum is subject to our analysis. If, however, the equations of condition of *both* minima are being combined in a single set of normal equations, the factor $1-\lambda$ is likely to be different for primary and secondary minima and must therefore be applied. It is this factor which will render the weight of the equations of condition pertaining to the secondary minimum small in comparison with those based on the primary minimum if the latter is deep and the former shallow.

The remainder of the terms in the denominator on the right-hand sides of Equations (2.14) can be easily evaluated by setting K for k, $p(K, \alpha)$ for $p(k, \alpha)$, and replacing the partial derivative $\partial p/\partial\alpha$ by a ratio of first tabular differences taken from the appropriate table in the neighbourhood of the point in question. It may be mentioned that this derivative is negative in the entire domain (which reverses the minus sign of the right-hand sides of Equations (2.14) into plus) and becomes infinite at the moment of either contact* – thus reducing the weight there to zero. A division by the instantaneous fractional luminosity l of the system will not be of much consequence if the amplitude of the light changes is small, but may become an important – and, in fact, controlling – element of the weight if the amplitude becomes large. An amplitude of 2.5 magnitudes can evidently cause $\sqrt{w}$ to vary from ten to one between the inner and outer contacts of the eclipse; an amplitude of five magnitudes would exaggerate this

* Except at the inner contact of a transit eclipse (i.e., commencement of the annular phase).

range a hundred times. This alone makes it obvious that whenever the errors of observation are constant on the magnitude scale (and, therefore, the absolute errors diminish with diminishing light), the weight of a determination of the elements of the eclipse will rest predominantly on observations made near the inner contact and in advanced partial phases, with the early stages of the eclipse contributing relatively little to the weight of the whole solution.

The expressions for $\sqrt{w}$ as defined by Equations (2.14) represent – we repeat – the intrinsic weights of the individual equations of condition, arising solely from the way in which the observational errors affect their absolute terms. If, in addition, the individual equations are also observationally of unequal weights, the *total weight* of the respective equation should be a *product* of the *empirical times intrinsic weight*; its square-root should then be taken and the respective equations of condition multiplied by it before the normal equations are formed. In what follows, however, $\sqrt{w}$ will continue to stand for the intrinsic weight as defined by Equations (2.14) alone; a multiplication of each equation by the square-root of its observational weight (whenever the latter differs from equation to equation) will be taken for granted. With regard to the numerical accuracy to which $\sqrt{w}$ should be evaluated, two significant figures are desirable and three figures are ample; while more than three figures in $\sqrt{w}$ would as yet be entirely superfluous.

B. DIFFERENTIAL CORRECTIONS FOR λ AND U

When the equations of condition of the form (4.8) have been formed for each observed normal point and multiplied by square-roots of their proper observational and intrinsic weights, we are almost ready to proceed with a solution for the elements of the eclipse, but not yet quite so. For the whole process of solution, as described in the preceding parts of this chapter, still makes use of two constants, the values of which had to be read off approximately from the observed data at the outset: namely, the brightness U of our system between minima (our unit of light), as well as the brightness λ of the system at the moment of internal tangency of the total or annular phase. The value of U is usually taken as the appropriate mean of all individual observations of the brightness of the system made in full light; and unless the minima are broad, of (what happens much more frequently) the number of observations made between minima is inadequate, its determination may carry considerable weight – much greater than that with which λ can as a rule be found. If the eclipse under consideration is an occultation which ultimately becomes total, λ_a remains constant during the whole interval of totality; and if its duration is appreciable, enough observations of minimum brightness can be secured to determine λ_a with fair accuracy. If, on the other hand, the eclipse happens to be a transit, the brightness of the system is apt to vary during annular phase, thus rendering a determination of λ_b at the moment of internal tangency inadvertently more difficult.

Suppose, however, that we estimated both U and λ from the available observational data as well as we could, and have proceeded with the iteration outlined earlier in this section. If the solution of properly weighted equations of condition has been performed by the method of least-squares, the most probable values of C_1, C_2 and C_3 have been obtained; but any error in the adopted values of U or would affect them all systematically. Moreover, if the errors within which the most probable values of C_1, C_2 and C_3 are defined by the available observational data were evaluated from the residuals of such a solution, they could faithfully reflect the dispersion of the observed normals between the outer and inner contacts, but would fail to take any cognizance of the uncertainty with which U as well as λ could be inferred from the observations and, in consequence, the real errors of all computed elements of the eclipse would be underestimated. This, if unremedied, would constitute indeed a serious defect of the method of solution which we proposed to follow. It has, however, been shown by Piotrowski (1948a) that we can improve the provisionally adopted values of U and λ *simultaneously* with a determination of the most probable values of C_1, C_2 and C_3, and obtain the differential corrections ΔU and $\Delta\lambda$ which are required to ensure the best possible fit. Piotrowski's procedure which makes this possible is as simple as it is elegant and can be summarized as follows.

Let the fundamental Equation (2.2) be rewritten as

$$\sin^2\theta = H(C_1, C_2, C_3; U, \lambda), \tag{2.15}$$

where

$$H \equiv (p^2-1)C_1 + 2(p+1)C_2 + C_3, \tag{2.16}$$

$p \equiv p(k, \alpha)$ and

$$\alpha = \frac{U-l}{U-\lambda}. \tag{2.17}$$

Let, furthermore,

$$U = 1 + \Delta U, \qquad \lambda = \lambda_0 + \Delta\lambda, \tag{2.18}$$

where U, λ are the true values of the respective quantities; 1, λ_0, their estimated values; and ΔU, $\Delta\lambda$, the requisite corrections which we shall now seek to determine. In order to do so, we shall expand $H(C_1, C_2, C_3, 1+\Delta U, \lambda_0+\Delta\lambda$, in a Taylor series in ascending powers of ΔU and $\Delta\lambda$, and anticipate these quantities to be small enough for their squares and higher powers to be negligible. Our fundamental Equation (2.15) generalized in this way then takes the form

$$H_0 + \left(\frac{\partial H}{\partial U}\right)_0 \Delta U + \left(\frac{\partial H}{\partial \lambda}\right)_0 \Delta\lambda + \cdots = \sin^2\theta, \tag{2.19}$$

where we have abbreviated $H_0 \equiv H(C_1, C_2, C_3; 1, \lambda_0)$.

Now

$$\frac{\partial H}{\partial U} = \frac{\partial H}{\partial \alpha}\frac{\partial \alpha}{\partial U}, \quad \frac{\partial H}{\partial \lambda} = \frac{\partial H}{\partial \alpha}\frac{\partial \alpha}{\partial \lambda}; \tag{2.20}$$

while from Equation (2.17) it follows that

$$\frac{\partial \alpha}{\partial U} = \frac{1-\alpha}{U-\lambda} \quad \text{and} \quad \frac{\partial \alpha}{\partial \lambda} = \frac{\alpha}{U-\lambda}. \tag{2.21}$$

Moreover, the square-root $\sqrt{w}$ of the intrinsic weight of our equations of condition can, with a sufficient accuracy, be approximated by

$$\sqrt{w} = -(1-\lambda)\frac{\mathrm{d}\alpha}{\mathrm{d}H_0}. \tag{2.22}$$

Combining this with the preceding relations we easily establish that

$$\sqrt{w}\left(\frac{\partial H}{\partial U}\right)_0 = -(1-\alpha) \tag{2.23}$$

and

$$\sqrt{w}\left(\frac{\partial H}{\partial \lambda}\right)_0 = -\alpha. \tag{2.24}$$

When we insert this in Equation (2.19) properly weighted, our generalized equation of condition then takes the neat form

$$\sqrt{w}(p^2-1)C_1 + 2\sqrt{w}(p+1)C_2 + \sqrt{w}C_3 - \alpha\Delta\lambda - (1-\alpha)\Delta U = \sqrt{w}\sin^2\theta, \tag{2.25}$$

and involves five unknowns to be simultaneously determined; the correction $\Delta\lambda$ pertaining always to the depth of the minimum under investigation – whether this be an occultation or a transit.

An inspection of the coefficients of the foregoing equation (2.25) discloses that $\alpha \equiv n$, as defined by Equation (0.12), is a quantity which follows directly from the observations, while $p(K, \alpha)$ as well as $\partial p/\partial\alpha$ (in $\sqrt{w}$) can be excerpted from the appropriate tables for the adopted value of K and each one of the observed values of it should be added to k in Equation (2.25) as it stands, the errors of the underlying observations are assumed to be constant on the intensity scale, and $\sqrt{w}$ to be given by Equation (2.14a). Should, on the other hand, the photometric errors be equal on the magnitude scale and Equation (2.14b) used to define $\sqrt{w}$, the coefficients of ΔU as well as of $\Delta\lambda$ in Equation (2.29) would still have to be divided by l. It was also mentioned earlier that, of all factors involved in Equations (2.14), C_2 in the denominator will usually not be known to us beforehand and we shall, therefore, have to eliminate it from $\sqrt{w}$ by multiplying all terms of Equation (2.25) by C_2. Since, however, the corrections and $\Delta\lambda$ in Equation (2.25) are not multiplied by $\sqrt{w}$, it follows that, after a pre-multiplication of all equations of condition of the form (2.25) by C_2, *the last two unknowns will be $C_2\Delta U$ and $C_2\Delta\lambda$ rather than the Δ-corrections themselves* – a fact to be borne in mind when we shall proceed later to combine such equations with other parts of the observational evidence.

The new equations of condition of the form (2.25) represent a very powerful tool for investigating the geometrical elements of the eclipse, and would permit us to determine the maximum or minimum brightness of our system even if no direct observations of these phases were available. One could simply estimate them – no matter how crudely – from the form of the light curve during partial phases, and refine them subsequently, by means of Equation (2.25), to the degree of accuracy attainable from the given data. In practice, however, a resort to such extreme measures will seldom be necessary; as a matter of fact, for well-covered light curves the corrections ΔU as well as $\Delta\lambda$ – particularly the former – may come out insignificant. Even should it be so, however, the effort entailed in including ΔU and $\Delta\lambda$ in our solution of partial phases of the eclipse was not misspent; for it will enable us later to ascertain the extent to which the uncertainty of U and λ will contribute to the errors of the computed elements of the eclipse.

In practical cases – even though the observations made during partial phases of the eclipse will contribute their share to our knowledge of U and λ – the main burden of their determination will naturally have to rest on the observations made during full light and during the total or annular phase. The question arises then as to the proper way in which this additional evidence should be combined with the observations of partial phases to compound the final result. If (as will usually be the case) our adopted unit of light represents an appropriate mean of all observations secured between minima and λ_a, the mean of all observations made during totality, the additional equations of condition bearing on their determination will be of the form

$$\sqrt{w_U}\Delta U = 0 \tag{2.26}$$

and

$$\sqrt{w_\lambda}\Delta\lambda_a = 0; \tag{2.27}$$

respectively. Any reader who may be inclined to regard them as trivial should remember that these are additional *equations of condition* which need not be fulfilled rigorously; how closely ΔU or $\Delta\lambda_a$ will actually approach zero will depend on the *weights* $\sqrt{w_U}$ and $\sqrt{w_\lambda}$ with which these equations will join the rest of our overdeterminate system. These weights are, in turn, specified by the relative uncertainty of the values adopted for U and λ_0 – or, in more specific terms, by the standard deviations σ_U and σ_λ of the individual observations from their respective means – in accordance with the equations

$$\sqrt{w_U} = \epsilon/\sigma_U \quad \text{and} \quad \sqrt{w_\lambda} = \epsilon/\sigma_\lambda, \tag{2.28}$$

where ϵ denotes, as before, the corresponding error of a single normal point, within minima, of unit weight. If, in particular, the observations between minima are grouped into N normals equivalent in weight to those within minima, then obviously $w_U = N$; and, similarly, if N such normals are available during totality, $w_\lambda = N/\lambda_a^2$ if they are equal on the magnitude scale.

This completes a survey of the equations which will be fundamental to our analysis; for we are now in a position to take simultaneously into account the relative contribution of each part of the whole light curve to a solution for the elements of the eclipse. Every normal point formed from the observations made during partial phases of the primary or secondary minimum will furnish us with an equation of condition of the form (2.25) for five unknowns $C_1, C_2, C_3, \Delta U$ and $\Delta\lambda_a$ if this minimum is an occultation, and $C_1, C_2, C_3, \Delta U$ and $\Delta\lambda_b$ if it is a transit. The observations between minima will furnish a single equation of the form (2.26) which will participate in the determination of ΔU, while the total phase of an occultation eclipse furnishes an equation of the form (2.27) adding its weight to the determination of $\Delta\lambda_a$. However, in any attempt to solve Equations (2.26) and (2.27) together with (2.25) for the most probable values of the unknowns we should remember that, if we pre-multiplied all equations of condition of the form (2.25) by C_2 in order to eliminate that constant from $\sqrt{w}$, the last two unknowns in Equation (2.25) became $C_2\Delta U$ and $C_2\Delta\lambda$. Therefore, before we can combine all equations of condition of the form (2.25) with (2.26) and (2.27) into a single set of normal equations, we have to rewrite the latter two equations as

$$(\sqrt{w_U}/C_2)C_2\Delta U = 0 \tag{2.26a}$$

and

$$(\sqrt{w_\lambda}/C_2)C_2\Delta\lambda_a = 0. \tag{2.27a}$$

In order to evaluate the coefficients of these equations, a preliminary estimate of C_2 is clearly a prerequisite. Ultimately, we should mention that Equations (2.26) or (2.27), as they stand, apply without qualification only to total eclipses and their depths. During the annular phase of a transit eclipse of an uniformly bright star, Equations (2.26) or (2.27) continue to hold good, with the sole change that λ_a is to be replaced by λ_b. Should, however, the larger component be darkened at the limb to an arbitrary degree, Equation (2.25) in the form appropriate for the partial phase of a transit continues to hold good during the annular phase as well, provided only that the requisite values of $p(K, \alpha)$ are taken from Tsesevich's 'annular' tables.

C. COMBINATION OF BOTH MINIMA

Now we are in a position to construct a set of normal equations, appropriate for either minimum, and iterate their solutions for k. A combination of the normal equations based on both minima into a single set should, however, be postponed until the final stage of intermediary analysis; for before arriving at it we must make sure that

(a) the type of eclipse at either minimum has been correctly identified, and that

(b) both minima yield a consistent set of the geometrical elements.

The first point can be settled conclusively by the success (or failure) of the successive approximations for k inherent in our process. It is obvious that *our iterations will converge to the true value of k if, and only if, the type of the eclipse under investigation has been correctly identified*, and the correct set of p-tables used to evaluate the coefficients of C_1 and C_2 (as well as the weights $\sqrt{w}$) in the equations of condition (2.25). If this has been done, the iterative process converges so rapidly that more than two cycles will but seldom be required. If, on the contrary, the difference between K and k after the second iteration fails to be substantially smaller than it was after the first one, that indicates that *a wrong type of the eclipse has been adopted* and that, instead of (say) an occultation, we are dealing with a transit. The alternative set of p-tables should then be used to set up a new system of our equations of condition, new normal equations formed, and their iterations repeated until k has at last been stabilized. In this way we shall always be able to find out *a posteriori* whether or not we were correct in our original assumption as to the type of the eclipse giving rise to the primary minimum, which we may have had to conjecture if no secondary minimum was observed (or if both minima were sensibly equal in depth).

If both minima have been adequately observed, we can go one step further. The foregoing discussion has made it clear that, in systems exhibiting total and annular eclipses, both minima can be subject to orbital analysis quite independently and on their own merits. It is, however, evident that not only should the solution of each minimum converge, but both should converge to the *same* set of geometrical elements. Let us fix our attention on the ratio k of the radii as the most sensitive one of such elements. If the values of the 'shape' k as deduced from the form of the light curve during partial phases of the occultation and transit eclipse fail to coincide within the limits of their uncertainty – while the value of the 'depth' k based on the depths of both minima and resulting from Equation (1.22) comes out intermediate between the two – the evidence is unmistakable that the degrees of limb-darkening of both components have been estimated incorrectly and, therefore, improper p-tables used to calculate the coefficients of our equations of condition. If, on the other hand, the 'depth' k happens to agree sensibly with the 'shape' k, this indicates that the degree of darkening of the smaller star only has been incorrectly estimated.

In general, if the 'shape' k as deduced from an analysis of an *occultation* eclipse comes out *smaller* than the 'depth' k, this indicates that the degree of darkening of the *smaller* star has been *underestimated*; while if the 'shape' k as deduced from a *transit* is *less* than the 'depth' k, the darkening of the *larger* star was assumed *too large.* In such cases, the recalcitrant coefficients of darkening of one (or both) stars should be altered, by trial and error, until both 'shape' k's and the 'depth' k's are in agreement at least within the limits of their uncertainties. Thus, in totally eclipsing systems, an initially *separate* treatment of both minima should permit us, not only *to distinguish an occultation from a transit eclipse* (by a success or failure of the iterative process based on the respective type of tables), but also *to check upon the correctness of the assumed degree of limb-darkening of both stars* (by the requirement that both minima should yield the same set of geometrical elements). *It is not until all this has been done that the equations of condition of the form* (2.25), *based upon the observations of both the primary and the secondary minima, should be combined together with Equations* (2.26) *and* (2.27) *into a single system of normal equations, which will yield the final set of the intermediary elements.*

The foregoing procedure tacitly assumes that the shape of both minima has been adequately observed. The same possibilities are, however, open to a lesser extent even if the secondary minimum is too shallow to enable us to make much use of its light record during partial phases; provided that at least its amplitude can be safely established. It is true that if the secondary minimum is very shallow, the factor $1 - \lambda_2$ in Equation (2.14) will reduce the weight of the equations of condition of the form (2.25) based upon it to a relative insignificance. Even though the *form* of a shallow secondary minimum by itself can disclose to us at this time very little, its *depth* λ_2, combined with λ_1 in Equation (1.22) can furnish a relatively accurate value of the 'depth' k. This 'depth' k should now be compared with the 'shape' k as deduced from the primary minimum. If the type of eclipse giving rise to this minimum has been correctly estimated, these two k's again should not differ significantly. If they do, the alternative type of eclipse should be adopted for the primary minimum; or the degree of darkening of the primary star adjusted until any significant discrepancy between the 'shape' and 'depth' k's has been removed.

The need of establishing a reliable value of the 'depth' k emphasizes the need of an accurate determination of the amplitude of the secondary minimum, even if the latter is too shallow to render the details of its shape to be of any avail. The difficulty of getting anything useful out of its shape arises from the fact that, for shallow eclipses, $1-\lambda$ becomes a small divisor in Equation (1.3), which will greatly magnify the errors in α due to the uncertainty of the observed values of $1-l$.

If, however, we knew the shape of the secondary minimum, we could obviously evaluate $1-l$ from each normal point observed within partial phases of the secondary minimum by means of the equation $1-\lambda=(1-l)/\alpha$, which is a direct consequence of Equation (1.3). In actual practice, the proper shape of the secondary minimum will, of course, not be known to us exactly. *It can*, however, *be approximated with an adequate precision from the shape of the primary (deep) minimum.* Barring the presence of complications of unknown origin, the alternate minima should differ in shape only on account of possibly unequal degree of darkening of the two components – a difference whose consequences are likely to be minor. Suppose that we ignore this difference and *assume the α's at corresponding moments of the alternate minima to be equal.* On the other hand, an insertion of Equations (218) in (2.17) discloses that

$$\alpha[1-\lambda_0+\Delta(U-\lambda)]=1-l+l\Delta U; \tag{2.29}$$

which by a substitution for α from Equation (1.3) can be rewritten as

$$(n-1)\Delta U-n\Delta\lambda=1-l-n(1-\lambda_0). \tag{2.30}$$

If the moment of the mid-secondary minimum and, therefore, the phase angle θ is known to us with a reasonable accuracy, we can replace, in the foregoing equation, the uncertain values of $\alpha(\pi+\theta)$ by those of $\alpha(\theta)$ which should be known much better from the well-observed light changes of the primary minimum. If, moreover, λ denotes the estimated fractional light of the system at the moment of internal tangency of the secondary minimum and l, that observed at any phase $\pi+\theta$, the only remaining unknown quantities in Equation (2.30) are ΔU and $\Delta\lambda$. *As many equations of the form* (2.30) *can obviously be set up as there are normal points observed within the secondary minimum, and solved for ΔU and $\Delta\lambda$* by the method of least squares; in many cases, we may be justified in putting $\Delta U=0$ in advance, which will leave us with $\Delta\lambda$ as the only unknown. The quantity $\lambda+\Delta\lambda$ then represents the best approximation obtainable for the fractional light of the system at the moment of internal tangency of the secondary minimum, and should be used whenever the latter is needed.

If is not until this has been done that we are in a position to consider combining our knowledge of the depth of the secondary minimum, and the shape of the primary minimum in a single solution for the elements of the eclipse. The reader should recall that Equation (2.6) permits us to write

$$C_1 - kC_2 = 0, \tag{2.31}$$

where k can be identified with our 'depth' k. Since a determination of the latter is subject to errors arising from the uncertainty of λ_a and λ_b, Equation (2.31) does not represent an exact relation between C_1 and C_2, but merely an additional *equation of condition* to be solved simultaneously with Equations (2.25), (2.26) and (2.27). The only point which still requires clarification is the relative weight $\sqrt{w_k}$ of Equation (2.31) as compared with that of other equations of our overdeterminate system; and this can be found from the following considerations. An error $\pm\Delta k$ in the adopted value of the 'depth' k will evidently produce an error of $\pm C_2\Delta k$ on the right-hand side of Equation (2.31). Now, differentiating Equation (1.22), we find that

$$\Delta k = \pm(1-\lambda_b)^{-1}\{(k^2\Delta\lambda_a)^2 + (\Delta\lambda_b)^2\}^{1/2}, \tag{2.32}$$

where we have set $Y = 1$, and where the uncertainties $\Delta\lambda_a$ and $\Delta\lambda_b$ of depths of the two minima can be estimated from the available observational data. If, as before, ϵ denotes the respective error of a single normal point, within minima, of unit weight, the square-root of the relative weight $\sqrt{w_k}$ of Equation (2.31) will again be given by

$$\sqrt{w_k} = (\epsilon/C_2)\Delta k; \tag{2.33}$$

and should be evaluated (or estimated as well as we can; for an accurate value of C_2 may not be known in advance) to multiply Equation (2.31) before the latter joins the rest of our overdeterminate system. The relative influence of this equation will depend on the ratio of uncertainties of the respective 'shape' and 'depth' k's. The reader should expect that, if the depths of both minima can be reliably estimated, the relative weight of the 'depth' k and, in consequence, of the additional equation (2.31) will always be considerable. If the primary minimum is deep, the 'shape' k based upon it may be of comparable weight; but *if*, on the other hand, *both minima are shallow, Equation* (2.31) *properly weighted may exert an important* – even controlling – *effect on the determination of all elements.*

D. EVALUATION OF THE ELEMENTS

Once the value of the ratio of the radii of both components has been stabilized by repeated iterations and the most probable values of the auxiliary constants C_1, C_2 and C_3 obtained, we are in a position to complete our analysis by evaluating the geometrical elements of the system under investigation. Since (cf. Equations (2.3)–(2.5))

$$C_1 = r_a^2 \csc^2 i, \quad C_2 = r_a r_b \csc^2 i, \quad C_3 = (r_b - r_a)^2 \csc^2 i - \cot^2 i, \tag{2.34}$$

and the constants on the left-hand sides are now known, a solution of these equations yields

$$r_a^2 = \frac{C_1^2}{(1 - C_3)C_1 + (C_2 - C_1)^2}, \tag{2.35}$$

$$r_b^2 = \frac{C_2^2}{(1 - C_3)C_1 + (C_2 - C_1)^2}, \tag{2.36}$$

$$\sin^2 i = \frac{C_1}{(1 - C_3)C_1 + (C_2 - C_1)^2}. \tag{2.37}$$

Since, moreover, during totality the remaining light of the system is due to that of the larger star, the fractional luminosities of the two components readily follow from

$$L_a = 1 - \lambda_a, \quad L_b = \lambda_a; \tag{2.38}$$

The ratio of the mean surface brightnesses J_a/J_b of the two components, defined by the relation

$$\frac{L_a}{L_b} = k^2 \frac{J_a}{J_b}, \tag{2.39}$$

follows from

$$\frac{J_a}{J_b} = \frac{1 - \lambda_a}{k^2 \lambda_a} \tag{2.40}$$

by Equations (1.22) and (2.38). Moreover, the theoretical angles θ_1 and θ_2 of the outer and inner contacts of the eclipse (and thus the theoretical durations of the minima) can ultimately be obtained by inserting in Equation (2.1) the appropriate values of r_a, r_b and i as deduced from Equations (2.35)–(2.37).

It should be stressed that, in Equations (2.37) and (2.39), λ_a and λ_b denote the fractional intensities corrected for $\Delta\lambda_a$ and $\Delta\lambda_b$, and pertaining to the corrected unit of light. A least-squares solution for the most probable values of C_1, C_2, C_3, $\Delta\lambda_a$, $\Delta\lambda_b$ and ΔU should, however, permit us to do more than to translate its results into the most probable values of the elements we seek to determine. The *uncertainty* of the auxiliary constants $C_1, C_2, \ldots$ etc., caused by the limited accuracy of the underlying observational data and implied in our least-squares solution, should permit us to determine also the corresponding *uncertainties of all elements*, which constitute indeed an inseparable part of our solution. Before we come, however, to discuss in Chapter VII the ways in which this can be done, we propose first to outline the process of solution for the elements to *partially*-eclipsing systems; for the methods of error analysis in either case will turn out to be so similar, and to have so much in common, that it is expedient to treat them jointly later (see Section VII-1).

IV-3. Iterative Methods: Partial Eclipses

If the light changes exhibited by an eclipsing system are found to be *continuous* throughout *both* minima, the eclipses giving rise to them are necessarily *partial*; and a determination of the geometrical elements of such eclipses from an analysis of their light curves becomes somewhat more involved. This is due to the fact that the fractional loss of light α due to eclipse will now be given by Equation (0.12) where n is given by Equation (2.17), and the maximum fractional loss of light $\alpha(k, p_0)$ is no longer unity as it was for total eclipses, but becomes an additional unknown to be determined by appropriate analysis. Such an analysis can, in general, be performed along the lines followed in the preceding section, but will possess certain particular features which we now wish to discuss.

In order to do so, let us depart from the fundamental equation (0.14) of this chapter, which at the moment of conjunction ($\theta = 0°$) reduces to Equation (1.10). Subtracting Equation (1.10) from (0.14) we find that

$$(p^2 - p_0^2)C_1 + 2(p - p_0)C_2 = \sin^2 \theta, \tag{3.1}$$

where (in accordance with Equation (1.11)) the constants C_1 and C_2 continue to be given by

$$C_1 = r_a^2 \csc^2 i \tag{3.2}$$

and

$$C_2 = r_a r_b \csc^2 i. \tag{3.3}$$

Suppose first that only *one* minimum has been observed (so that Equations (1.19) or (1.20) are not available). Is it possible, in such a case, to determine the empirical value of α_0 in $p_0 \equiv p(k, \alpha_0)$ simultaneously with r_a, r_b and i by an appropriate analysis of the observed light changes? To begin with, we usually do not know whether the minimum under investigation is due to occultation or transit eclipse; and the values of k as well as α_0 must be estimated before a system of equations of the form (3.1) can be set up for every observed point and solved for the unknown constants C_1 and C_2. Once this has been done, the initially assumed value of K can be checked by a comparison with the ratio C_1/C_2; but is there any way to verify the correctness of the assumed value of K (which depends not only on the geometrical elements r_a, r_b and i, but also on the extent of limb-darkening of the star undergoing eclipse)?

This should indeed be possible in principle if we regard

$$(p^2 - p_0^2)C_1 + 2(p - p_0)C_2 \equiv H(C_1, C_2, \alpha_0 + \Delta\alpha_0), \tag{3.4}$$

and expand the H-function in a Taylor series of the form

$$H_0 + \frac{\partial H_0}{\partial \alpha_0}\Delta\alpha_0 + \cdots = \sin^2 \theta, \tag{3.5}$$

where $H_0 \equiv H(C_1, C_2, \alpha_0)$. Before the equations of condition of the form (3.4) can be solved for the most probable values of the unknowns, they should (apart from the possibly different observational weights) be multiplied by square-roots $\sqrt{w}$ of their intrinsic weights, which can again be derived by the method of the preceding section and, for partial eclipses, are found to take the explicit form*

$$\sqrt{w} \equiv \frac{\mathrm{d}l}{\mathrm{d}\sin^2\theta} = -\frac{1-\lambda}{2\alpha_0 C_2(1+kp)(\partial p/\partial\alpha)}, \tag{3.6}$$

which differs from Equation (2.14a) only by the presence of α_0 (which formerly was equal to one) in the denominator. Differentiating H_0 with respect to α_0 and multiplying this derivative by $\sqrt{w}$ we readily establish that

$$\sqrt{w}\,\frac{\partial H_0}{\partial \alpha_0} = \frac{1-\lambda}{\alpha_0}\left\{\sqrt{\frac{w}{w_0}} - n\right\}, \tag{3.7}$$

where $\sqrt{w_0}$ denotes the intrinsic weight at the moment of maximum eclipse.

When the expression (3.7) is inserted in (3.5), an equation results containing C_1, C_2 and $\Delta\alpha_0$ as unknowns. If this equation could actually be iterated for all these unknowns, this would imply that no matter how crude a guess at α_3 we start from, it could be improved by successive approximations until the correct value is obtained. This is indeed possible in theory but, unfortunately, *not* in practice, because the coefficient of $\Delta\alpha_0$ in Equation (3.5) is not only numerically very small,† but – to make matters worse – its variation simulates that of the coefficient of C_1 to such an extent as to render a simultaneous determination of C_1 and $\Delta\alpha_0$ from the same set of equations a virtual impossibility. An inclusion of $\Delta\alpha_0$ as an additional unknown to be determined simultaneously with C_1 and C_2 be a least-squares solution of the equations of condition of the form (3.5) is found to entail so drastic a loss of weight of the whole solution that such a procedure would appear to be futile even if based on the best light curves now available. This demonstrates, therefore, that a *determination of a complete set of geometrical elements of partially eclipsing systems from one minimum alone remains as yet impracticable.*

A. COMBINATION OF THE ALTERNATE MINIMA

When, however, at least the *depths* of *both* minima have been observed, the situation is thoroughly altered; for then the quantities k and α_0 are connected by Equation (1.19) if the eclipse is an occultation, or by Equation (1.20) if it is a transit. For total eclipses ($\alpha_0' = 1$) Equation (1.19) led to a determination of the 'depth' k by Equation (1.22); but now – with α_0 as an additional unknown – Equations (1.19) or (1.20) furnish us with new relations between k and α_0 which are independent of the shape of the light curve, and which should permit us to ascertain the maximum obscuration α_0 corresponding to any assumed value of k.

* Whenever, in what follows, the values of λ are used without subscripts, it is understood that they apply equally to either occultations or transits. The same will be true of α_0 when used without primes.

† Vanishing as it does not only at the beginning of the eclipse (when $\sqrt{w} = n = 0$) and at the moment of maximum obscuration (when $\sqrt{w/w_0} = n = 1$), but also for $p = -p_0$ during partial phase.

If, therefore, the depths of both minima of a partially eclipsing system are known, a determination of the elements of our system can take a different course.* As the first step toward such a solution, we must estimate a *provisional value of* k – let us denote it again by K – and settle the question of the *types of eclipse* giving rise to the two minima – i.e., decide whether $\lambda_1 = \lambda_e a$ or λ_b. In contrast with the case of total eclipses, for partial eclipses there is – alas – no 'depth' or even 'duration' k available now to guide us at the beginning of our work; so that unless our system was already subject to previous investigations and some preliminary elements are available, our choice of k may have to be an almost outright guess. Spectroscopic evidence, if available, may sometimes enable us to specify k with some accuracy independently of the form of the light curve, or may indicate whether or not the component of greater surface brightness is also likely to be greater in size. If, however, in the absence of any preliminary elements or other pertinent information we are completely in the dark as to the type of eclipses giving rise to the primary or secondary minima, the best strategy is to start the solution by assuming $K = 1$ – in which case a distinction between occultations and transits loses any meaning.

Having estimated K and deduced from the observations the values of λ_a and λ_b (as well as our unit of light) to the best of our knowledge, our next step should be to evaluate α_0 from Equations (1.19) or (1.20). The reader should note that, in order to set up the explicit forms of these relations, a knowledge (or estimate) of the extent of limb-darkening of *both* components is prerequisite. And once the value of $\alpha_0 \equiv \alpha(K, p_0)$ has thus been established, the fractional loss of light at any moment during eclipse will be given by $\alpha \equiv n\alpha_0$; from which the corresponding value of $p(k, \alpha)$ can be extracted from Tsesevich's tables for the appropriate type of the eclipse and the degree of darkening adopted in Equations (1.19) or (1.20). An equation of condition of the form (3.1) can now be set up for every normal point based on observations secured within minima. With regard to the numerical accuracy to which the coefficients of these equations should be evaluated, the remarks made in connection with our treatment of total eclipses in the preceding section continue to hold good – i.e., three decimals in the coefficients and four decimals in the absolute terms should be ample in all but very exceptional cases. A least-squares solution of the set of such equations, weighted in accordance with Equation (3.6), should then yield the most probable values of C_1 and C_2 based on the assumed value of K. A check on the latter is again provided by the ratio $C_1/C_2 = k$. If K and k differ significantly, the iterative process should be repeated – each time with the previously improved value of k – until the assumed and resulting values of the ratio of the radii of both components are consistent.

* In order to forestall any possible misunderstanding, let it be stressed that this does not require the secondary minimum to be deep; what matters is that an absolute (not proportional) error in its determination be small. A knowledge that the depth of the secondary minimum is quite negligible, with an upper limit of 0.01 magnitude, is just as valuable in fixing the value of α_0 as if its depth were (say) ten times as large. It is only when we know *nothing whatever* of the secondary minimum – whether it is shallow or deep – that the determination of elements of partially eclipsing systems observed by the present methods becomes indeterminate.

Our ability to achieve this presumes tacitly that the iterations converge; and this requires that the types of eclipse giving rise to the primary and secondary minima have been correctly identified at the outset. If this was not so and, for this reason, we started the iterative process by assuming $K = 1$, the ratio C_1/C_2 resulting from the first iteration will very likely differ significantly from unity; so that a decision as to which one of the two minima is an occultation or a transit can no longer be deferred. At this time we are, however, already in possession of a value of $k = C_1/C_2$ which should not be too far from the true value of this ratio; and with this value of k *two* new parallel solutions should be started: one on the assumption that the primary minimum is due to an occultation eclipse, the other assuming it to be a transit. Only one of these alternatives can obviously be true and *only that one will converge to a definite answer.* Since the quantity K with which the two alternative solutions were started cannot be too far from truth, the outcome of one (the correct one) should closely confirm it; while, on the other hand, the ratio k resulting from the other solution should prove recalcitrant and refuse to stabilize in the course of repeated iterations. To iterate for k on the assumption of a wrong type of eclipse (i.e., using improper p-tables to calculate the coefficients of C_1 and C_2 in our equations of condition) is indeed as hopeless as chasing one's own shadow: for the ratio $C_1/C_2 = k$ will then be always different from K no matter how many times we put it through the mill. It is thus again the convergence (or divergence) of our iterations which will disclose to us *a posteriori* whether or not the types of our eclipses have been correctly identified.

B. GENERALIZED EQUATIONS OF CONDITION: DIFFERENTIAL CORRECTIONS

Before we actually embark on the iterative process – or at least before the final iteration is performed – systematic errors produced by possible errors in the adopted values of minimum brightness of the system at the moments of either conjunction, or in the adopted maximum brightness of the system (i.e., in our unit of light) should again be considered. The whole process of solution, as outlined in the preceding sections, utilizes *three* constants whose values had to be approximately read off the observed data: namely, λ_a, λ_b and U; and their provisional values may, therefore, be subject to an error. The question then arises again as to the effect of these errors on the correctness of our solution for the elements of the eclipse, and the way in which the provisional values of all three constants might be improved simultaneously with a determination of the most probable values of the geometrical elements. This way is so closely parallel with that followed in the preceding section in the course of our treatment of total eclipses that a brief outline will now be sufficient.

Let, as before, Equation (3.2) be written symbolically as

$$H(C_1, C_2, 1 + \Delta U, \lambda_0 + \Delta\lambda, \alpha_0 + \Delta\alpha_0) = \sin^2\theta, \tag{3.8}$$

where λ_0 and α_0 pertain to the minimum under investigation and $\Delta\lambda$, $\Delta\alpha_0$ denote the corrections to the respective quantities which we seek to determine. If we expand Equation (3.8) again in a Taylor series retaining only the first powers of the requisite corrections, we obtain

$$H_0 + \left(\frac{\partial H}{\partial U}\right)_0 \Delta U + \left(\frac{\partial H}{\partial \lambda}\right)_0 \Delta\lambda + \left(\frac{\partial H}{\partial \alpha_0}\right)_0 \Delta\alpha_0 + \cdots = \sin^2\theta, \tag{3.9}$$

where $H_0 \equiv H(C_1, C_2, 1, \lambda_0, \alpha_0)$. Before this equation can be used for a determination of the unknowns, it should again be multiplied by a square-root $\sqrt{w}$ of its intrinsic weight as defined by Equation (3.6). Moreover, in evaluating its coefficients we find that, exactly as in the case of a total eclipse,

$$\sqrt{w}\left(\frac{\partial H}{\partial U}\right)_0 = n - 1, \tag{3.10}$$

$$\sqrt{w}\left(\frac{\partial H}{\partial \lambda}\right)_0 = -n, \tag{3.11}$$

and $(\partial H/\partial\alpha)_0$ continues to be given by Equation (3.7); but in virtue of Equations (1.19)–(1.20) and of the fact that

$$\Delta(1-\lambda) = \Delta U - \Delta\lambda, \tag{3.12}$$

with the aid of Equation (1.19) we may now express $\Delta\alpha_0'$ in terms of the underlying errors of ΔU, $\Delta\lambda_a$ and $\Delta\lambda_b$ as

$$\Delta\alpha_0' = \Delta U - \Delta\lambda_a + \frac{\Delta U - \Delta\lambda_b}{k^2 Y(a, c_0)} \tag{3.13}$$

if the eclipse happens to be an occultation; and, by Equation (1.20),

$$\Delta\alpha_0'' = \Delta U - \Delta\lambda_b + k^2 Y(a, c_0)(\Delta U - \Delta\lambda_a) \tag{3.14}$$

if it is a transit. If we insert Equations (3.10)–(3.14) in (3.9) and remember that the function $Y(a, c_0)$ continues to be defined by Equation (1.21), our complete equation of condition for the determination of all unknown constants takes the explicit form

$$\begin{aligned}\sqrt{w}(p^2 - p_0^2)C_1 + 2\sqrt{w}(p - p_0)C_2 + \Big\{n - 1 + \frac{1-\lambda_a}{\alpha_0'} \\ \times\left[\sqrt{\frac{w}{w_0}} - n\right]\left[1 + \frac{1}{k^2 Y}\right]\Big\}\Delta U - \left\{n + \frac{1-\lambda_a}{a_0'}\left[\sqrt{\frac{w}{w_0}} - n\right]\right\}\Delta\lambda_a \\ - \frac{1-\lambda_a}{\alpha_0'}\left\{\sqrt{\frac{w}{w_0}} - n\right\}\frac{\Delta\lambda_b}{k^2 Y} = \sqrt{w}\sin^2\theta\end{aligned} \tag{3.15}$$

if the eclipse is an occultation, and

$$\begin{aligned}\sqrt{w}(p^2 - p_0^2)C_1 + 2\sqrt{w}(p - p_0)C_2 + \Big\{n - 1 + \frac{1-\lambda_b}{\alpha_0''}\left[\sqrt{\frac{w}{w_0}} - n\right] \\ \times(1 + k^2 Y)\Big\}\Delta U - \left\{n + \frac{1-\lambda_b}{\alpha_0''}\left[\sqrt{\frac{w}{w_0}} - n\right]\right\}\Delta\lambda_b \\ - \frac{1-\lambda_b}{\alpha_0''}\left\{\sqrt{\frac{w}{w_0}} - n\right\}k^2 Y\Delta\lambda_a = \sqrt{w}\sin^2\theta\end{aligned} \tag{3.16}$$

if it is a transit.* It should again be added that if, in order to remove the constant

* It may be of interest to note that, by Equations (1.17) and (1.18), $(1-\lambda_a)/\alpha_0' = L_a$ and $(1-\lambda_b)/\alpha_0'' = L_b$, where L_a, L_b are the fractional luminosities of the two components.

C_2 from the denominator of the Equation (3.6) defining $\sqrt{w}$, we pre-multiplied all equations of condition of the above forms by C_2, the last three unknown quantities automatically become equal to $C_2\Delta U$, $C_2\Delta\lambda_2$ and $C_2\Delta\lambda_b$. If, in addition, we found it expedient to multiply the w's by any other arbitrary factor c (applied, for instance, to keep their numerical values within reasonable bounds), it goes without saying that the values of the last three unknowns resulting from our solution would be equal to c times the respective differential correction.

Whatever is the case, the most general form of the equations of condition of the form (3.15) or (3.16) is found to contain five unknowns to be determined simultaneously by an appropriate least-squares solution. This is the same number of unknowns as we encountered earlier in connection with the case of total eclipses; the only difference being that C_3 is now by definition zero and that the corrections to the depths of both minima occur explicitly, with different coefficients, in the equations of condition pertaining to each type of eclipse. The reason is easy to understand; for whereas an error in the assumed depth of the minimum under investigation affects systematically all values of n as well as α_0, an error in depth of the other minimum will affect α_0 alone. As inspection of the numerical magnitude of the coefficients of ΔU, $\Delta\lambda_a$ and $\Delta\lambda_b$ in Equations (3.15) and (3.16) reveals that, although the rigorous forms of these coefficients are rather complicated, that of the correction to the depth of the minimum under investigation is again essentially equal to n, and that of ΔU is sensibly equal to $n-1$, as in the case of total eclipses; the numerical values of both coefficients will, therefore, range roughly from zero to one. The coefficient of the correction to the depth of the alternate minimum, proportional to the difference $(\sqrt{w/w_0}-n)$, is, on the other hand, apt to be numerically very small for reasons previously discussed (see p. 130). These facts disclose that the correction to the depth of the minimum under investigation will, in general, be a well-determined quantity, while the correction to the depth of the other minimum will remain largely indeterminate for the same reasons which made a determination of $\Delta\alpha_0$ well-nigh impossible.* If, however, we combine the equations of condition of the form (3.15) and (3.16) pertaining to both minima into a single set of normal equations – as we should always do at least in the final iteration – well-determined corrections to the depths of both minima should be obtained.

The light changes of a partially eclipsing system exhibit, by definition, no phase of constant light at the bottom of either minimum. In consequence, Equation (2.27) of the case of total eclipses will have no analogy in the present problem; but Equation (2.26) respecting the contribution of the observations made between minima to the determination of ΔU continues to hold good. Ultimately, we may add that if, in the foregoing equations, we set $p_0=-1$ and thus let the eclipse become a grazing total (or annular) one, α_0' or α_0'' becomes

* These reasons will be more readily apparent if we re-state them in the following terms. Since an error in the adopted depth of the minimum under investigation affects n as well as α_0, the corresponding error in n will affect all p's except p_0; its effect on the differences $p^2-p_0^2$ or $p-p_0$ constituting the coefficients of C_1 and C_2 may, therefore, become appreciable. An error in the adopted depth of the other minimum, will, however, affect only α_0 and, through it, both p and p_0 alike – so that by forming their differences we can minimize its consequences to a large extent.

exactly equal to unity (i.e., $\Delta\alpha_0' = \Delta\alpha_0'' = 0$); and, in consequence, Equations (3.13) or (3.14) provide a closed relation between ΔU and $\Delta\lambda_a$ or $\Delta\lambda_b$ of the form

$$\Delta\lambda_b - \Delta\lambda_a + \frac{\Delta U - \Delta\lambda_b}{k^2 Y} = k^2 Y(\Delta U - \Delta\lambda_a). \tag{3.17}$$

If we insert Equation (3.17) in (3.15) or (3.16), the latter equations reduce indeed to Equation (2.25) appropriate for grazing eclipses (i.e., when $C_3 = 0$). In such a case, Equations (3.15) and (3.16) would then become completely independent, and the corrections $\Delta\lambda_a$ and $\Delta\lambda_b$ could be solved separately from the data pertaining to either minimum. For partially eclipsing systems this is, however, not the case.

C. EVALUATION OF THE ELEMENTS

Once the final least-squares solution of a system of full-dress equations of condition of the form (3.15) and (3.16) has been obtained, and the value of C_1/C_2 found not do differ significantly from the adopted value K, we are in a position to reap the fruits of our investigation and evaluate at last the geometrical elements of our eclipsing system. In order to do so, we first establish the definitive values of the maximum obscuration α_0' and α_0'' at the bottom of each minimum from Equations (1.19) and (1.20), by inserting in them the corrected values of λ_a and λ_b (expressed in terms of our corrected unit of light) and the final value of k. With the aid of the definitive values of k and α_0 enter next the Tsesevich table appropriate for the accepted degree of darkening of the component undergoing eclipse at the respective minimum, to extract the corresponding value of $p(k, \alpha_0)$.

The equations defining the geometrical elements of partially eclipsing systems in terms of our auxiliary constants are

$$C_1 = r_a^2 \csc^2 i, \quad C_2 = r_a r_b \csc^2 i, \quad p_0 = (\cos i - r_b)/r_a; \tag{3.18}$$

and solving them we obtain

$$r_a^2 = \frac{C_1^2}{C_1 + (p_0 C_1 + C_2)^2}, \tag{3.19}$$

$$r_b^2 = \frac{C_2^2}{C_1 + (p_0 C_1 + C_2)^2}, \tag{3.20}$$

$$\sin^2 i = \frac{C_1}{C_1 + (p_0 C_1 + C_2)^2}; \tag{3.21}$$

or

$$r_a = \sqrt{C_1} \sin i, \tag{3.22}$$

$$r_b = (C_2/\sqrt{C_1}) \sin i, \tag{3.23}$$

$$\cot i = (p_0 C_1 + C_2)/\sqrt{C_1}. \tag{3.24}$$

The fractional luminosities $L_{a,b}$ of the two components then follow as a solution of Equations (0.3) and (1.17)–(1.18) from

$$\frac{L_a}{L_b} = k^2 Y(a, c_0) \frac{1-\lambda_a}{1-\lambda_b} \quad \text{and} \quad L_a + L_b = 1, \tag{3.25}$$

where again the corrected values of λ_a and λ_b referred to the corrected unit of light should be used. The ratio of the mean surface brightnesses of the two components is defined by

$$\frac{J_a}{J_b} = Y(a, c_0) \frac{1-\lambda_a}{1-\lambda_b}; \tag{3.26}$$

while the theoretical angle θ_1 of the first contact of a partial eclipse (and thus the computed duration of the minimum) can be obtained from Equation (2.1) by insertion of appropriate values of the elements r_a, r_b and i as deduced from Equations (3.19)–(3.24).

Before we conclude the present section, some retrospective considerations regarding the determination of elements of totally and partially eclipsing systems may be pointed out. The reader has undoubtedly noticed that the inclusion of α_0 among the unknown elements of partially eclipsing systems has not only complicated the procedure, but also diminished substantially the weight of the whole solution. Whereas, for total eclipses, three points of a light curve were in principle sufficient to specify the geometrical elements, all points within one minimum due to a partial eclipse were found inadequate to do so in practice, and in order to accomplish our aim we were compelled to draw further information from the secondary minimum as well. In solving for the elements of a totally eclipsing system we had to know the degree of darkening of one (the smaller) component only; whereas a knowledge (or estimate) of the darkening of both components was found to be prerequisite for deducing the elements of partially eclipsing systems. All this is bound to render the determination of elements of partially eclipsing systems unavoidably less exact. Yet, according to the laws of chance, a great majority of known eclipsing variables is likely to exhibit partial eclipses. An interpretation of their light curves therefore represents an important, though perhaps less attractive and certainly more laborious, field of double-star astronomy; and as such it is bound to commend itself to the attention of the students of our subject.

IV-4. Effects of Orbital Eccentricity

The whole discussion of the analysis of the light changes of eclipsing binary systems in the time-domain, as contained in the preceding sections of this chapter, has been based on the assumption that the components mutually eclipsing each other revolve around their common centre of gravity in *circular* orbits; and that, consequently, their changes of light within minima due to eclipses are *symmetrical* with respect to the moments of conjunctions. In reality, of course, the circular orbits represent only a mathematical abstraction – never to be met exactly in any practical case. Indeed, the orbits of all known eclipsing variables will be characterized by a finite eccentricity e; though observations disclose that, for a very large majority of such systems, their eccentricity may not produce any noticeable effect. Such is, for instance, the case with *close* binaries, where the space between the components is not large enough to leave room for variation of the radius-vector. On the other hand, among well-separated systems (or longer periods), eccentric orbits are encountered not too infrequently; but again if such systems are to exhibit eclipses, the inclinations of their orbital planes to the line of sight (i.e., $\cos i$) must remain very small (though this need not be true of e).

As is well known, the effects of orbital eccentricity on the light curves of eclipsing variables manifest themselves in three distinct ways. First (in accordance with Kepler's second law of elliptic motion) the times of the alternate minima of light will generally not be separated by exactly half the orbital period, but will be *displaced* relative to each other by an amount depending on both the orbital eccentricity e and the orientation of the apsidal line with respect to the line of sight. Other consequences of the same law will also render both minima to be *asymmetric*, and of *unequal duration.*

A. CENTRAL ECLIPSES

In order to ascertain the extent to which this will be the case, let

$$R = \frac{a(1-e^2)}{1+e\cos v} \tag{4.1}$$

denote the radius-vector of the relative orbit of our binary, of period P and semi-major axis a, in which the position of one component with respect to the other is specified by the true anomaly v measured from the longitude of periastron ω. If, moreover, the line of sight of a distant observer lies in the plane of the orbit (i.e., its inclination to the celestial sphere $i = 90°$), the extent of the three photometric phenomena arising from orbital eccentricity can be ascertained exactly and in a closed form.

Thus if $t_{1,2}$ denote the times of the mid-primary and secondary minima of light, and $n \equiv 2\pi/P$ stands for the mean daily motion, then, according to Kepler's second law,

$$n(t_2 - t_1) = (1-e^2)^{3/2} \int_{90°-\omega}^{270°-\omega} \frac{dv}{(1+e\cos v)^2} = \Psi - \sin\Psi, \tag{4.2}$$

where we have abbreviated

$$\Psi = \pi + 2\tan^{-1}\frac{e\cos\omega}{\sqrt{1-e^2}}. \tag{4.3}$$

Equation (4.2) represents a particular case of Kepler's celebrated equation, from which the quantity Ψ as defined by Equation (4.3) follows as the eccentric anomaly, in a parabolic orbit, corresponding to the mean anomaly $n(t_2 - t_1)$. The methods of solution of Kepler's equation are too numerous to be reviewed here in full. According to the late W. E. Brown, more than 100 approximate procedures to that end have been offered in the past three centuries, until an exact solution was obtained by Sievert and Burniston (1972). If – as will often be true – the displacement $n(t_2 - t_1)$ of the minima could be ascertained from the observations with only a moderate precision, the following graphical method should be sufficient. Consider a rectangular system of $y - \Psi$ coordinates, and construct graphs of the equations

$$y = \sin \Psi \quad \text{and} \quad y = \Psi - n(t_2 - t_1). \tag{4.4}$$

The sine curve can be drawn once for all; while a line inclined to the Ψ-axis by 45° and subtending an intercept $n(t_2 - t_1)$ needs alone to be constructed in each individual case. By virtue of Equation (4.2), the abscissa of the point of intersection of the curves (4.4) is the required value of Ψ.

Should a higher accuracy be desired than one attainable by graphical construction, recourse may be had to existing tables of the solution of Kepler's equation for parabolic orbits (cf. e.g., Crawford, 1930; Appendix C). However, whichever procedure is adopted, it should be borne in mind that the observed differences $t_2 - t_1$ do not specify e and ω separately, but only their combination $e(1-e^2)^{-1/2} \cos \omega$, which, if quantities of the order of e^3 and higher are ignored, reduces to the *tangential* component of the eccentricity $e \cos \omega$.

The durations of the ascending and descending branches of either minimum can be determined as follows. Let $t'_{1,2}$ and $t''_{1,2}$ denote the moments of the first and last contact of the primary and secondary eclipse, and let $\phi'_{1,2}$ be the corresponding phase angles measured from the respective conjunction at the time $t_{1,2}$. The differences

$$t_{1,2} - t'_{1,2} \equiv \tau'_{1,2} \quad \text{and} \quad t''_{1,2} - t_{1,2} \equiv \tau''_{1,2} \tag{4.5}$$

then yield the durations of the descending and ascending branches of each minimum.

According to the law of areas,

$$n\tau'_{1,2} = \frac{1}{a^2\sqrt{1-e^2}} \int_{v_{1,2}-\phi'_{1,2}}^{v_{1,2}} R^2 \, dv \tag{4.6}$$

and

$$n\tau''_{1,2} = \frac{1}{a^2\sqrt{1-e^2}} \int_{v_{1,2}}^{v_{1,2}+\phi''_{1,2}} R^2 \, dv, \tag{4.7}$$

where R continues to represent the radius-vector as given by Equation (4.1), $v_1 = 90° - \omega$, $v_2 = 270° - \omega$, and the angles of contacts ϕ' and ϕ'' are given by $\sin^{-1}(a_1 + a_2)/R$, where $a_{1,2}$ denote the radii of the two components; or, more explicitly, by the equations

$$\begin{aligned} \sin \phi'_{1,2} &= \alpha\{1 \pm e \sin(\omega + \phi'_{1,2})\}, \\ \sin \phi''_{1,2} &= \alpha\{1 \pm e \sin(\omega - \phi''_{1,2})\}, \end{aligned} \tag{4.8}$$

where we have abbreviated

$$\alpha \equiv \frac{a_1 + a_2}{a(1-e^2)} = r_1 + r_2, \tag{4.9}$$

in which $r_{1,2}$ denote the fractional radii of the respective components expressed in terms of the semi-latus rectum $a(1-e^2)$ of their relative orbit.

A solution of Equations (4.8) yields

$$\begin{aligned}\sin \phi'_{1,2} &= (\alpha/U^2_{1,2})(1 \mp \alpha e \cos \omega \pm e\sqrt{U^2_{1,2}-\alpha^2} \sin \omega),\\ \sin \phi''_{1,2} &= (\alpha/U^2_{2,1})(1 \pm \alpha e \cos \omega \pm e\sqrt{U^2_{2,1}-\alpha^2} \sin \omega),\end{aligned} \tag{4.10}$$

where

$$U^2_{1,2} = 1 \mp 2\alpha e \cos \omega + \alpha^2 e^2. \tag{4.11}$$

If we insert now for R and ϕ in Equations (4.6)–(4.7) and integrate between appropriate limits, we obtain

$$\begin{aligned}n\tau'_{1,2} &= 2 \tan^{-1} \sqrt{\frac{1-e}{1+e}} \tan \tfrac{1}{2}\phi'_{1,2}\\ &\quad \pm \sqrt{1-e^2}\left\{\frac{e \cos(\omega+\phi'_{1,2})}{1 \pm e \sin(\omega+\phi'_{1,2})} - \frac{e \cos \omega}{1 \pm e \sin \omega}\right\}\end{aligned} \tag{4.12}$$

and

$$\begin{aligned}n\tau''_{1,2} &= 2 \tan^{-1} \sqrt{\frac{1-e}{1+e}} \tan \tfrac{1}{2}\phi''_{1,2}\\ &\quad \pm \sqrt{1-e^2}\left\{\frac{e \cos(\omega-\phi''_{1,2})}{1 \pm e \sin(\omega-\phi''_{1,2})} - \frac{e \cos \omega}{1 \pm e \sin \omega}\right\},\end{aligned} \tag{4.13}$$

where the angles $\phi'_{1,2}$ and $\phi''_{1,2}$ are given by Equations (4.10) above. The sums $\tau'_{1,2}+\tau''_{1,2}$ then specify the *durations* of both minima, while the differences $\tau'_{1,2}-\tau''_{1,2}$ furnish their *asymmetry*.

An inspection of the right-hand sides of Equation (4.11) and (4.12) discloses that, for $e > 0$, the differences $\tau'_{1,2}-\tau''_{1,2}$ vanish only if $\omega = 90°$ or $270°$ – i.e., if the apsidal line is parallel with the line of sight – and attain a maximum for $\omega = 0°$ or $180°$ (i.e., at the time when the displacement of the minima is greatest). If ω lies between 270° and 90°, then $\tau'_1 > \tau''_1$ and $\tau''_2 < \tau'_2$; while for $90° < \omega < 270°$ the opposite is true. The branch of each minimum which lies nearest to the neighbouring minimum is always the steeper of the two. On the other hand, the durations $\tau'_1+\tau''_1$ and $\tau'_2+\tau''_2$ of both minima turn out to be equal if $\omega = 0°$ or $180°$ and differ most when the apsidal line is parallel to the line of sight. Thus the difference in durations of the minima and their displacement or asymmetry can never vanish both at the same time.

B. INCLINED ORBITS

All results given in the preceding section hold good exactly only if the plane of the eclipsing orbit happens to be normal to the celestial sphere, and the eclipses are consequently central. Should this cease to be true, however, the geometry of our problem becomes considerably more complicated on account of the fact that, if $e > 0$ and $i < 90°$, *the moments of the minima of light will fail to coincide with the times of conjunctions* and, in consequence, the areas of each component eclipsed at corresponding phases of either minimum will generally be unequal. In point of fact, if a total (or annular) eclipse occurs in the neighbourhood of

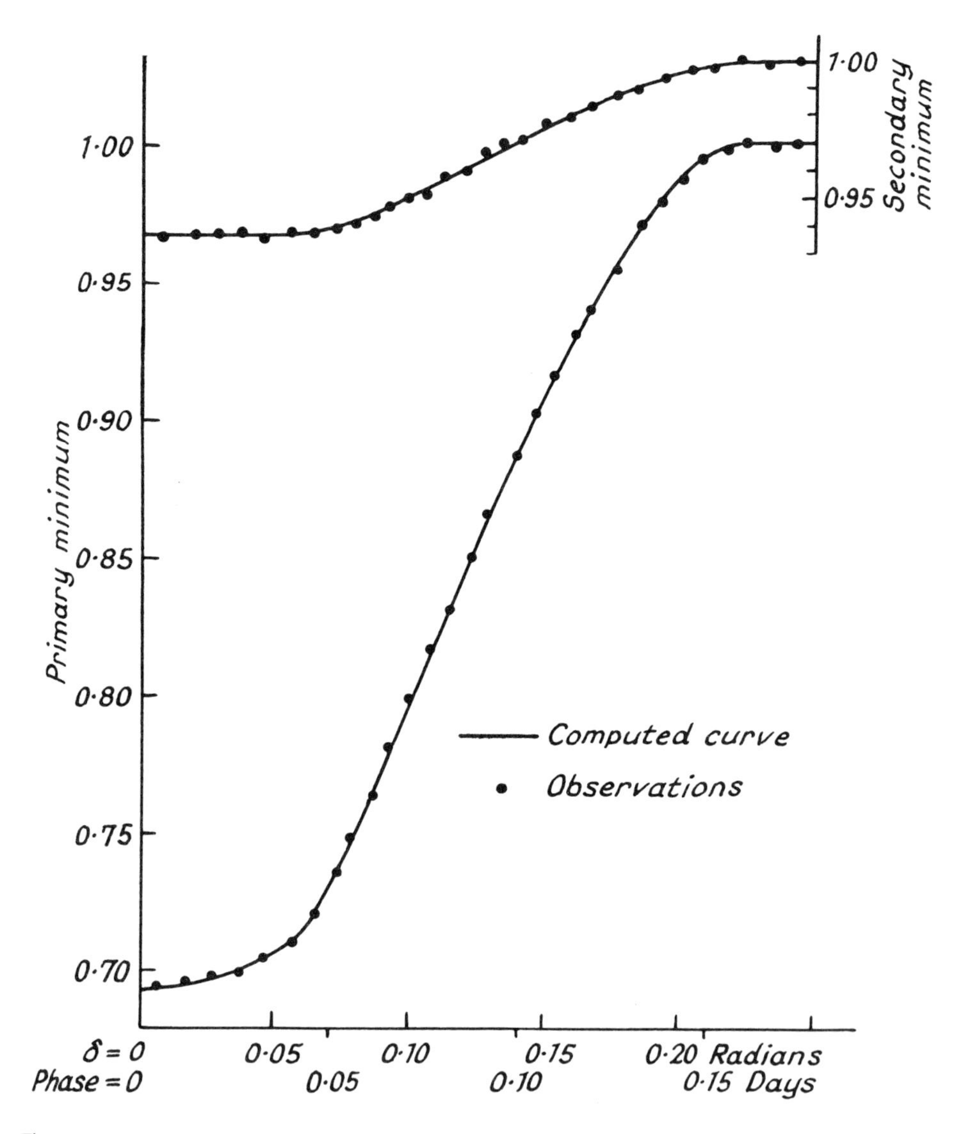

Fig. 4-1. The observed and theoretical light curves of the eclipsing system YZ Cassiopeiae, after G. E. Kron (1939).

The primary (deeper) minimum is due to an annular eclipse (of a limb-darkened star); while the eclipse giving rise to the secondary minimum (above) is total.

Abscissae: the fractional intensity of the system; ordinates, the phase (in days) and the phase angle (in radians).

periastron, the other may be partial, and vice versa. In extreme cases, an eclipse may occur at periastron and none at apastron at all!

In eccentric eclipsing systems the minima of light occur, by definition, when the centres of the apparent discs of both components are closest – i.e., when their projected separation δ becomes a minimum. For elliptical orbits,

$$\delta^2 = R^2(\sin^2\theta \sin^2 i + \cos^2 i) = \frac{a^2(1-e^2)^2(1-\sin^2 i \cos^2\theta)}{[1-e\sin(\theta-\omega)]^2}, \tag{4.14}$$

where (cf. Figure 4-2) the phase angle is

$$\theta = v + \omega - 90°. \tag{4.15}$$

A necessary condition for δ to be a minimum is the requirement that $d\delta/d\theta$ vanishes at that point. Differentiating Equation (4.14) we find this to happen for the values of θ which satisfy

$$\{1-e\sin(\theta-\omega)\}\sin^2 i \sin 2\theta + 2e(1-\cos^2\theta\sin^2 i)\cos(\theta-\omega) = 0. \tag{4.16}$$

When $i = 90°$, this equation admits of two roots at $\theta = 0$ and π, for which $\delta = 0$. For $i < 90°$ let us, therefore, seek the roots of Equation (4.16) which are in the neighbourhood of 0 and π. Their sines will obviously be quantities of the same order of magnitude as the orbital eccentricity e. If the third and higher powers of such quantities are ignored, the roots of Equation (4.26) which render δ a minimum are found to be

$$\begin{aligned} \theta_1 &= -e(1-e\sin\omega\csc^2 i)\cos\omega\cot^2 i + \cdots, \\ \theta_2 &= \pi + e(1+e\sin\omega\csc^2 i)\cos\omega\cot^2 i + \cdots; \end{aligned} \tag{4.17}$$

and the true anomalies at the moments of deepest eclipses become

$$v_{1,2} = 90° - \omega + \theta_{1,2}. \tag{4.18}$$

According to Kepler's law of areas, the relative *displacement* $n(t_2 - t_1)$ of the minima will be obtained from

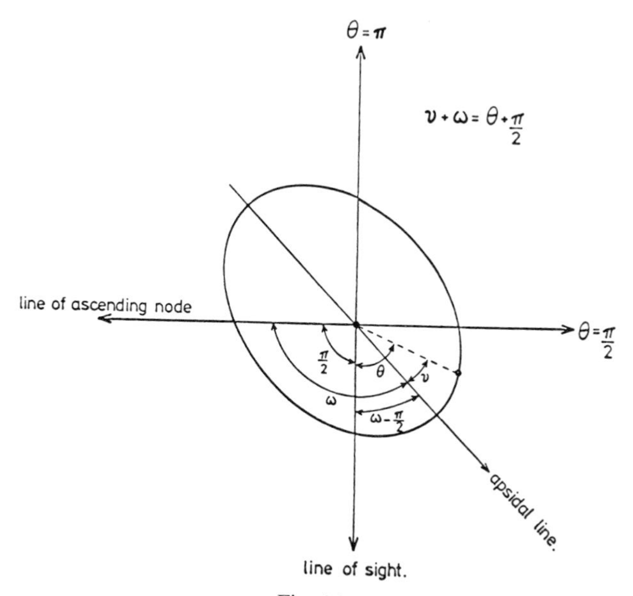

Fig. 4-2.

$$n(t_2 - t_1) = \frac{1}{a^2\sqrt{1-e^2}} \int_{v_1}^{v_2} R^2\,\mathrm{d}v = \pi + 2e(1 + \csc^2 i)\cos\omega + \cdots, \tag{4.19}$$

which, for $i = 90°$, coincides to the first order in e with a Fourier expansion of the right-hand side of Equation (4.2).

The *durations* as well as *asymmetry* of the alternative minima of eccentric eclipsing systems with orbits inclined to the line of sight can be established in a way similar to that followed in the preceding section. For inclined orbits ($i < 90°$) Equation (4.8) assume the forms

$$\begin{aligned} \sqrt{1 - \sin^2 i \cos^2 \phi'_{1,2}} &= \alpha\{1 \pm e \sin(\omega + \phi'_{1,2})\}, \\ \sqrt{1 - \sin^2 i \cos^2 \phi''_{1,2}} &= \alpha\{1 \pm e \sin(\omega - \phi''_{1,2})\}. \end{aligned} \tag{4.20}$$

Moreover, if we introduce an auxiliary angle ϕ_0, defined by

$$\alpha^2 = 1 - \sin^2 i \cos^2 \phi_0, \tag{4.21}$$

we find that, correctly to terms of the first order in e, Equation (4.20) can be solved to yield

$$\phi'_{1,2} = \phi_0 \pm 2e(\sin^2 \phi_0 + \cot^2 i)\sin(\omega + \phi_0)\csc 2\phi_0 + \cdots \tag{4.22}$$

and

$$\phi''_{1,2} = \phi_0 \pm 2e(\sin^2 \phi_0 + \cot^2 i)\sin(\omega - \phi_0)\csc 2\phi_0 + \cdots. \tag{4.23}$$

If we integrate now (to the same order of approximation) $R^2\,\mathrm{d}v$ from the moment of deepest eclipse to that of its first or last contact, the fractional *durations* of the two minima become

$$\begin{aligned} d_{1,2} &= n(\tau'_{1,2} + \tau''_{1,2}) \\ &= 2\phi_0 \mp 2e(1 - \cot^2 i \csc^2 \phi_0)\sin\omega \sin\phi_0; \end{aligned} \tag{4.24}$$

while their *asymmetries* are given by

$$n(\tau'_{1,2} - \tau''_{1,2}) = \pm 2e(\csc^2 i - \cot\phi_0)(\sec\phi_0 - 1)\cos\omega. \tag{4.25}$$

Lastly, the intervals between the time of maximum eclipse (i.e., minimum of light) and the moments of conjunctions are equal to $\frac{1}{2}(\tau'_{1,2} - \tau''_{1,2})$.

In order to demonstrate the effects of orbital inclination on the duration of the alternate minima in eccentric eclipsing systems, let us eliminate from Equation (4.23) the auxiliary angle ϕ_0 by means of Equation (4.21). In particular, since

$$1 - \cos^2 i \csc^2 \phi_0 = \frac{\alpha^2 - 2\cos^2 i}{\alpha^2 - \cos^2 i}, \tag{4.26}$$

Equations (4.24) make it evident that the effect of diminishing inclination i upon the duration $d_{1,2}$ of the alternate minima will depend essentially on whether $\cos i \lesseqgtr \alpha/\sqrt{2}$: when $\cos i < \alpha/\sqrt{2}$, the minimum nearest apastron will be of longer duration; while for $\alpha > \cos i > \alpha/\sqrt{2}$, the converse will be true. A decrease in i will, therefore, tend at first to diminish a disparity in duration of the two minima until the equality $\cos i = \alpha/\sqrt{2}$ is reached; but beyond this limit the disparity will tend to increase again. When $\cos i = \alpha/\sqrt{2}$, both minima should be of equal duration (as far as terms of the order of e^2 and higher remain negligible); the light curve should then simulate very closely one produced by a system of stars for which $e \sin \omega = 0$, and the eclipses should be total or partial – depending on whether the ratio of the radii of its components is smaller or greater than $(2-\sqrt{2})/(2+\sqrt{2})$. If, ultimately, $\cos i$ approaches α, the denominator on the right-hand side of Equation (4.26) will tend to zero and so, consequently, will the angle ϕ_0. In such a case, the expansion on the right-hand side of Equation (4.24) will cease to converge – no matter now small e may be. This situation explains why, near this limit, no eclipse may occur any longer in the proximity of the apastron while one persists near periastron.

A sum of the durations of both minima follows from Equation (4.24) as

$$d_2 + d_2 = 4\phi_0, \tag{4.27}$$

and is (within the scheme of our accuracy) independent of e or ω, but diminishes steadily with i. On the other hand, a difference

$$d_1 - d_2 = -4e(1 - \cot^2 i \csc^2 \phi_0) \sin \omega \cos \phi_0 + \cdots, \tag{4.28}$$

so that (if ϕ_0 is not too large) the ratio

$$\frac{d_1 - d_2}{d_1 + d_2} = -e(1 - \cot^2 i \csc^2 \phi_0) \sin \omega, \tag{4.29}$$

which by Equation (4.26) yields

$$e \sin \omega = \frac{d_2 - d_1}{d_2 + d_1} \left\{ \frac{\alpha^2 - \cos^2 i}{\alpha^2 - 2\cos^2 i} \right\}. \tag{4.30}$$

A determination of the *radial* component $e \sin \omega$ of orbital eccentricity from the unequal durations of both minima is, therefore, possible in principle provided that the orbital inclination is known (or, if unknown, is such that $\cos i \ll \alpha$). In practice, however, such a determination will usually be less precise than that of the tangential component $e \cos \omega$ from Equation (4.19). In this latter case, an error of the respective determination is proportional to a ratio of the displacement $t_2 - t_1$ to the orbital period P, and in the former case to a ratio of $d_2 - d_1$ to $d_1 + d_2$ – both of which (especially the difference $d_2 - d_1$) are very much smaller than $\frac{1}{2}P$, so that, unlike $e \cos \omega$, $e \sin \omega$ obtains as a ratio of factors which can be numerically small.

When both light minima are reasonably deep, the inclination i cannot be far from 90° and a separation of e and ω can, therefore, be effected from purely photometric data. Shallow minima are, however, practically useless for this purpose, so that the theoretically interesting case when $\cos i > \alpha/\sqrt{2}$ is of no practical importance. All we can deduce then from photometric evidence alone is the tangential component of orbital eccentricity, and to obtain e and ω separately we have to wait for a spectroscopic determination of the absolute orbit or for a definite indication of apsidal motion.

Of all photometric effects produced by orbital eccentricity, the asymmetry of the minima turns out, by far, to be the least conspicuous. This fact can be demonstrated by a comparison of the asymmetry with displacement: dividing $n(\tau'_{1,2} - \tau''_{1,2})$ as given by Equation (4.25) by $n(t_2 - t_1) - \pi$ as given by Equation (4.19), it follows that

$$\frac{\tau'_{1,2} - \tau''_{1,2}}{t_2 - t_1 - \frac{1}{2}P} = \pm \frac{(\sec \phi_0 - 1)(\csc^2 i - \cos \phi_0)}{1 + \csc^2 i}, \tag{4.31}$$

which for $i = 90°$ reduces to

$$\frac{\tau'_{1,2} - \tau''_{1,2}}{t_2 - t_1 - \frac{1}{2}P} = \pm 2 \sec \phi_0 \sin^4 \tfrac{1}{2}\phi_0. \tag{4.32}$$

The ratio on the right-hand side of Equation (4.31) is always smaller than that in Equation (4.32) – a fact demonstrating that, for a given e and ω, the asymmetry becomes a maximum for central eclipses and diminishes as $i < 90°$.

A ratio of the displacements of the conjunctions with respect to the minima of light is equal to one-half of the amounts given by Equations (4.31) or (4.32) and in either case it remains very small. If, for example, we set $\phi_0 = 45°$ and $i = 70°$ (corresponding, by Equations (4.9) and (4.21), to a sum $r_1 + r_2 = 0.783$, which is well outside the range for most contact systems), the ratio (4.31) becomes equal to 0.15; and if ϕ_0 is diminished to 30° (corresponding to $r_1 + r_2 = 0.648$), this ratio reduces to 0.04. The asymmetry of the light minima of typical eclipsing systems which can (on account of their separation) exhibit noticeable eccentricity will, therefore, never amount to more than a few per cent of the displacement of the minima – which means that, in most cases, it is likely to remain observationally insignificant. This constitutes a fortunate circumstance, of which advantage will be taken in a subsequent part (Section V-3) of this book.

Bibliographical Notes

IV-1. An interpretation of the light changes of eclipsing binary systems in the time-domain – i.e., an analysis of a plot of the observed brightness against the time (or phase angle) – happened to be the first road adopted for the solution of the observed data for the elements (geometrical, or physical) of the respective system; and as it happened often before in the history of science, the first road followed was not necessarily the best that could have been adopted for the purpose.

Perhaps it was so adopted because it appeared intuitively simplest. But apparent simplicity is not always the best guide towards the desired returns; and even so almost a hundred years had to elapse between Goodricke's famous surmise that ... "if it were not perhaps too early to hazard even a conjecture on the cause of this variation" (i.e., of Algol, the periodicity of which was discovered by Goodricke in 1782) "I should imagine it could hardly be accounted for otherwise than ... by the interposition of a large body revolving around Algol" (J. Goodricke, 1783), and the first attempt to place this qualitative hypothesis on a quantitative basis (E. C. Pickering, 1880).

Pickering's work inaugurated the first era in theoretical investigations of eclipsing variables, which enrolled first but slowly (cf., J. Harting, 1889; G. W. Myers, 1896; E. Hartwig, 1900; A. Pannekoek, 1902; A. W. Roberts, 1903; S. Blazhko, 1911; or R. S. Dugan, 1911). In 1912, most part of this early work was systematized and superseded by a series of investigations by H. N. Russell (1912a, b), followed by H. N. Russell and H. Shapley (1912a, b), which together with extensive applications to practical cases carried out by H. Shapley (1915) inaugurated a new era in the study of eclipsing systems. Subsequently, small modifications to their method were offered by H. Vogt (1919), J. Fetlaar (1923), J. Stein (1924), S. Scharbe (1925), B. W. Sitterly (1930), V. A. Krat (1934, 1935, 1936),

J. Ellsworth (1936), S. L. Piotrowski (1937) or O. E. Brown (1938). Many years later, H. N. Russell and J. E. Merrill (1952) wrote up the early 1912 work with new tables and many small modifications, but without injecting any essentially new ideas into the problem.

That this could not have been done was since made evident by criticisms stemming from many sources (cf., e.g., Z. Kopal, 1941), which pointed out that the results obtained in this way hardly deserved the name of a 'solution' of the problem – as a part of it (i.e., the location of 'fixed points' on the light curve) had to be anticipated in advance; and the 'direct' methods contained no built-in safeguards for systematic improvements of their positions – or, indeed, for any kind of error analysis which could indicate their uncertainty. These facts alone were sufficient to bring any further development along these lines to a standstill.

IV-2. In order to circumvent difficulties arising from these sources, Z. Kopal (1941) proposed to exchange direct methods of analysis of the light changes of eclipsing variables for an *iterative* process, which permitted the entire solution to be based on the actually observed data (rather than of smooth curves drawn by free hand to follow the course of individual observations), and dispensing with the use of any fixed points altogether.

Iterative methods for treating different types of eclipses were first comprehensively described in Chapter III of the writer's *Introduction to the Study of Eclipsing Variables* (Z. Kopal, 1946). Their important feature was the facility with which they lent themselves for an error analysis of the outcome – an analysis developed later by S.L. Piotrowski (1947, 1948a) as well as Z. Kopal (1948a). A comprehensive summary of the entire procedure has subsequently been summarized by the present writer in two sources: Chapter III of the *Computation of the Elements of Eclipsing Binary Systems* (Z. Kopal, 1950), and Chapter VI of *Close Binary Systems* (Z. Kopal, 1959). A presentation of the subject as given in Sections 2 and 3 of this chapter follows (with some modifications) that given in Kopal's 1959 treatise.

Methods expounded in these sections continue to remain adequate for an analysis of the light changes of eclipsing binary systems in the time-domain as long as their components can be regarded as spherical, and photometric proximity arising from ellipticity and reflection can be ignored. This can, however, be true (within the limits of present-day observational errors) only for relatively wide pairs, whose likelihood of discovery is much smaller than that of close pairs exhibiting wider minima, in which the photometric proximity effects are not only noticeable, but may become conspicuous. If they do, no way has been discovered in the time-domain so far for a meaningful separation of photometric proximity effects from those that may arise from the eclipses; and it is this difficulty on which the time-domain approach to the solution of our problem – by any method – ran on the rocks.

As will be shown in subsequent Chapters V and VI of this book, this shipwreck of our efforts can, however, be avoided by a translation of the entire problem from the time- to the frequency-domain; and if so, the question is bound to obtrude on our minds: why was this not done to begin with? Surely not for a lack of knowledge of the underlying mathematics: the Fourier theorem was discovered by its author already in 1807; and all that we shall need in Chapters V and VI of the present volume to 'rectify' the light changes in the frequency-domain was well known, for instance, to Gauss. Why did not the problem at issue attract his genius to carry out (and better than ourselves) what we shall do in Chapters V–VI – and why has (for that matter) the loss of light arising from eclipses not been identified with the Hankel transform (3.34) of Chapter I before our time?

We do not, of course, know the answer. One possibility may be that, in Gauss's days, the problem at issue did not appear to be observationally significant enough to attract the attention of great minds of the time. It is true that photometric observations which could be made in the first half of the 19th century were seldom more accurate than to 0.1 of a magnitude – a precision which increased to 0.01 magn. only around 1900, and to 0.001 magn. by the middle of the 20th century. Even today, the difficulties besetting an analysis of the light changes of eclipsing variables based on the best available observations are still comparable with those which would confront a computer of planetary or asteroidal orbits if the positional measurements at his disposal were inaccurate to 0.001 of a quadrant (i.e., some 20 minutes of arc!). In dynamical astronomy this would correspond to quite a pre-historic degree of accuracy; not meriting as yet much attention.

On the other hand, the more inaccurate the observations, the greater number of them is required to lessen the cumulative effects of their accidental errors; and the method of 'least-squares' (co-discovered by Gauss in 1809) would have been very much more justified to be put to this task than to re-discover the asteroid Ceres. Perhaps the best explanation may be that the *physics* underlying phenomena, exhibited by Algol and other eclipsing systems, is much more complicated than that governing the motions of planets or asteroids; and it was an increase in understanding of the physical basis of the observed phenomena which provided incentive for the current efflorescence of our subject.

IV-3. The effects of orbital eccentricity on the light changes of eclipsing variables received an early attention of the investigators; for the classical aspects of it cf. N. C. Dunér (1900) or Ch. André (1900); a comprehensive account of this early work can be found on pp. 212–217 of the treatise on *Die Veränderlichen Sterne* by J. G. Stein (1924).

The present version of the subject as given in this section follows closely Section VI.9 of the writer's book on *Close Binary Systems* (Z. Kopal, 1959); cf. also Z. Kopal and H. M. Al-Naimyi (1978).

CHAPTER V

ANALYSIS OF THE LIGHT CHANGES IN THE FREQUENCY-DOMAIN: SPHERICAL STARS

In the preceding chapter of this book we gave an outline of the iterative solutions of the light curves of eclipsing variables in the time-domain for the elements of the eclipses of the two stars. In retrospect, we note that while the problem at issue does not admit of any closed (or even algebraic) solution on account of its non-linearity, it can be solved by successive approximations (the convergence of which will be discussed more fully in Chapter VII). But whichever method we employed in our efforts to decipher the light changes of eclipsing variables in the time-domain so far, we found one auxiliary function consistently in our way: namely, the 'geometrical depth' $p(k, \alpha)$ of the eclipse as defined by Equation (0.6). It is this function – rather than the fractional loss of light $\alpha(k, p)$ – which played directly so fundamental a role in *all* methods of interpretation of the light changes caused by eclipses in the time-domain. Unfortunately, a transcendental character of its dependence on k and α has made it impossible to construct a mathematical solution of the underlying problem otherwise than by successive approximations; and no one succeeded so far to express this function analytically in an algebraic or differential form – no one has, indeed, seen it in any form other than of numerical tables!

This fact is doubly unfortunate; for a relationship between p and α depends, for any constant k, on the distribution of brightness over the apparent disc of the star undergoing eclipse; and must be recomputed for every adopted degree of limb-darkening; moreover, even for the same degree of darkening separate tables are required for occultations and transits. Tsesevich (1939, 1940) produced such tables (accurate to four decimal places) for the case of linear law of limb-darkening specified by the coefficients $u = 0.0(0.2)1.0$; but nothing is known about the behaviour of functions $p(k, \alpha)$ if the limb-darkening becomes non-linear; and no mathematical machinery (other than inverse interpolation of the $\alpha(k, p)$-tables) is known to obtain them.

To an astronomer concerned mainly with more utilitarian aspects of the case, such considerations may perhaps seem specious. Unjustly so; for they may be also symptomatic of an improper approach to the problem, making the difficulties encountered in the preceding chapter self-imposed? Do the time-domain methods outlined in Chapter IV represent the only way for the solution of our problem; and if not, what are the alternatives; and what assurance do we possess that we have followed in Chapter IV the right road? The aim of the present chapter will be to demonstrate that the answers to these questions are manifestly in the negative: other methods exist which allow us to reach our goal in a completely different way – namely, *by transposing our problem from the time- to the frequency-domain* – in other words, not to analyze for the elements of the eclipse a function represented by a plot of the observations versus the time (i.e., the 'light curve' in the time-domain), but its *Fourier transform*.

V-1. Introduction: Uniformly Bright Stars

In order to embark on this task in simplest possible terms, let us consider an eclipsing system which consists of two spherical stars revolving around the common centre of gravity in circular orbits, and appearing in projection on the celestial sphere as uniformly bright circular discs. As in Chapter IV, the task confronting us is to ascertain from the light changes exhibited during eclipses the 'geometrical elements' of the system – i.e., the fractional dimensions $r_{1,2}$ of its components, and the inclination i of their orbit. In order to do so, let us – unlike in Chapter IV – de-emphasize our interest in the location of the individual observed points (l, θ) in the time-domain, and focus instead our attention on the *area* subtended by the light curve in the $l - \sin^{2m}\theta$ coordinates $(m = 1, 2, 3, \ldots)$, as shown in Figure 5-1. The areas A_{2m} between the lines $l = 1$, $\sin^{2m}\theta = 0$, and the actual shape of the light curve are then evidently given by the integrals

$$A_{2m} = \int_0^{\theta_1} (1 - l)\, \mathrm{d}(\sin^{2m}\theta), \tag{1.1}$$

hereafter referred to as *moments of the eclipse*, of index m, where θ_1 denotes the angle of the first contact of the eclipse. For $\theta > \theta_1$, by definition $l = 1$ for spherical stars, and thus any extension of the limits of integration on the right-hand side of Equation (1.1) would no longer affect the requisite areas.

The empirical values of A_{2m} can be readily ascertained by quadratures, which

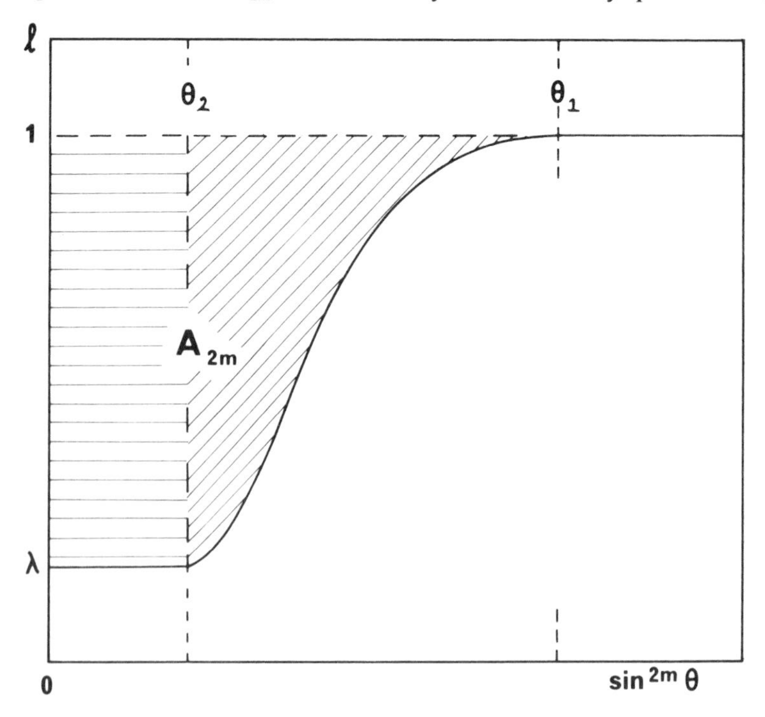

Fig. 5-1. Light curve of a total eclipse.

may assume no more elaborate form than a 'counting of the squares' of the area cross-hatched on Figure 5-1 – limited on one side by the ensemble of the observed points. In performing this planimetry, we may find it of advantage to represent this boundary again by a smooth curve drawn by free hand to represent the course of observed points.* In doing so we are, however, not relying on any particular points of this curve as we did in Chapter IV, but propose to give all a simultaneous representation.

On the other hand, by virtue of Equation (0.4) of Chapter IV, the integrand $1-l$ on the right-hand side of Equation (1.1) can be theoretically expressed as

$$1-l=\alpha(k,p)L_1; \tag{1.2}$$

and from Equation (0.2) of Chapter IV it follows that the element of integration

$$\mathrm{d}(\sin^{2m}\theta)=m(\delta^2-\delta_0^2)^{m-1}(1-\delta_0^2)^{-m}\,\mathrm{d}\delta^2, \tag{1.3}$$

where we have abbreviated

$$\delta_0=\cos i \tag{1.4}$$

to denote the minimum distance between the centres of both components at the time of maximum eclipse (i.e., when $\theta=0$).

If, moreover, $\delta_{1,2}$ denote the separations of their centres at the moments of the first and second contacts of the eclipse, it follows that (for the model adopted)

$$\delta_1=r_1+r_2 \tag{1.5}$$

for any type of eclipse; while

$$\delta_2=r_2-r_1 \quad \text{or} \quad r_1-r_2, \tag{1.6}$$

depending on whether the eclipse is total ($r_2>r_1$) or annular ($r_1>r_2$).

A. TOTAL (ANNULAR) ECLIPSES

For total (or annular) eclipses

$$\delta_1>\delta_2>\delta_0; \tag{1.7}$$

while if the eclipse is partial,

$$\delta_1>\delta_0>\delta_2. \tag{1.8}$$

In view of these results, the areas A_{2m} whose empirical values can be obtained by quadratures of the form (1.1) can, by Equations (1.2) and (1.3) be also theoretically expressed as

$$A_{2m}=mL_1\csc^{2m}i\int_{\delta_2^2}^{\delta_1^2}(\delta^2-\delta_0^2)^{m-1}\alpha\,\mathrm{d}\delta^2$$

* Later in this book – in the discussion of error analysis (Section VII-3) – a systematic method will be developed for a purely analytical determination of the areas A_{2m}, making direct use of individual observations and replacing planimetry by least-squares techniques.

$$= mL_1 \csc^{2m} i \left\{ \int_{\delta_2^2}^{\delta_1^2} (\delta^2 - \delta_0^2)^{m-1} \alpha \, d\delta^2 + \int_{\delta_0^2}^{\delta_2^2} (\delta^2 - \delta_0^2)^{m-1} \alpha \, d\delta^2 \right\}. \tag{1.9}$$

for total (annular) eclipses, where the first integral on the right-hand side stands for the contribution arising from the partial phases of the eclipse ($\delta_1 > \delta > \delta_2$); and the second, from the total (annular) phase ($\delta_2 > \delta > \delta_0$).

In the case of total eclipses ($r_2 > r_1$), $\alpha = 1$ during totality; while during annular phase ($r_2 < r_1$) it is equal to $(r_2/r_1)^2$ – in each case, a constant. Therefore,

$$\int_{\delta_0^2}^{\delta_2^2} (\delta^2 - \delta_0^2)^{m-1} \alpha \, d\delta^2 = m^{-1} (\delta_2^2 - \delta_0^2)^m \tag{1.10}$$

for the case of total eclipse (when $\delta_2 = r_2 - r_1$); while if the eclipse is annular, $\delta_2 = r_1 - r_2$ and the right-hand side of Equation (1.10) should be multiplied by $(r_2/r_1)^2$.

In the course of partial phases of either type of eclipses – be it an occultation or transit – the fractional loss of light α – equal, for uniformly bright discs, to the fractional area eclipsed, varies with δ for $\delta_1 > \delta > \delta_2$ in accordance with Equations (2.14) of Chapter I, as

$$\begin{aligned} \pi r_1^2 \alpha &= 2r_1^2 \int_0^{\phi_1} \sin^2 \phi \, d\phi + 2r_2^2 \int_0^{\phi_2} \sin^2 \phi \, d\phi \\ &= r_1^2 \{\phi_1 - \tfrac{1}{2} \sin 2\phi_1\} + r_2^2 \{\phi_2 - \tfrac{1}{2} \sin 2\phi_2\}, \end{aligned} \tag{1.11}$$

where the angles $\phi_{1,2}$ (cf. Figure 1-3) are defined by the equations

$$\phi_1 = \cos^{-1} \frac{\delta^2 + r_1^2 - r_2^2}{2\delta r_1} = \cos^{-1} \frac{s}{r_1} \tag{1.12}$$

and

$$\phi_2 = \cos^{-1} \frac{\delta^2 - r_1^2 + r_2^2}{2\delta r_2} = \cos^{-1} \frac{\delta - s}{r_2}, \tag{1.13}$$

such that

$$r_1 \sin \phi_1 = r_2 \sin \phi_2. \tag{1.14}$$

In order to evaluate the first integral on the right-hand side of Equation (1.9) between the limits $\delta_{1,2}^2$ it is, however, not necessary to make use of Equation (1.11) for α as it stands. For remembering that $\alpha(\delta_1) = 0$ while $\alpha(\delta_2) = 1$ during totality and $(r_2/r_1)^2$ during annular phase, by partial integration we find that, in the case of occultation eclipses ($r_1 < r_2$ and $\delta_2 = r_2 - r_1$),

$$\int_{\delta_2^2}^{\delta_1^2} (\delta^2 - \delta_0^2)^{m-1} \alpha \, d\delta^2 = -m^{-1} (\delta_2^2 - \delta_0^2)^m - m^{-1} \int_{\delta_2}^{\delta_1} (\delta^2 - \delta_0^2)^m \frac{\partial \alpha}{\partial \delta} \, d\delta; \tag{1.15}$$

while if the eclipse is a transit ($r_1 > r_2$ and $\delta_2 = r_1 - r_2$), the first term the right-hand side of Equation (1.15) should be multiplied by $(r_2/r_1)^2$.

By differentiation of Equation (1.12) with respect to δ we find that, for uniformly bright discs,

$$\frac{\partial\alpha}{\partial\delta} = -\frac{1}{\pi r_2}\left(\frac{r_2}{r_1}\right)^2 \sqrt{\frac{(\delta_1^2-\delta^2)(\delta^2-\delta_2^2)}{\delta^2 r_2^2}}, \tag{1.16}$$

an expression which should be inserted on the right-hand side of Equation (1.15) before the integration can be performed. In order to facilitate this integration, let us introduce a new variable ϕ related with δ by means of the equation*

$$\delta^2 = r_1^2 - 2r_1r_2\cos\phi + r_2^2, \tag{1.17}$$

in terms of which Equation (1.16) can be rewritten as

$$\frac{\partial\alpha}{\partial\delta} = -\frac{2}{\pi}\left(\frac{r_2}{r_1}\right)\frac{\sin\phi}{\delta}. \tag{1.18}$$

Since, moreover, by a differentiation of Equation (1.17)

$$\mathrm{d}\delta = (r_1r_2/\delta)\sin\phi\,\mathrm{d}\phi, \tag{1.19}$$

the theoretical areas A_{2m} as defined by Equations (1.9)–(1.10) and (1.15) can be rewritten in the contracted form as

$$A_{2m} = -L_1\int_{\delta_2}^{\delta_1}\left(\frac{\delta^2-\delta_0^2}{1-\delta_0^2}\right)^m \frac{\partial\alpha}{\partial\delta}\,\mathrm{d}\delta, \tag{1.20}$$

expressible in terms of partial integrals of the type

$$\int_{\delta_2}^{\delta_1}\delta^{2j}\frac{\partial\alpha}{\partial\delta}\,\mathrm{d}\delta = \frac{2r_2^2}{\pi}\int_0^{\pi}(r_1^2-2r_1r_2\cos\phi+r_2^2)^{j-1}\sin^2\phi\,\mathrm{d}\phi \tag{1.21}$$

for $j = 0, 1, 2, \ldots$. For $j > 0$ the right-hand side of the foregoing equation consists of trigonometric integrals whose evaluation is trivial (those involving odd powers of $\cos\phi$ all vanish on integration between 0 and π); while when $j = 0$, the left-hand side makes it immediately obvious that

$$\int_{\delta_2}^{\delta_1}\frac{\partial\alpha}{\partial\delta}\,\mathrm{d}\delta = \alpha(\delta_1) - \alpha(\delta_2) = -1 \quad\text{or}\quad -k^2, \tag{1.22}$$

depending on whether the eclipse under consideration is occultation ($r_1 < r_2$) or transit ($r_1 > r_2$).

In making use of all foregoing results we find it easy to establish that, for $m = 1, 2, 3, \ldots$,

$$A_2 = L_1C_3, \tag{1.23}$$

$$A_4 = L_1(C_3^2 + C_2^2), \tag{1.24}$$

* We may note from the triangle O_1O_2Q on Figure 1-3 that the angle so defined is equal to $\pi - \phi_1 - \phi_2$.

$$A_6 = L_1(C_3^3 + 3C_2^2C_3 + C_1C_2^2), \tag{1.25}$$

where we have abbreviated

$$C_1 = r_1^2 \csc^2 i, \tag{1.26}$$

$$C_2 = r_1 r_2 \csc^2 i, \tag{1.27}$$

$$C_3 = r_2^2 \csc^2 i - \cot^2 i; \tag{1.28}$$

and where

$$L_1 = 1 - \lambda; \tag{1.29}$$

λ denoting the fractional luminosity of the system observed during totality. The reader may note that the constants C_1 and C_2 as defined by Equations (1.26) and (1.27) are closely analogous to those introduced previously by Equations (2.34) or (3.18) of Chapter IV – identical with them if $r_1 \equiv r_a$. The constant C_3 as defined by Equation (1.28) above is, however, different from $C_3 \equiv \sin^2 \theta_{1,2}$ used in Chapter IV; in terms of the present notations, $\sin^2 \theta_{1,2} = C_3 \pm 2C_2 + C_1$.*

It should also be added that the foregoing Equations (1.26)–(1.29) are valid as they stand only if the eclipse under consideration is an occultation (i.e., $r_2 < r_1$) ending in totality (i.e., if $\cos i < r_2 - r_1$). If the eclipse is a transit ($r_1 > r_2$) culminating in an annular phase, Equations (1.26)–(1.28) continue to hold good provided that the fractional radii r_1 and r_2 are interchanged, and the fractional luminosity of the eclipsed component is replaced by $(r_2/r_1)^2 L_1$.

By this a solution for the elements of the totally-eclipsing systems under restrictions set forth in this section has really been completed. To recapitulate it, first specify (by planimetry or otherwise) the 'moments' A_{2m} ($m = 1, 2, 3$) of the light curve plotted in the $l - \sin^{2m} \theta$ coordinates. Next insert these 'empirical' values of these moments on the left-hand sides of Equations (1.23), (1.24) and (1.25) to solve for the corresponding values of the auxiliary constants $C_{1,2,3}$; by Equation (1.23) A_2 yields C_3 directly; and on insertion in Equation (3.24) the latter yields C_2; lastly, with C_2 and C_3 already known, Equation (1.25) can be used to evaluate C_1. And once this has been done, an inversion of Equations (1.26)–(1.28) yields the desired elements of the system in the form

$$r_{1,2}^2 = \frac{C_{1,2}^2}{(1 - C_3)C_1 + C_2^2} \tag{1.30}$$

and

$$\sin^2 i = \frac{C_1}{(1 - C_3)C_1 + C_2^2}, \tag{1.31}$$

by which the solution of the restricted problem set forth in this section has been completed.

We may only add that the foregoing expressions (1.30) and (1.31) for the geometrical elements $r_{1,2}^2$ and $\sin^2 i$ of the eclipse in terms of the auxiliary

* The reader may note that, with C_3 as given by Equation (1.28) above, Equation (2.2) of Chapter IV could be rewritten, more symmetrically, as $p^2C_1 + 2pC_2 + C_3 = \sin^2 \theta$.

constants $C_{1,2,3}$ are convenient, but not necessary. If we prefer to express them in terms of the moments A_{2m} of the light curves, an elimination of the C_j's in favour of the A_{2m}'s by means of Equations (1.23)–(1.25) permits us to rewrite Equation (1.31) for $\sin^2 i$ in terms of the A_{2m}'s as a (terminating) continuous fraction of the form

$$\sin^2 i = \cfrac{A_0}{A_0 - \frac{2}{3}A_2 - \cfrac{A_0 A_4}{3A_2 - \cfrac{A_0 A_6}{A_4 + \cfrac{A_0 A_2 A_6}{3(A_2^2 - A_0 A_6)}}}}; \tag{1.32}$$

while

$$r_1 = \left\{\frac{A_0 A_6}{A_0 A_4 - A_2^2} - \frac{3A_2}{A_0}\right\}^{1/2} \sin i \tag{1.33}$$

and

$$r_2 = \left\{\frac{A_6}{A_0} - \frac{3A_2}{A_0}\left(\frac{A_4}{A_0} - \frac{A_2^2}{A_0^2}\right)\right\} \sin i, \tag{1.34}$$

where we have abbreviated $A_0 \equiv L_1 \equiv 1 - \lambda$.

B. PARTIAL ECLIPSES

Having succeeded in constructing a closed algebraic solution for the elements of totally-eclipsing systems in the frequency domain, let us turn our attention to *partial eclipses* – be these occultations of transits. In such a case, the first part of Equation (1.9) for computation of the theoretical moments A_{2m} continues to hold good; but on partial integration it transforms into

$$A_{2m} = -L_1 \csc^{2m} i \int_{\delta_0}^{\delta_1} (\delta^2 - \delta_0^2)^m \frac{\partial \alpha}{\partial \delta}\, d\delta, \tag{1.35}$$

where the derivative $\partial\alpha/\partial\delta$ continues to be given by Equation (1.16).

In order to proceed with the evaluation of the integrals on the right-hand side of the foregoing equation, let us change over from δ as the variable of integration to the angular variable ϕ introduced by Equation (1.17) introduced already in the preceding section of this chapter: by use of Equations (1.18) and (1.19) we find that, for partial eclipses,

$$A_{2m} = \frac{2L_1 r_2^2}{\pi} \int_{\phi_0}^{\pi} \left(\frac{\delta^2 - \delta_0^2}{1 - \delta_0^2}\right)^m \left(\frac{\sin\phi}{\delta}\right)^2 d\phi, \tag{1.36}$$

where

$$\phi_0 = \cos^{-1} \frac{r_1^2 + r_2^2 - \delta_0^2}{2r_1 r_2}. \tag{1.37}$$

In order to normalize further the limits of integration on the right-hand side of

Equation (1.36), introduce a new variable u related with ϕ by the equation

$$u = \frac{1+\cos\phi}{1+\cos\phi_0} \tag{1.38}$$

in such a way that, at first contact of the eclipse ($\delta = \delta_1$), $\phi = \pi$ corresponds to $u = 0$; while at maximum eclipse ($\delta = \delta_0$), $\phi = \phi_0$ and $u = 1$. If so, it can be shown that

$$\delta^2 = \delta_1^2\{(1-h^2) + h^2(1-u)\} \tag{1.39}$$

and

$$\sin^2\phi\,\mathrm{d}\phi = -4\kappa^3\sqrt{u(1-\kappa^2 u)}\,\mathrm{d}u, \tag{1.40}$$

where we have abbreviated

$$h^2 = 1 - \left(\frac{\delta_0}{\delta_1}\right)^2 \tag{1.41}$$

and

$$\kappa^2 = \frac{\delta_1^2 - \delta_0^2}{\delta_1^2 - \delta_2^2}. \tag{1.42}$$

Introduce further a family of auxiliary integrals of the form

$$2\pi J^0_{\beta,\gamma}(\kappa) = (2\kappa)^{\beta+\gamma+2}\int_0^1 u^{\beta/2}(1-\kappa^2 u)^{\beta/2}(1-u)^{\gamma/2}\,\mathrm{d}u, \tag{1.43}$$

well-known from the theory of the light curves of distorted eclipsing systems (cf. Equation (3.33) of Chapter III). If so, and if

$$\alpha(k, \delta_0) = \frac{2r_2^2}{\pi}\int_{\phi_0}^{\pi} \frac{\sin^2\phi}{\delta^2}\,\mathrm{d}\phi \equiv \alpha_0 \tag{1.44}$$

denotes the fractional loss of light suffered at maximum eclipse, it is possible to express the moments A_{2m} as defined by Equation (1.36) in the form

$$A_2 = L_1\{2(C_2^2/C_1)J^0_{1,0} - \alpha_0\cot^2 i\}, \tag{1.45}$$

$$A_4 = L_1\{2(C_2^2/C_1)(C_2 J^0_{1,2} - 2J^0_{1,0}\cot^2 i) + \alpha_0\cot^4 i\}, \tag{1.46}$$

$$A_6 = L_1\{2(C_2^2/C_1)(C_2^2 J^0_{1,4} - 3C_2 J^0_{1,2}\cot^2 i + 3J^0_{1,0}\cot^4 i) - \alpha_0\cot^6 i\}, \tag{1.47}$$

etc., where the constants $C_{1,2}$ continue to be given by Equations (1.26)–(1.27), and

$$2\pi J^0_{1,0} = \cos^{-1}\mu - \mu\sqrt{1-\mu^2}, \tag{1.48}$$

$$3\pi J^0_{1,2} = (2+\mu^2)\sqrt{1-\mu^2} - 3\mu\cos^{-1}\mu, \tag{1.49}$$

$$6\pi J^0_{1,4} = 3(1+4\mu^2)\cos^{-1}\mu - \mu(13+2\mu^2)\sqrt{1-\mu^2}, \tag{1.50}$$

where the auxiliary parameter μ is related with the modulus (1.42) by the equation

$$\mu = 1 - 2\kappa^2. \tag{1.51}$$

If the eclipse becomes grazing – i.e., when $\delta_0 = \delta_2$ and, consequently (by Equations 1.41 and 1.51) $\kappa^2 = 1$ and $\mu = -1$, it follows from Equations (1.48)–(1.50) that $J^0_{1,0} = \frac{1}{2}$, $J^0_{1,2} = 1$, $J^0_{1,4} = \frac{5}{2}, \ldots$, in which case the moments A_{2m} as defined by Equations (1.45)–(1.47) become identical with Equations (1.23)–(1.25).

By this our task of evaluating the different moments A_{2m}, subtended by the light curves due to partial eclipses, in terms of the geometrical elements of the respective system has been completed. Unlike in the case of total eclipses, the theoretical expressions for the A_{2m}'s are not algebraic in terms of the elements of the eclipse, but involve also inverse trigonometric functions. Therefore, although the direct problem of expressing the A_{2m}'s in terms of the elements of the eclipse has thus again been solved in a closed form, the inverse problem of solving for the elements from known values of the A_{2m}'s cannot be solved (as it was in the case of total eclipses) algebraically, but must proceed by successive approximations. These approximations are facilitated by tables of the $J^0_{1,n}$-functions occurring on the right-hand sides of Equations (1.45)–(1.47), which are given in the Appendix VI as functions of the argument

$$\nu = 2 \sin^{-1} \kappa, \tag{1.52}$$

such that

$$\mu = 1 - 2 \sin^2 \tfrac{1}{2}\nu, \tag{1.53}$$

to an ample degree of accuracy.

The strategy for an analysis of the light changes due to partial eclipses in the frequency-domain can be outlined as follows:

1. Adopt a trial value of the ratio of the radii k – if we do not know whether the eclipse under scrutiny is an occultation or a transit, depart from $K = 1$.

2. Establish the corresponding value of the maximum obscuration α_0 with the aid of Equations (1.19) or (1.20) of Chapter IV.

3. By inversion (numerical) of the function $\alpha_0 \equiv \alpha(K, p_0)$ ascertain the value of the maximum geometrical depth p_0 of the eclipse; and, hence (by Equation (0.9) of Chapter IV), $\cos i = r_b(1 + Kp_0)$; then establish the value of ϕ_0 corresponding to these parameters by Equation (1.34).

4. The fractional luminosities $L_{1,2}$ (or $L_{a,b}$) can now be determined by a recourse to Equations (3.22) of Chapter IV.

5. Turn now to Equation (1.45) relating A_2 with the elements of the eclipse, which now contains only one unknown: namely, r_2. Solve for it with the aid of tables given in Appendix VI.

6. Turn next to Equation (1.46) for A_4 which contains an additional unknown: namely, r_1. Solve for it; then form the ratio r_1/r_2 to specify the type of eclipse ($r_1/r_2 < 1$ corresponds to an occultation; $r_2/r_1 < 1$, to a transit) and ascertain the extent to which the resulting ratio r_1/r_2 differs from that adopted at the outset. If it does so by a margin greater than we are willing to tolerate, adopt its new value, and repeat the preceding cycle 1–6 as many times as may be necessary to bring the adopted and resulting values of k into satisfactory agreement. Values of A_{2m} for $m > 2$ contain no more new unknowns to be determined, and can be used as a check on the correctness of our numerical work.

As is evident, unlike in the case of total (annular) eclipses discussed first a solution for the elements of partially-eclipsing systems can proceed towards its goal only by successive approximations – a strategy which forced itself on us already by an iterative approach outlined in Chapter IV. However – unlike iteration in the time-domain – successive approximations in the frequency-domain, leading to the goal through our 6-step cycle, do not call for any least-squares solutions to be performed at each cycle (which constituted their most laborious feature) and therein rests the computational advantage of the present scheme.

C. RELATION WITH FOURIER SERIES

Having progressed thus far with an outline of the analysis of the light changes of eclipsing variables in the frequency-domain, let us compare its salient features with the time-domain methods expounded in Chapter IV. While the basic data of the time-domain methods have been individual observations of $l(\theta)$, in the frequency-domain these data enter only collectively – through the moments A_{2m} of the light curves, representing areas subtended by the light curves in the $l - \sin^{2m}\theta$ coordinates. Secondly, the location of the observed points in the time-domain can be related with the geometrical elements r_1, r_2, i of the respective system only through *non-linear* equations of the form (0.14) of Chapter IV, involving auxiliary *transcendental* functions $p(k, \alpha)$ which exist only in tabular form.

In the frequency-domain, on the other hand, the basic data are the moments A_{2m} corresponding to different values of m; in terms of which the solution for the geometrical elements becomes very much simpler. If the eclipses are total (or annular), the A_{2m}'s can be related with the geometrical elements by simple *algebraic* equations of the form (1.30)–(1.34). The latter hold good, as they stand, only for eclipses of the stars exhibiting uniformly bright discs; but (as we shall show in the next section) are capable of generalization – for total (not annular) eclipses – to an arbitrary degree of the law of limb-darkening over the apparent disc of the eclipsed star. If, however, the eclipse becomes partial, the solution can be obtained algebraically, but not in a closed form.

Before we extend the procedure exemplified in the present section of a treatment of uniformly bright discs to more general cases, the question remains still to be answered; to what extent do the developments expounded in this section constitute a Fourier analysis, and where did the concept of *frequency* enter into our work?

The answer will become obvious when we consider more closely the significance of the moments A_{2m} of the light curves. As is well known (cf. e.g., Oberhettinger, 1973; p. 31*)

* Equation (1.54) continues to hold good for *non-integral* values of m (which will be needed later) as well, if $0 \leq \theta \leq \pi$, provided that the factorials involving m are replaced by the respective gamma-functions, and the summation is extended to infinity.

$$\sin^{2m}\theta = \frac{(2m)!}{4^m}\sum_{j=0}^{m}\frac{(-1)^j\epsilon_j\cos 2j\theta}{(m+j)!(m-j)!}, \tag{1.54}$$

where $\epsilon_0 = 1$ while, for $j > 0$, $\epsilon_j = 2$. Take a differential of both sides of the foregoing equation, multiply by the loss of light $1 - l$, and integrate between $(0, \theta_1)$; the outcome discloses that

$$\int_0^{\theta_1}(1-l)\,\mathrm{d}(\sin^{2m}\theta) = \frac{(2m)!}{4^{m-1}}\sum_{j=1}^{m-1}\frac{(-1)^{j+1}j}{(m+j)!(m-j)!} \times \int_0^{\theta_1}(1-l)\sin 2j\theta\,\mathrm{d}\theta. \tag{1.55}$$

The integral of the left-hand side of Equation (1.55) is obviously identical with our moments A_{2j} of the light curves. In order to identify those on the right-hand side of Equation (1.55), let us return to Equations (4.8) and (4.9) of Chapter I which define the real and imaginary parts $F_{1,2}(\nu)$ of the Fourier transform $F(\nu)$ of the function $f(\theta)$. Let us hereafter assume – in contrast with Equation (4.13) of Chapter I – that

$$f(-\theta) = -f(\theta). \tag{1.56}$$

Such an assumption annihilates the real part of $F(\nu)$; while the imaginary part

$$F_2(\nu) = 2\int_0^{\theta_1} f(\theta)\sin(2\pi\nu\theta)\,\mathrm{d}\theta \tag{1.57}$$

for $f(\theta) = 1 - l$ and $\nu = j/\pi$ assumes the form

$$\int_0^{\theta_1}(1-l)\sin 2j\theta\,\mathrm{d}\theta = \tfrac{1}{2}F_2\left(\frac{j}{\pi}\right), \tag{1.58}$$

which is identical with the integrals on the r.h.s. of Equation (1.55).

Moreover, as $1 - l = 0$ for $\theta > \theta_1$, the foregoing equation will remain unaltered if the upper limit θ_1 of integration is increased (say) to $\frac{1}{2}\pi$. If so, however, the Fourier series (4.11) of Chapter I can be particularized to its sine part

$$1 - l = \sum_{j=1}^{\infty} b_{2j}\sin 2j\theta, \tag{1.59}$$

where (by the second one of Equations (4.12) of Chapter I),

$$b_{2j} \equiv \frac{2}{\pi}F_2\left(\frac{j}{\pi}\right) = \frac{4}{\pi}\int_0^{\pi/2}(1-l)\sin 2j\theta\,\mathrm{d}\theta. \tag{1.60}$$

Inserting ultimately Equation (1.60) in (1.55) we find that

$$A_{2m} = \frac{(2m)!}{4^m}\sum_{j=1}^{m}\frac{\pi(-1)^{j+1}j}{(m+j)!(m-j)!}b_{2j}, \tag{1.61}$$

i.e., that *the moments A_{2m} of the light curves can be expressed as weighted means of m coefficients b_j of the expansion of a function defined by*

$$f(\theta)=1-l,\quad f(-\theta)=-f(\theta), \tag{1.62}$$

in a Fourier sine series valid within $\pm\frac{1}{2}\pi$.

In particular,

$$A_2=\frac{\pi}{4}\,b_2, \tag{1.63}$$

$$A_4=\frac{\pi}{4}\,(b_2-\tfrac{1}{2}b_4), \tag{1.64}$$

$$A_6=\frac{\pi}{4}\,(\tfrac{15}{16}b_2-\tfrac{3}{4}b_4+\tfrac{3}{16}b_6), \tag{1.65}$$

etc.; and, conversely,

$$b_2=\frac{4}{\pi}\,A_2, \tag{1.66}$$

$$b_4=\frac{8}{\pi}\,(A_2-A_4), \tag{1.67}$$

$$b_6=\frac{4}{3\pi}\,(9A_2-24A_4+16A_6). \tag{1.68}$$

If, moreover, we insert for the A_{2m}'s from Equations (1.23)–(1.25) we find that, for totally-eclipsing systems, the b_{2j}'s are related with the geometrical elements r_1, r_2 and i, constituting the constants $C_{1,2,3}$ as defined by Equations (1.26)–(1.28) by the equations

$$\pi b_2=4L_1C_3, \tag{1.69}$$

$$\pi b_4=8L_1(C_3-C_3^2-C_2^2), \tag{1.70}$$

$$\pi b_6=\tfrac{4}{3}L_1\{(4C_3-3)^2C_3+8(6C_3+2C_1-3)C_2^2\}, \tag{1.71}$$

etc. These appear, however, to be more complicated than the corresponding relations (1.23)–(1.25) for the moments of the light curves; so that it seems preferable henceforward to continue our analysis in terms of the A_{2m}'s rather than of the b_{2j}'s.

The significance of the latter is, however, clear. The coefficients b_{2j} correspond to discrete frequencies of the terms of the r.h.s. of the expansion (1.59); and as they are expressible as linear combinations of the moments A_{2m}, the index m of the latter is, therefore, proportional to the *frequencies* in terms of which our light curves are being analyzed. In the case of total eclipses, an analysis in terms of *four* fundamental frequencies (i.e., $m=0, 1, 2, 3$) has proved sufficient for a complete solution of our problem; but the properties of a *continuous* spectrum of m will be considered at a later stage of our analysis.

V-2. General Case: Limb-Darkened Stars

A. EVALUATION OF THE MOMENTS A_{2m} OF ECLIPSES

The method of integration used in the preceding section to evaluate the moments A_{2m} of the light curves is applicable only to particular cases in which the star undergoing eclipse can be regarded as exhibiting to a distant observer an uniformly bright disc. In general this will, however, not be the case. An assumption of uniformly bright discs may offer a very good approximation for an interpretation of observations made in the infrared ($\lambda > 10^4$ Å); but *not* those made at optical frequencies; for if atomic absorption in the atmospheres of very hot stars (of O- and early B-type) would make such an assumption tolerable, it would be largely nullified by the effects of electron scattering. Therefore, in order to provide foundations which could be applied with confidence to an analysis of the light changes observed at optical frequencies, our procedures as developed in the preceding section must be generalized to allow for the darkening (not necessarily linear) of the eclipsed disc towards the limb; and in the present section such an analysis will be given for any type of eclipse.

In order to do so, let us return to Equation (1.9) which defines the theoretical values of the moments A_{2m} in terms of the δ-variable as

$$A_{2m} = mL_1 \csc^{2m} i \int_{\delta_0^2}^{\delta_1^2} (\delta^2 - \delta_0^2)^{m-1} \alpha \, d\delta^2 \tag{2.1}$$

where, in accordance with Equation (2.7) of Chapter I,

$$\alpha = \sum_{l=0}^{\infty} C^{(l)} \alpha_l^0, \tag{2.2}$$

in which the functions $C^{(l)}$ of the limb-darkening coefficients continue to be given by Equations (2.8)–(2.9) of Chapter I; while the associated α-functions of the type α_l^0 have been defined in Chapter I as the Hankel transforms of the form (3.34) or (3.55), and approximated in Section I-3 by several alternative expansions in terms of the elements of the eclipse. By use in Equation (2.2) above of (say) the expansion (3.91) of Chapter I, and insertion in Equation (2.1) we find that

$$\begin{aligned} A_{2m} = mL_1 \csc^{2m} i &\sum_{l=0}^{\infty} \frac{C^{(l)}}{\nu^2 \Gamma(\nu)} \sum_{n=0}^{\infty} (-1)^n (\nu + 2n + 2) \\ &\times \frac{\Gamma(\nu + n + 1)}{(n+1)!} \{G_{n+1}(\nu, \nu + 1, a)\}^2 \\ &\times \int_{\delta_0^2}^{\delta_1^2} (\delta^2 - \delta_0^2)^{m-1} (1 - c^2)^{\nu+1} G_n(\nu + 2, 1, c^2) \, d\delta^2, \end{aligned} \tag{2.3}$$

where a and c are defined by Equations (3.56)–(3.58) of that chapter.

In order to evaluate the remaining integral on the r.h.s. of the foregoing equation, let us avail ourselves of the transformation (4.19) already used in

Chapter I to normalize its limits from (δ_1^2, δ_0^2) to $(1, 0)$. This can indeed be accomplished if we set

$$u = \frac{\delta^2 - \delta_0^2}{\delta_1^2 - \delta_0^2} = \frac{c^2 - c_0^2}{1 - c_0^2}, \tag{2.4}$$

in terms of which

$$\int_{\delta_0^2}^{\delta_1^2} (\delta^2 - \delta_0^2)^{m-1}(1 - c^2)^{\nu+1} G_n(\nu + 2, 1, c^2)\, d\delta^2$$
$$= (\delta_1^2 - \delta_0^2)^m (1 - c_0^2)^{\nu+1} \int_0^1 u^{m-1}(1-u)^{\nu+1} G_n(\nu+2, 1, c^2)\, du, \tag{2.5}$$

where

$$c^2 = 1 - (1 - c_0^2)(1 - u); \tag{2.6}$$

and, therefore (cf. Jurkevich and Heard, 1977; or Demircan, 1978a), the Jacobi polynomial

$$G_n(\nu + 2, 1, c^2) = n! \sum_{i=0}^{n} \frac{(n + \nu + 2)_i}{i!(n-i)!} \sum_{k=0}^{i} \frac{(-1)^{i+k}}{k!(i-k)!} (1 - c_0^2)^k (1-u)^k. \tag{2.7}$$

If we insert Equation (2.7) in (2.5), the integrals on the r.h.s. of Equation (2.5) prove again to be of the hypergeometric type; and their term-by-term integration eventually discloses that

$$A_{2m} = \Gamma(m+1) L_1 \sin^{2m}\theta_1 \sum_{l=0}^{\infty} \frac{C^{(l)}}{\nu} \frac{(1 - c_0^2)^{\nu+1}}{\Gamma(\nu+1)\Gamma(\nu+m+1)}$$
$$\times \sum_{n=0}^{\infty} \frac{\nu + 2n + 2}{n+1} (n + \nu + 1) \left\{ \frac{\Gamma(\nu+n+1)}{\Gamma(n+1)} G_{n+1}(\nu, \nu+1, a) \right\}^2$$
$$\times G_n(\nu + 2, \nu + m + 2, 1 - c_0^2), \tag{2.8}$$

where θ_1 denotes the phase angle of the first contact of the eclipse (when $\delta = \delta_1$).

The series on the r.h.s. of the foregoing equation does not, by any means, represent a unique approximation to the moments A_{2m}. Thus if we use Equation (3.94) of Chapter I in place of Equation (3.91) in (2.2) we find (in the same way) that

$$A_{2m} = b^2 L_1 \sin^{2m}\theta_1 \sum_{l=0}^{\infty} \frac{C^{(l)}}{\nu} \sum_{n=0}^{\infty} (2n + \tfrac{1}{2}) P_{2n}(0)$$
$$\times F^{(4)}(-n + \tfrac{1}{2}, n + 1; \nu + 1, 2; a^2, b^2)$$
$$\times {}_3F_2(-n, n + \tfrac{1}{2}, 1; \tfrac{1}{2}, m + 1; 1 - c_0^2), \tag{2.9}$$

where $b = 1 - a$. Or, again, if we use Demircan's expansion (3.99) of Chapter I for α_l^0, we obtain

$$A_{2m} = b^2 L_1 \sin^{2m} \theta_1 \sum_{l=0}^{\infty} \frac{C^{(l)}}{\nu} \sum_{n=0}^{\infty} (-1)^n (2n+1)$$
$$\times F^{(4)}(-n, n+1; \nu+1, 2; a^2, b^2) G_n(1, m+1, 1-c_0^2). \qquad (2.10)$$

Equations (2.8)–(2.10) for the moments A_{2m} converge at different rates in different domains of a and c_0; and hold good for *any* (not necessarily integral) values of $m \geqslant 0$; but should $m < 0$, the series on the right-hand sides of Equations (2.8)–(2.10) diverge. If $m = 0$,

$$A_0 = L_1 \alpha(a, c_0) \equiv L_1 \alpha_0. \qquad (2.11)$$

Furthermore, the foregoing expansions (2.8)–(2.10) for the moments A_{2m} of the light curves hold good for *any* type of eclipse – be these partial, total, or annular; occultations or transits. As the reader may recall from Equations

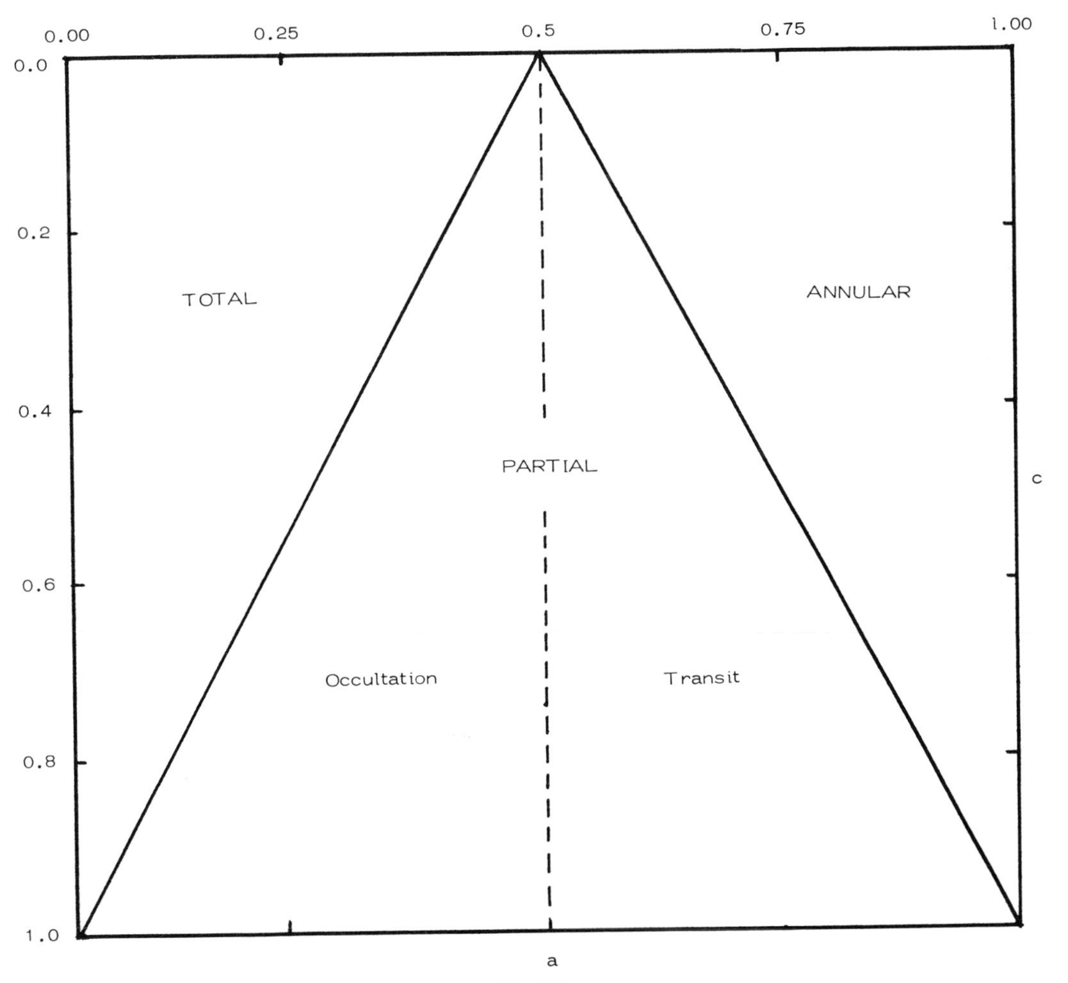

Fig. 5-2.

(3.56)–(3.58) of Chapter I, their arguments a and c_0 have been normalized to remain within the (closed) interval (0, 1). For $0 < a < \frac{1}{2}$, the corresponding eclipses will be occultations; while for $\frac{1}{2} < a < 1$ they become transits. As regards c_0, the value of $c_0 = 1$ corresponds to the moment of first contact; and $c_0 = 0$, to that of central eclipse. If $1 > c_0 > |a - b|$ – or, since $a + b = 1$, $1 > c_0 > |2a - 1|$ – the eclipses will be partial (be they occultations or transits); for $c_0 \leq 1 - 2a$ they become total; while if $c_0 \leq 2a - 1$, they will be annular.

A connection between the values of the parameters a, c_0 and the nature of the eclipses is diagramatically illustrated in Figure 5-2. Each eclipse starts at the bottom abscissa (corresponding to $c = 1$); its ordinate corresponding to the value of the ratio of the radii of the respective system. As the eclipse progresses, $c < 1$ and the location of the point c_0 will rise along the ordinate corresponding to a given value of a – unit for $c_0 = 0$ the eclipse becomes central (occultation or transit, depending on whether $a \lesseqgtr \frac{1}{2}$).

How rapid is the convergence of the Fourier expansions on the right-hand sides of Equations (2.8)–(2.10) – or, rather, what are their asymptotic properties? Extensive numerical work has disclosed that if a knowledge of the values of A_{2m} is required correctly to one part in a thousand, a summation of ten terms on the r.h.s. of Equations (2.8)–(2.10) will generally yield the desired result; but to attain an accuracy of the order of 1 part in 10^4, we must sum 80–100 terms. To do so by hand would be impracticable; but for automatic computers this constitutes no problem. For example, a CDC 7600 computer can sum up one hundred terms on the r.h.s. of Equations (2.4) or (2.6) – and print out the answer – in approximately 0.7 of a second – i.e., faster than we could write it by hand if we knew the answer! Machines less powerful than the CDC 7600 may take longer to accomplish this task; but the requisite computing time should still be of the order of 1 sec. Therefore, within the contemporary 'state of the art', the use of Fourier expansions for A_{2m} of the form (2.8)–(2.10) is entirely practicable for anyone having access to automatic computers.

B. ALTERNATIVE EVALUATION OF THE MOMENTS A_{2m}

The expansions for A_{2m} set up in the first part of this section hold good for any type of eclipses – be these occultations or transits; partial, total, or annular. Before, however, such expansions are actually put to task the question can be raised: is it possible – at least in certain cases – to sum up the series on the right-hand sides of Equations (2.8)–(2.10) in a closed form? In the first section of this chapter we found indeed this to be possible if the disc of the star undergoing eclipse were uniformly bright: can our foregoing results be extended to an arbitrary degree of limb-darkening?

In order to investigate such a possibility, let us return to Equation (2.1) defining the moments of the light curves, and set out to evaluate them, not (as before), by inserting for α the expansions established in Section I-3 and integrating term-by-term; but by direct use of closed Hankel transforms for α_l^0 as given by Equation (3.34) of Chapter I. Since only one Bessel function behind the integral sign of Equation (3.34) involves δ (through h), the limits of integration can obviously be interchanged – so that

$$A_{2m} = mL_1 \csc^{2m} i \sum_{l=0}^{\infty} C^{(l)} 2^{\nu+1} \Gamma(\nu) \int_0^{\infty} (2\pi q r_1)^{-\nu} J_\nu(2\pi q r_1)$$

$$\times J_1(2\pi q r_2) \left\{ \int_{\delta_0}^{\delta_1} (\delta^2 - \delta_0^2)^{m-1} J_0(2\pi q \delta) \delta \, \mathrm{d}\delta \right\} \mathrm{d}(2\pi q r_2). \tag{2.12}$$

By successive application of the differential recursion formula

$$(2\pi q)\delta^j J_{j-1}(2\pi q \delta) = \frac{\mathrm{d}}{\mathrm{d}\delta} \{\delta^j J_j(2\pi q \delta)\}, \tag{2.13}$$

valid for Bessel functions of arbitrary order we readily establish that, for $m = 1$,

$$\int_{\delta_0}^{\delta_1} J_0(2\pi q \delta)\delta \, \mathrm{d}\delta = \{h_1^2 x^{-1} J_1(h_1 x) - h_0^2 x^{-1} J_1(h_0 x)\} r_2^2; \tag{2.14}$$

while for $m = 2$,

$$\int_{\delta_0}^{\delta_1} (\delta^2 - \delta_0^2) J_0(2\pi q \delta)\delta \, \mathrm{d}\delta = \{(h_1^2 - h_0^2)(h_1/x) J_1(h_1 x) - 2(h_1/x)^2 J_2(h_1 x)$$
$$+ 2(h_0/x)^2 J_2(h_0 x)\} r_2^2; \tag{2.15}$$

for $m = 3$,

$$\int_{\delta_0}^{\delta_1} (\delta^2 - \delta_0^2)^2 J_0(2\pi q \delta)\delta \, \mathrm{d}\delta = \{(h_1^2 - h_0^2)^2 (h_1/x) J_1(h_1 x)$$
$$- 4(h_1^2 - h_0^2)(h_1/x)^2 J_2(h_1 x)$$
$$+ 8(h_1/x)^3 J_3(h_1 x)$$
$$- 8(h_0/x)^3 J_3(h_0 x)\} r_2^6 , \tag{2.16}$$

where $x \equiv 2\pi q r_2$ and

$$h_1 = \frac{\delta_1}{r_2} = \frac{r_1 + r_2}{r_2} = 1 + k, \tag{2.17}$$

while

$$h_0 = \frac{\delta_0}{r_2} = \frac{\cos i}{r_2}. \tag{2.18}$$

If, furthermore, we abbreviate

$$2^{\nu+1} \Gamma(\nu) h_j^m \int_0^{\infty} \frac{J_\nu(kx)}{(kx)^\nu} J_1(x) \frac{J_m(h_j x)}{x^m} \, \mathrm{d}x \equiv \Pi_m^{(\nu)}(h_j), \tag{2.19}$$

where $j = 0$ or 1, it follows from Equation (2.1) that

$$A_2 = L_1 r_2^2 \csc^2 i \sum_{l=0}^{\infty} C^{(l)} \{\Pi_1^{(\nu)}(h_1) - \Pi_1^{(\nu)}(h_0)\}, \tag{2.20}$$

$$A_4 = 2L_1 r_2^4 \csc^4 i \sum_{l=0}^{\infty} C^{(l)}\{(h_1^2 - h_0^2)\Pi_1^{(\nu)}(h_1) - 2\Pi_2^{(\nu)}(h_1) + 2\Pi_2^{(\nu)}(h_0)\}, \quad (2.21)$$

$$A_6 = 3L_1 r_2^6 \csc^6 i \sum_{l=0}^{\infty} C^{(l)}\{(h_1^2 - h_0^2)^2 \Pi_1^{(\nu)}(h_1) - 4(h_1^2 - h_0^2)\Pi_2^{(\nu)}(h_1) + 8\Pi_3^{(\nu)}(h_1) - 8\Pi_3^{(\nu)}(h_0)\}, \quad (2.22)$$

etc.

In order to evaluate the Π-integrals on the right-hand sides of the preceding equations, consider first those with the argument h_1, referring to the first contact of the eclipse. Since, at that moment, $h_1 = 1 + k$, the use of Bailey's equation (3.59) of Chapter I discloses that

$$\Pi_m^{(\nu)}(h_1) = \frac{1}{\nu \Gamma(m)} \left(\frac{h_1^2}{2}\right)^{m-1} F_4(1 - m, 1; 2, \nu + 1; h_1^{-2}, k^2/h_1^2) \quad (2.23)$$

for *any* type of eclipse, when $m = 1, 2, 3, \ldots$ and, therefore, Appell's generalized hypergeometric series $F^{(4)}$ on the r.h.s. of Equation (2.23) reduces to a polynomial – regardless of the degree of the law of limb-darkening.

It is only the Π's with the argument h_0 (corresponding to the moment of maximum eclipse) which will be different for different types of eclipse; and the following three cases must be distinguished. If the eclipse is *total*, $\delta_0 < r_2 - r_1$ and, therefore, $1 > h_0 + k$. In such a case, Bailey's formula (3.59) continues to apply (the series $F^{(4)}$ reducing to its first term) and yields

$$\Pi_m^{(\nu)}(h_0) = \frac{2}{m!\nu} \left(\frac{h_0^2}{2}\right)^m; \quad (2.24)$$

which on insertion in Equations (2.20)–(2.22) together with (2.23) verifies the expressions

$$A_2 = L_1 \bar{C}_3, \quad (2.25)$$

$$A_4 = L_1(\bar{C}_3^2 + \bar{C}_2^2), \quad (2.26)$$

$$A_6 = L_1(\bar{C}_3^3 + 3\bar{C}_2^2 \bar{C}_3 + \bar{C}_1 \bar{C}_2^2), \quad (2.27)$$

where

$$\bar{C}_3 = (r_2^2 \csc^2 i - \cot^2 i) \sum_{l=0}^{\infty} \frac{1!C^{(l)}}{\nu}, \quad (2.28)$$

$$\bar{C}_2^2 = r_1^2 r_2^2 \csc^4 i \sum_{l=0}^{\infty} \frac{2!C^{(l)}}{\nu(\nu+1)}, \quad (2.29)$$

$$\bar{C}_1 \bar{C}_2^2 = r_1^2 r_2^4 \csc^6 i \sum_{l=0}^{\infty} \frac{3!C^{(l)}}{\nu(\nu+1)(\nu+2)}, \quad (2.30)$$

The foregoing equations (2.25)–(2.27) represent generalization of Equations (1.23)–(1.25), deduced previously for mutual eclipses of uniformly bright discs, to discs of arbitrary limb-darkening. They remain of the same form; and reduce indeed to Equations (1.23)–(1.25) if all coefficients of limb-darkening vanish identically. If this is not the case, the above Equations (2.25)–(2.27) must be used in place of Equations (1.23)–(1.25); and the $\bar{C}_j$'s as defined by Equations (2.28)–(2.30) should replace the C_j's given by Equations (1.26)–(1.28). One additional point should, moreover, be stressed: namely, the fact that – consistent with Equations (2.8)–(2.9) of Chapter I,

$$\sum_{l=0}^{\infty} \frac{1!C^{(l)}}{\nu} = \sum_{l=0}^{\infty} \frac{2C^{(l)}}{l+2} = 1, \tag{2.31}$$

it follows that

$$\bar{C}_3 \equiv C_3 \tag{2.32}$$

i.e., that *the quantity C_3 is invariant with respect to the limb-darkening* – a fact not, however, true of $C_{1,2}$.

If, next, the eclipse happens to be *annular* – i.e., if $r_1 > \delta_0 + r_2$ and, therefore, $k > h_0 + 1$, Bailey's Equation (3.59) of Chapter I continues to hold good; and its application to Equation (2.23) yields

$$\Pi_m^{(\nu)}(h_0) = \frac{2}{m!k^2}\left(\frac{h_0^2}{2}\right)^m F^{(4)}(1-\nu, 1; 2, m+1; k^2, h_0^2/k^2), \tag{2.33}$$

as a result of which

$$A_2 = L_1\bar{C}_3 + L_1 \cos^2 i \left\{1 - (r_2/r_1)^2 \sum_{l=0}^{\infty} C^{(l)} F^{(4)}(1-\nu, 1; 2, 2; r_2^2/r_1^2, \delta_0^2/r_1^2)\right\}, \tag{2.34}$$

$$A_4 = L_1(\bar{C}_3^2 + \bar{C}_2^2) - L_1 \cos^4 i \left\{1 - (r_2/r_1)^2 \sum_{l=0}^{\infty} C^{(l)} \times F^{(4)}(1-\nu, 1; 2, 3; r_2^2/r_1^2, \delta_0^2/r_1^2)\right\}, \tag{2.35}$$

$$A_6 = L_1(\bar{C}_3^3 + 3\bar{C}_2^2\bar{C}_3 + \bar{C}_1\bar{C}_2^2) + L_1 \cos^6 i \left\{1 - (r_2/r_1)^2 \sum_{l=0}^{\infty} C^{(l)} F^{(4)}(1-\nu, 1; 2, 4; r_2^2/r_1^2, \delta_0^2/r_1^2)\right\}, \tag{2.36}$$

can again be expressed in terms of Appell's $F^{(4)}$-series which reduces to polynomials if ν happens to be a negative integer, but not otherwise; and the infinite series for $F^{(4)}$ converge none too rapidly for practical use.

Should, lastly, the eclipses become *partial*, $r_1 + r_2 > \delta_0 > |r_1 - r_2|$ and, in consequence, none of three factors h_0, k and 1 becomes equal to (or larger than) a sum of the other two. As a result, Bailey's Equation (3.59) of Chapter I will no longer help us to evaluate the $\Pi_m^{(\nu)}(h_0)$'s. However, an application of Equations (3.81)–(3.90) used in Section I-3 to evaluate the α_l^0's will disclose that, for partial eclipses,

$$\Pi_m^{(\nu)}(h_0) = \frac{2^{1-m}h_0^{2m}}{\nu\Gamma(m+1)} \sum_{n=0}^{\infty} (-1)^n B(n+1, \nu+1) \frac{\nu+2n+2}{\nu+n+1} \times \{G_{n+1}(\nu, \nu+1, a)\}^2 {}_2F_1(-n-\nu-1, n+1; m+1; c_0^2), \tag{2.37}$$

valid for any value of $m \geqslant 0$, where (in accordance with Equations (3.56) and (3.58) of Chapter I)

$$a = \frac{r_1}{r_1 + r_2} \quad \text{and} \quad c_0 = \frac{\cos i}{r_1 + r_2} = ah_0. \tag{2.38}$$

The foregoing result represented by Equation (2.37) can be simplified by use of Sonine's well-known expansion

$$\frac{J_m(x)}{x^m} = \frac{\Gamma(m+1)}{2^{m-1}} \sum_{j=0}^{\infty} \frac{J_{2j}(x)}{\Gamma(m-j+1)\Gamma(m+j+1)} \tag{2.39}$$

(cf., e.g., Watson, 1945; p. 139), which for integral values of m reduces to a sum of m terms. By insertion of Equation (2.39) in (2.19) the latter can be rewritten as

$$\Pi_m^{(\nu)}(h_0) = \frac{\Gamma(m+1)}{2^{m-1}} h_0^{2m} \sum_{j=0}^{\infty} \frac{\mathfrak{P}_{2j}^{(\nu)}(h_0)}{\Gamma(m-j+1)\Gamma(m+j+1)}, \tag{2.40}$$

where we have abbreviated

$$\mathfrak{P}_{2j}^{(\nu)}(h_0) = 2^\nu \Gamma(\nu) \int_0^\infty (kx)^{-\nu} J_\nu(kx) J_1(x) J_{2j}(h_0 x)\, \mathrm{d}x. \tag{2.41}$$

differing from Equation (2.19) by the fact that the order of the Bessel function of $h_0 x$ is now an even integer.

Accordingly,

$$\Pi_1^{(\nu)}(h_0) = h_0^2\{\mathfrak{P}_0^{(\nu)}(h_0) + \mathfrak{P}_2^{(\nu)}(h_0)\}, \tag{2.42}$$

$$\Pi_2^{(\nu)}(h_0) = h_0^4\{\tfrac{1}{4}\mathfrak{P}_0^{(\nu)}(h_0) + \tfrac{1}{3}\mathfrak{P}_2^{(\nu)}(h_0) + \tfrac{1}{12}\mathfrak{P}_4^{(\nu)}(h_0)\}, \tag{2.43}$$

$$\Pi_3^{(\nu)}(h_0) = h_0^6\{\tfrac{1}{24}\mathfrak{P}_0^{(\nu)}(h_0) + \tfrac{1}{16}\mathfrak{P}_2^{(\nu)}(h_0) + \tfrac{1}{40}\mathfrak{P}_4^{(\nu)}(h_0) + \tfrac{1}{240}\mathfrak{P}_6^{(\nu)}(h_0)\}, \tag{2.44}$$

etc. If, moreover, we recall that

$$\mathfrak{P}_0^{(\nu)}(h_0) = \alpha_l^0(h_0), \tag{2.45}$$

we find that Equations (2.20)–(2.22) for partial eclipses assume the forms

$$A_0 = L_1 \alpha_0, \tag{2.46}$$

$$A_2 = L_1\{\bar{C}_3 + (1-\alpha_0)\cos^2 i\} - L_1 \cos^2 i \sum_{l=0}^{\infty} C^{(l)} \mathfrak{P}_2^{(\nu)}(h_0), \tag{2.47}$$

$$\begin{aligned} A_4 = L_1\{&\bar{C}_3^2 + \bar{C}_2^2 - (1-\alpha_0)\cos^4 i] \\ &+ L_1 \cos^4 i \sum_{l=0}^{\infty} C^{(l)}\{\tfrac{4}{3}\mathfrak{P}_2^{(\nu)}(h_0) + \tfrac{1}{3}\mathfrak{P}_4^{(\nu)}(h_0)\}, \end{aligned} \tag{2.48}$$

$$\begin{aligned} A_6 = L_1\{&\bar{C}_3^3 + 3\bar{C}_2^2\bar{C}_3 + \bar{C}_1\bar{C}_2^2 + (1-\alpha_0)\cot^6 i\} \\ &- L_1 \cos^6 i \sum_{l=0}^{\infty} C^{(l)}\{\mathfrak{P}_2^{(\nu)}(h_0) + \tfrac{3}{5}\mathfrak{P}_4^{(\nu)}(h_0) + \tfrac{1}{10}\mathfrak{P}_6^{(\nu)}(h_0)\}, \end{aligned} \tag{2.49}$$

where

$$\alpha_0 \equiv \sum_{l=0}^{\infty} C^{(l)} \alpha_l^0(h_0) \tag{2.50}$$

denotes the maximum obscuration of the component undergoing eclipse.

A further discussion of the foregoing equations (cf. Kopal, 1976, 1977), too long to be reproduced here, has disclosed that the $\mathfrak{P}$-integrals are expressible in a *closed* form in terms of a class of integrals of the form $I^0_{\beta,\gamma}$ defined by Equation (2.28) of Chapter II; the properties of which were extensively discussed in Section III-3. Indeed, one can show (cf. Kopal, 1976a) that Equation (2.47) can be rewritten as

$$A_2 = \alpha_0 L_1 \bar{C}_3 + \mathfrak{A}_2, \tag{2.51}$$

where

$$\begin{aligned} \mathfrak{A}_2 &= L_1 r_2^2 \csc^2 i \sum_{l=0}^{\infty} \frac{C^{(l)}}{\nu} \left(\frac{r_2}{r_1}\right)^{2\nu} \{I^0_{-1,2\nu} + h_0 I^1_{-1,2\nu}\} \\ &= L_1 \csc^2 i \sum_{l=0}^{\infty} \frac{C^{(l)}}{\nu} \left(\frac{r_2}{r_1}\right)^{2\nu} \{r_2^2 I^0_{-1,2\nu} + 2\nu\delta_0^2 I^0_{1,2(\nu-1)}\}; \end{aligned} \tag{2.52}$$

and similarly for the moments A_{2m} of other indices, for which the reader is referred to the sources already quoted.

C. DIFFERENTIAL PROPERTIES OF THE MOMENTS A_{2m}

The moments A_{2m} of the light curves introduced in this chapter constitute a fundamental tool for a determination of the elements of eclipsing binary systems by an analysis of their light changes in the frequency-domain; and, for this reason, we wish to explore further their more general properties before putting them to the actual task in Section 3 of this chapter.

In order to do so, let us return to the definition (1.1) of the A_{2m}'s, where the fractional loss of light $1-l$ can be expressed in accordance with Equation (1.2) valid for any type of eclipses, as

$$1 - l = L_1 f(r_1, r_2, \delta); \tag{2.53}$$

f representing a homogeneous function of its parameters of zero order. Moreover, earlier in this chapter we have shown (cf. Equations (1.23)–(1.25) or (2.25)–(2.27)) that the functional dependence of the moments A_{2m} on the geometrical elements $r_{1,2}$ and i of the eclipse assumes a particularly simple form in terms of the auxiliary parameters $C_{1,2,3}$ defined by Equations (1.26)–(1.28). Of these, the first two ($C_{1,2}$) are bound to be positive regardless of the type of eclipses; but C_3 may be positive or negative, depending on whether $r_2 \gtrless \cos i$.

Let us, therefore, in what follows regard the constants $C_{1,2,3}$ as 'generalized coordinates' of our problem; and consider, in particular, the differential operator

$$\frac{\partial}{\partial C_3} \equiv \sin^2 i \left\{ r_1^2 \frac{\partial}{\partial r_1^2} + r_2^2 \frac{\partial}{\partial r_2^2} + \sin^2 i \frac{\partial}{\partial \sin^2 i} \right\}, \tag{2.54}$$

which – unlike those taken with respect to $C_{1,2}$ – is symmetrical with respect to $r_{1,2}$ and, therefore, valid as it stands for any type of eclipse.

On the other hand, since the function $f(r_1, r_2, \delta)$ on the right-hand side of Equation (2.53) is known (cf. Section III-3) to be homogeneous in terms of its parameters, Euler's theorem on homogeneous functions discloses that

$$r_1^2 \frac{\partial f}{\partial r_1^2} + r_2^2 \frac{\partial f}{\partial r_2^2} + \delta^2 \frac{\partial f}{\partial \delta^2} = 0. \tag{2.55}$$

A combination of Equations (2.54) and (2.55) then discloses that, for such functions,

$$\begin{aligned} \frac{\partial f}{\partial C_3} &= \sin^2 i \left\{ \sin^2 i \frac{\partial}{\partial \sin^2 i} - \delta^2 \frac{\partial}{\partial \delta^2} \right\} f \\ &= -(1-\delta_0^2) \left\{ \delta^2 \frac{\partial}{\partial \delta^2} - \delta_0^2 \frac{\delta}{\partial \delta_0^2} + \frac{\partial}{\partial \delta_0^2} \right\} f, \end{aligned} \tag{2.56}$$

where, as before, $\delta_0 \equiv \cos i$.

Let us differentiate now Equation (1.1) with respect to C_3 behind the integral sign. Since the loss of light $1-l$ commences at the moment of first contact, $f(\theta_1) = 0$ and, consequently,

$$\frac{\partial A_{2m}}{\partial C_3} = L_1 \int_0^{\theta_1} \frac{\partial f}{\partial C_3} \, \mathrm{d}(\sin^{2m} \theta); \tag{2.57}$$

or, changing over from θ to δ as the variable of integration we can rewrite Equation (2.57) as

$$\frac{\partial A_{2m}}{\partial C_3} = -mL_1 \csc^{2(m-1)} i \int_{\delta_0^2}^{\delta_1^2} \left\{ \delta^2 \frac{\partial f}{\partial \delta^2} - \delta_0^2 \frac{\partial f}{\partial \delta_0^2} + \frac{\partial f}{\partial \delta_0^2} \right\} (\delta^2 - \delta_0^2)^{m-1} \, \mathrm{d}\delta^2. \tag{2.58}$$

On the other hand, by lowering the index of A_{2m} from m to $m-1$, it also follows from Equation (1.1) that

$$\begin{aligned} A_{2(m-1)} &= L_1 \int_0^{\theta_1} f \, \mathrm{d}(\sin^{2(m-1)} \theta) \\ &= (m-1) L_1 \csc^{2(m-1)} i \int_{\delta_0^2}^{\delta_1^2} (\delta^2 - \delta_0^2)^{m-2} f \, \mathrm{d}\delta^2 \\ &= -L_1 \csc^{2(m-1)} i \int_{\delta_0^2}^{\delta_1^2} (\delta^2 - \delta_0^2)^{m-1} \frac{\partial f}{\partial \delta^2} \, \mathrm{d}\delta^2 \end{aligned} \tag{2.59}$$

by partial integration; so that, by a subtraction of Equations (2.58) and (2.59), we obtain

$$\frac{\partial A_{2m}}{\partial C_3} - mA_{2(m-1)} = mL_1 \csc^{2(m-1)} i \times \int_{\delta_0^2}^{\delta_1^2} \left\{(1-\delta^2)\frac{\partial f}{\partial \delta^2} - (1-\delta_0^2)\frac{\partial f}{\partial \delta_0^2}\right\}(\delta^2 - \delta_0^2)^{m-1}\,\mathrm{d}\delta^2. \tag{2.60}$$

From Equation

$$\delta^2 = 1 - \cos^2\theta \sin^2 i = \sin^2\theta + \delta_0^2 \cos^2\theta, \tag{2.61}$$

it follows, however, that

$$(1-\delta^2)\frac{\partial f}{\partial \delta^2} = \frac{\partial f}{\partial \delta^2}\sin^2 i \cos^2\theta, \tag{2.62}$$

while

$$(1-\delta_0^2)\frac{\partial f}{\partial \delta_0^2} = \sin^2 i \frac{\partial f}{\partial \delta^2}\frac{\partial \delta^2}{\partial \delta_0^2} = \frac{\partial f}{\partial \delta^2}\sin^2 i \cos^2\theta \tag{2.63}$$

as well. Therefore, the difference of Equations (2.62) and (2.63) in curly brackets of Equation (2.60) will be zero in the whole range of $\delta_0 \leqslant \delta \leqslant \delta_1$ – a fact sufficient to annihilate the entire right-hand side of this equation; and to leave us with an outcome disclosing that

$$\frac{\partial A_{2m}}{\partial C_3} = mA_{2(m-1)}, \tag{2.64}$$

which represents a fundamental differential recursion formula satisfied by our moments A_{2m} of the light curves, and valid for *any* value of $m > 0$ (not necessarily integral).

Before we proceed to discuss the significance of this remarkably simple relation, let us underline its extreme generality. The only condition required for its proof is, in fact, the validity of Equation (2.55) – *requiring* $f(r_1, r_2, \delta)$ *on the right-hand side of Equation* (2.53) *to be a homogeneous function of zero degree in its parameters.* But this is bound to be true for any type of eclipse – occultation or transit; partial, total or annular – and for any law of limb-darkening; since all associated α-functions are homogeneous in $r_{1,2}$ and δ. The reader can verify by direct differentiation that all moments A_{2m} deduced earlier in this chapter satisfy indeed Equation (2.64) which can, therefore, serve also as a check of our previous work.

It may be added that, by virtue of Equation (2.64), our moments A_{2m} of the light curves turn out to belong to a class of functions generated by that difference-differential equation, the properties of which were studied by Appell (1880) and, more recently, by Truesdell (1948).

What is the form of differential equation satisfied by the moments $A_{2m}(C_1, C_2, C_3)$ regarded as functions of C_3 as the independent variable? A repeated differentiation of Equation (2.64) with respect to C_3, and combination of previous members of this series, discloses A_{2m} to satisfy a linear partial differential equation

$$\frac{\partial^m A_{2m}}{\partial C_3^m} = m!A_0, \tag{2.65}$$

of m-th order, subject to the initial conditions

$$\left(\frac{\partial^j A_{2m}}{\partial C_3^j}\right)_0 = \frac{m!}{(m-j)!} A_{2(m-j)}(C_1, C_2; 0) \quad \text{for } j = 1, 2, 3, \ldots, m-1;$$

the general solution of which can be expressed as a power series of the form

$$A_{2m} = L_1 \sum_{j=0}^{m} \binom{m}{j} f_j(C_1/C_2) C_2^j C_3^{m-j}, \tag{2.67}$$

where the v_j's are polynomials in C_1 and C_2.

Let, moreover,

$$u(h) = \sum_{j=0}^{\infty} \frac{v_j}{j!} h^j \tag{2.68}$$

represent a function of a dummy variable h in terms of the polynomials v_j. If so, it can be shown (cf. Appell, 1880) that

$$u(h)\,e^{hC_3} = \sum_{j=0}^{\infty} \frac{h^j}{j!} A_{2j}, \tag{2.69}$$

which represents a 'generating function' of the moments A_{2j} of the light curve of even orders.

Lastly, what is the significance of the zero-th moment A_0 of the light curve, encountered on the right-hand side of Equation (2.65)? Since

$$\lim_{m\to 0} \{d(\sin^{2m}\theta)\} = \begin{cases} 0 & \text{for } \theta > 0, \\ 1 & \text{for } \theta = 0, \end{cases} \tag{2.70}$$

Equation (1.1) discloses that A_0 no longer represents any area subtended by the light curve, but its ordinate at $\theta = 0$. In other words, *the quantity A_0 is identical with the maximum fractional loss of light $1 - \lambda$ suffered by the system at zero phase.* For total eclipses, we evidently have $A_0 \equiv L_1$; while for partial or annular eclipses, $A_0 \equiv L_1\alpha_0$, where α_0 stands for the maximum obscuration of the eclipsed star.

D. ATMOSPHERIC ECLIPSES

So far in this chapter we have tacitly assumed that the light changes of eclipsing variables are caused by mutual eclipses of stars which appear in projection as opaque and sharp-edged discs. Although the density varies smoothly from inside out to approach zero well outside the visible confines of a star, the physical reason of sharp-edge illusion is well understood; and goes back to the fact that the optical depth τ along the line of sight is a rapidly-varying function of the distance from the centres: the photospheric structure of typical Main-Sequence Stars (like our Sun) is such that τ drops from 10 to 10^{-2} – a change entailing a transition from practically complete opacity to a complete transparency – within the span of a few hundred kilometres (i.e., within 1 part in 10^4 of its diameter).

Recent decades brought forth, however, discoveries of a number of eclipsing variables whose behaviour in this respect is likely to be anomalous: namely, of binary systems in which at least one component is surrounded by an *extended atmosphere*, the optical depth of which varies in more gradual manner. Classical examples of one type of such variables are the systems of ζ Aurigae or VV Cephei, in which a normal star of early spectral type undergoes occultation by an extended atmosphere of a late-type supergiant, and suffers an appreciable loss of light long before it peters completely out of sight. In other words, the bodily eclipses in these (and other similar) systems are preceded by atmospheric eclipses of their early-type components; and these two kinds of eclipse may gradually merge into each other. Another type of atmospheric eclipse is encountered in close binaries, one component of which happens to be a Wolf-Rayet star – as exemplified by the well-known variable V 444 Cygni. In such systems a semi-transparent envelope of one (the Wolf-Rayet) star is in a rapid state of expansion; and its mutual eclipses with its more normal mate give rise to minima of light of very unequal duration – a fact established for the first time in V 444 Cyg by Kron and Gordon (1943) – and, in such systems, the effects of atmospheric and bodily eclipses of stars of comparable dimensions are inextricably connected. Since, moreover, systems exhibiting atmospheric eclipses are likely to grow in number as well as astrophysical importance at an accelerated pace, it seems desirable to establish also a photometric theory of such eclipses along similar lines as was already accomplished earlier in this book for bodily eclipses of opaque stellar discs.

The aim of the present sub-section will, therefore, be to extend our previous work to incorporate the effects of finite transparency of the eclipsing disc. In the time-domain, the first steps in this direction were taken by the present writer many years ago (cf. Kopal, 1945; Kopal and Shapley, 1946; Kopal, 1959); and the effects of semi-transparency on the explicit form of Hankel transforms expressing the fractional loss of light have already been considered briefly in Section I-3, pp. 31 or 34 of this book. The more specific aim of this section will be to transpose our problem again in the frequency-domain, and to outline a theory of the moments A_{2m} of the light changes arising from atmospheric eclipses, for an arbitrary structure of the semi-transparent envelope responsible for such eclipses – along the lines initiated (for an interpretation of bodily eclipses of opaque stellar discs) by Smith (1976).

In order to do so, attention is invited to Figure 5-3, the shaded portion of which represents the actual area

$$f \equiv \pi r_1^2 \alpha(r_1, \rho; \delta), \tag{2.71}$$

where

$$\begin{aligned} &0 \leq r \leq r_1, \\ &0 \leq \rho \leq r_2, \\ &\delta_0 \leq \delta \leq r_1 + r_2. \end{aligned} \tag{2.72}$$

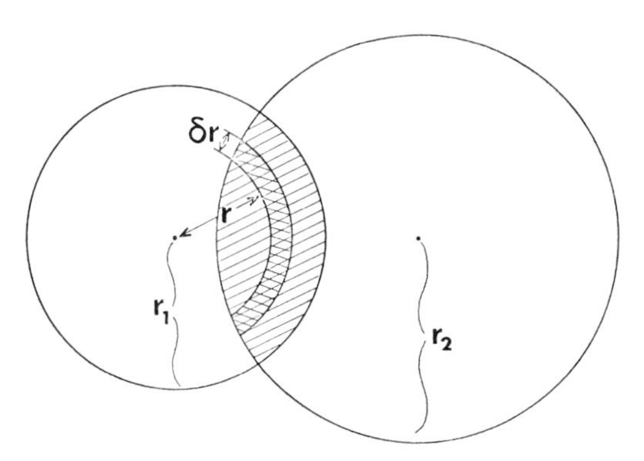

Fig. 5-3. Area of mutual eclipse is denoted by the shaded portion.

The loss of light represented by the cross-hatched part of the annulus will be given by

$$J(r)\,\mathrm{d}f \equiv J(r)\frac{\partial f}{\partial r}\,\mathrm{d}r = J(r)\frac{\partial}{\partial r}(\pi r^2\alpha)\,\mathrm{d}r, \tag{2.73}$$

where $J(r)$ denotes an arbitrary radially-symmetrical distribution of brightness over the apparent disc of the eclipsed star. The loss of light $1-l$ emitted by the total shaded area on Figure 5-3 should then be given

$$\begin{aligned}1-l &= L_1\alpha(r_1, r_2, \delta)\\ &= L_1\int_0^{r_1} J(r)\frac{\partial\alpha}{\partial r}\{\pi r^2\alpha(r, r_2, \delta)\}\,\mathrm{d}r.\end{aligned} \tag{2.74}$$

If the eclipsing disc were totally opaque up to a radius $\rho = r_2$, the explicit form of the function $\alpha_0^0 \equiv \alpha(r, r_2; \delta)$ is well known to us already from Chapter I. If, however, this disc happens to be semi-transparent, the function α in the integrand of Equation (2.74) should be replaced by

$$\alpha = \int_0^{r_2} \Lambda(\rho)\frac{\partial\alpha_0^0}{\partial\rho}\,\mathrm{d}\rho, \tag{2.75}$$

where $\Lambda(\rho)$ denotes a 'transparency function', generally deducible from physical considerations; and, on partial integration,

$$\alpha = [\Lambda(\rho)\alpha_0^0(r, \rho, \delta)]_0^{r_2} - \int_0^{r_2} \frac{\partial\Lambda}{\partial\rho}\,\alpha_0^0\,\mathrm{d}\rho. \tag{2.76}$$

If the eclipsing star were transparent, $\Lambda(\rho) \equiv 0$; accordingly, $\alpha = 0$ and the eclipse would give rise to no loss of light. If, on the other hand, this star were completely opaque up to the limb, $\Lambda(\rho) = 1$ for $0 \leqslant \rho \leqslant r_2$; and since $\alpha(r, 0; \delta) = 0$, it follows that $\alpha = \alpha_0^0(r, r_2; \delta)$. If, however, Λ represents a continuous function of ρ, Equation (2.56) reduces to

$$\alpha = \Lambda(r_2)\alpha_0^0(r, r_2, \delta) - \int_0^{r_2} \frac{\partial \Lambda}{\partial \rho} \alpha_0^0(r, \rho, \delta)\, \mathrm{d}\rho \tag{2.77}$$

which, if $\Lambda(r_2) = 0$, would reduce further to

$$\alpha = -\int_0^{r_2} \frac{\partial \Lambda}{\partial \rho} \alpha_0^0(r, \rho, \delta)\, \mathrm{d}\rho; \tag{2.78}$$

the negative sign of its right-hand side being offset by a generally negative sign of the derivative $\partial\Lambda/\partial\rho$.

The moments A_{2m} of the light curves arising from atmospheric eclipses then should – in accordince with Equation (1.1) – be given by

$$\begin{aligned}
\tilde{A}_{2m} &= L_1 \int_0^{\theta_1} \mathrm{d}(\sin^{2m}\theta) \int_0^{r_1} J(r) \frac{\partial}{\partial r}\left\{\pi r^2 \int_0^{r_2} \Lambda(\rho) \frac{\partial \alpha_0^0}{\partial \rho}\, \mathrm{d}\rho\right\} \mathrm{d}r \\
&= L_1\Lambda(r_2) \int_0^{\theta_1} \mathrm{d}(\sin^{2m}\theta) \int_0^{r_1} J(r) \frac{\partial}{\partial r}\{\pi r^2 \alpha_0^0(r, r_2, \delta)\}\, \mathrm{d}r \\
&\quad - L_1 \int_0^{\theta_1} \mathrm{d}(\sin^{2m}\theta) \int_0^{r_1} J(r) \frac{\partial}{\partial r}\left\{\pi r^2 \int_0^{r_2} \frac{\partial \Lambda}{\partial \rho} \alpha_0^0(r, \rho, \delta)\, \mathrm{d}\rho\right\} \mathrm{d}r.
\end{aligned} \tag{2.79}$$

The first part of the expression on the right-hand side of the foregoing equation we know already from Section 1 of this chapter to be identical with the moments $A_{2m}(r_1, r_2, i)$ of the light curves of bodily eclipses, as given by Equation (2.4) for any type of eclipse or law of limb-darkening; but the second is still to be evaluated. Since, the limits of integration with respect to r, ρ and $\sin^{2m}\theta$ are constant, they can be interchanged – a strategy which permits us to assert that

$$\begin{aligned}
\tilde{A}_{2m} &= \Lambda(r_2) A_{2m}(r_1, r_2, i) \\
&\quad - \pi \int_0^{r_1} J(r) \frac{\partial}{\partial r}\left\{r^2 \int_0^{r_2} A_{2m}(r, \rho, i) \frac{\partial A}{\partial \rho}\, \mathrm{d}\rho\right\} \mathrm{d}r
\end{aligned} \tag{2.80}$$

as the final result of our analysis, from which the moments A_{2m} can be evaluated in terms of the A_{2m}'s for the same geometry if the form of the opacity function $\Lambda(\rho)$ can be deduced from physical considerations.

To provide a specific example of the form which such a Λ-function may assume, consider that the physical cause of semi-transparency of an extended atmosphere stratified in spherically-symmetrical layers is atomic absorption (or electron scattering) specified by a coefficient $\kappa(r)$. If, moreover, ρ denotes new the density of its material and the mass-absorption coefficient

$$\kappa\rho = \kappa_0\, \mathrm{e}^{-nr}, \tag{2.81}$$

where κ_0 and n are constants (as would be the case in a neutral atmosphere in hydrostatic equilibrium), the equation

$$d\tau = \kappa\rho \, ds, \tag{2.82}$$

defining the element dr of the optical depth along the line of sight of the element ds, yields

$$\tau = 2\kappa_0 \int_0^\infty e^{-nr} \, ds, \tag{2.83}$$

where

$$s^2 = r^2 - \delta^2. \tag{2.84}$$

If, furthermore, we set $r = \delta x$ and replace ds by dx as the element of integration, the optical depth τ along the line of sight can be expressed as

$$\tau = 2\delta\kappa_0 K_1(a), \tag{2.85}$$

where

$$K_1(a) = \int_1^\infty \frac{x \, e^{-ax}}{\sqrt{x^2 - 1}} \, dx \tag{2.86}$$

constitutes an integral representation of the Bessel function of the second kind with imaginary argument, and $a \equiv \delta n$ as well as $\delta\kappa_0$ are non-dimensional parameters. The transparency function on the r.h.s. of Equation (2.80) can then be expressed as

$$\Lambda = 1 - e^{-\tau}. \tag{2.87}$$

At the distance $a = 0$ from the centre of the eclipsing component, $K_1(0) = \infty$ and $\tau = \infty$, in which case the foregoing Equation (2.87) yields $\Lambda = 1$ (corresponding to complete opacity of the eclipsing layers); while if $a = \infty$ and $K_1(\infty) = 0$ ($\tau = 0$), Equation (2.87) leads to $\Lambda = 0$ (corresponding to their complete transparency).

V-3. Solution for the Elements

In the preceding sections of this chapter we initiated a new approach to the problem of the analysis of the light changes of eclipsing variables for the elements of their eclipses; and the principal tool for this purpose has been found to be the moments of the light curves A_{2m}, in terms of which the desired elements can be evaluated by algebraic means. For one particular type of eclipses: namely, when the latter prove to be *total*, this task has already been accomplished – at least as long as both components of totally-eclipsing systems can be regarded as spherical (and complications arising from distortion will be treated in the next chapter).

To recapitulate the essential steps of the solution, a knowledge of the empirical values (obtainable from the observed light curves) of the first four moments A_0, A_2, A_4 and A_6 is sufficient for a determination of three auxiliary constants $\bar{C}_{1,2,3}$ related with the A_{2m}'s by simultaneous algebraic equations (2.25)–(2.27) where $L_1 = 1 - \lambda \equiv A_0$; and which (for stars of arbitrary degree of the law of limb-darkening) can be converted into purely geometrical auxiliary constants $C_{1,2,3}$ as defined by Equations (1.26)–(1.28), by means of Equations (2.28)–(2.30). The remaining desired geometrical elements $r_{1,1}$ and $\sin i$ then follow in terms of the reduced constants $C_{1,2,3}$ (freed from the effects of limb-darkening) from Equations (1.30) and (1.31). We may add that the same can, in principle, be accomplished by use of *any* four moments A_{2m} whatever the four m's may happen to be. The higher moments (being progressively smaller in absolute value) are, however more difficult to deduce from observations with sufficient significance; and the geometrical relations between the observed moments and the desired elements of the eclipse become progressively more complicated.

Suppose, however, that the type of eclipses which gave rise to the observed minima cannot be inferred in advance from an inspection of the light curve; or that the latter indicates an alternation of partial eclipses. That this will be true for a large majority of observed cases follows from the fact that partial eclipses can occur for much larger range of inclinations between their orbital planes and the line of sight than is necessary to bring about an alternation of total and annular eclipses. Therefore, in what follows a general procedure will be developed which should permit similar treatment of every type of eclipses (inclusive of those ending in totality), and which can be made amenable to solution by automatic computers.

In order to do so, let us return to Equations (2.8)–(2.10) representing a general expression for the moments A_{2m} of the light curve due to any type of eclipses, which we shall rewrite as

$$A_{2m} = L_1 \sin^{2m} \theta_1 \, f_{2m}(a, c_0). \tag{3.1}$$

Suppose, moreover, that the coefficients u_l in the law of limb-darkening (involved in the $C^{(l)}$'s) have been estimated from the theory of stellar atmospheres, and that our attention is focussed on the determination of the four elements r_1, r_2, i and L_1 (or, which is equivalent, L_1, $\sin^2 \theta_1$, a and c_0) from four moments A_{2m} of the light curve empirically determined for four distinct values of m (not necessarily integral, or equally-spaced).

All theoretical expressions (3.1) for A_{2m} are factored by the luminosity L_1 of the star undergoing eclipse. Hence, their *linear* ratios

$$\frac{A_{2m}}{A_{2\mu}} = \sin^{2(m-\mu)} \theta_1 \frac{f_{2m}(a, c_0)}{f_{2\mu}(a, c_0)} \tag{3.2}$$

depend only on θ_1, a_1 and c_0. If, moreover, we form the *quadratic* ratios

$$\frac{A_{2m}A_{2m'}}{A_{2\mu}A_{2\mu'}} = \sin^{2(m+m'-\mu-\mu')} \theta_1 \frac{f_{2m}(a, c_0) f_{2m'}(a, c_0)}{f_{2\mu}(a, c_0) f_{2\mu'}(a, c_0)}, \tag{3.3}$$

where m, m' and μ, μ' are all distinct but such that

$$m + m' = \mu + \mu', \tag{3.4}$$

the terms in Equation (3.3) involving $\sin^2 \theta_1$ cancel as well; and its right-hand side – let us denote it by $g_{m,m';\mu,\mu'}$ – depend on a and c_0 only.

Suppose that, in particular,

$$\frac{(A_2)^2}{A_0 A_4} \equiv \frac{A_2/A_0}{A_4/A_0} \equiv g_2(a, c_0) \tag{3.5}$$

and

$$\frac{(A_4)^2}{A_2 A_6} \equiv \frac{A_4/A_2}{A_6/A_4} \equiv g_4(a, c_0). \tag{3.6}$$

With the left-hand sides of the preceding equations established from the observations as ratios of the respective empirical moments of the light curves, $g_2(a, c_0)$ and $g_4(a, c_0)$ constitute two independent relations between the unknown parameters a and c_0; and can be solved for them (numerically or otherwise) with the aid of the expansions on the r.h.s. of Equations (2.8)–(2.10). An automatic iterative procedure to this end has been developed by Demircan (1979); the necessary and sufficient condition for the convergence of such iterations being the requirement that the Jacobian determinant

$$\frac{\partial(g_2, g_4)}{\partial(a, c_0)} \neq 0. \tag{3.7}$$

The expansions (2.8)–(2.10) can be also used in Equation (3.3) as a tool for the construction of tables of the functions $g_2 \equiv g_{2,2;0,4}$ or $g_4 \equiv g_{4,4;2,6}$ for any adopted distribution of brightness of the apparent disc of the eclipsed star. Demircan (1978) did so for a quadratic law of limb-darkening, characterized by the coefficients $u_1 = 0.650$ and $u_2 = -0.023$, which approximates closely the exact solution of the equations of radiative transfer in grey plane-parallel atmospheres within errors graphically shown on the accompanying Figure 5-4 (after Kopal, 1949); and a graphical representation of the g_2- and g_4-functions in terms of a and c_0 is exhibited on Figures 5-5 and 5-6.

If the outcome of the solution of a pair of the g-functions of the type (3.5) or (3.6) discloses that

$$0 < a < \tfrac{1}{2}, \tag{3.8}$$

the eclipse giving rise to the observed minimum is an *occultation* (signifying that $r_1 < r_2$); while if

$$\tfrac{1}{2} < a < 1, \tag{3.9}$$

this eclipse proves to be a *transit*. Moreover, if the value of c_0 transpires to be such that

$$1 > c_0 > |2a - 1|, \tag{3.10}$$

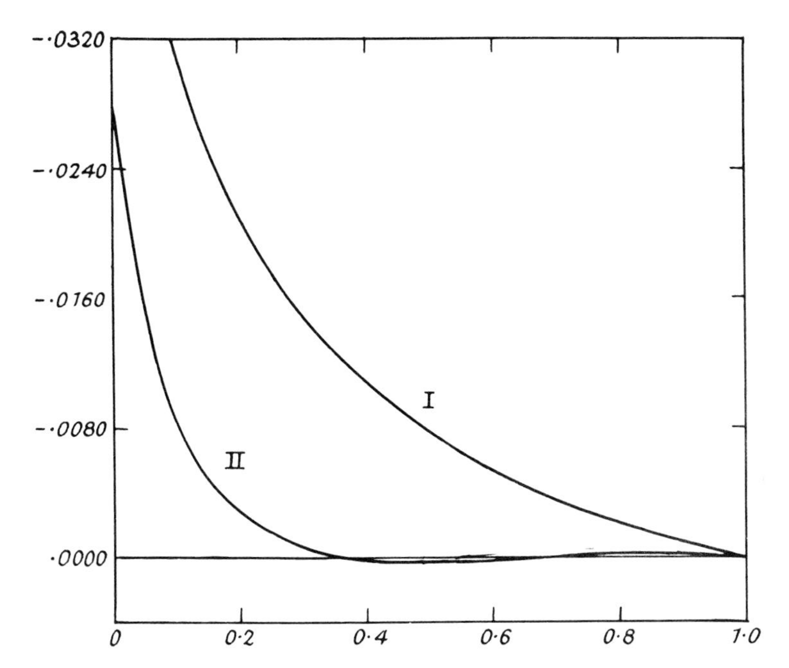

Fig. 5-4. Deviations of the linear (I) and quadratic (II) approximations of the law of limb-darkening from the exact solution of radiative transfer in grey plane-parallel atmospheres (abscissae), plotted against $\cos\gamma$ of the angle of foreshortening (ordinates). Curve I represents the deviations of the exact solution of the problem from a linear approximation of the form $J(\gamma) = 0.4 + 0.6\cos\gamma$; curve II, the corresponding deviations from the quadratic approximation $J(\gamma) = 0.373 + 0.650\cos\gamma - 0.023\cos^2\gamma$ (after Kopal, 1949).

the eclipse is bound to be *partial.* If

$$c_0 < 1 - 2a, \tag{3.11}$$

the eclipse becomes *total*; while if

$$c_0 < 2a - 1, \tag{3.12}$$

in turns out to be *annular.* A graphical representation of a family of functions defined by the preceding Equations (3.8)–(3.12) can be seen on Figure 5-2 on p. 161 and with its aid a knowledge of any pair of the parameters a and c_0 will at once specify the nature of the eclipse which gave rise to the respective minimum.

Moreover, once these two parameters have been satisfactorily determined, the actual geometrical elements L_1, r_1, r_2 and i of the system can be determined without difficulty in the following manner. Inasmuch as

$$\sin^2\theta_1 = (r_1 + r_2)^2 \csc^2 i - \cot^2 i = \frac{r_1^2(1 - c_0^2)}{a^2 - (r_1 c_0)^2}, \tag{3.13}$$

an insertion of Equation (3.13) in (3.2) particularized (say) for $\mu = 0$ and $m = 1$ furnishes an equation of the form

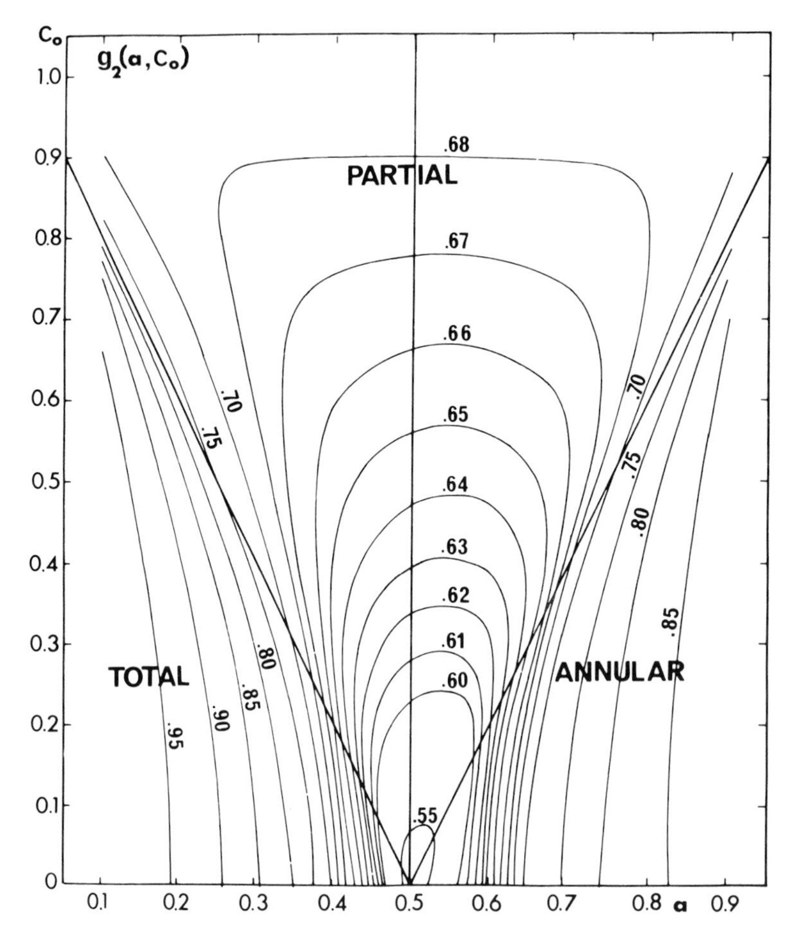

Fig. 5-5. A diagrammatic representation of the function $g_2(a, c_0)$ as defined by Equation (3.5) for every type of eclipse (after Kopal and Demircan, 1978).

$$\{(1-c_0^2)A_0 f_2(a, c_0) + c_0^2 A_2 f_0(a, c_0)\} r_1^2 = a^2 A_2 f_0(a, c_0), \tag{3.14}$$

which for known values of $A_0 \equiv 1 - \lambda$, A_2, a and c_0 contains r_1^2 as the only unknown. Once we have determined it, the values of r_2 and i follow from

$$a r_2 = (1-a) r_1 \tag{3.15}$$

and

$$\cos i = (c_0/a) r_1; \tag{3.16}$$

while the fractional luminosity L_1 of the component undergoing eclipse obtains, e.g., from Equation (3.1) particularized for $m = 0$ as

$$f_0(a, c_0) L_1 = A_0 = 1 - \lambda \tag{3.17}$$

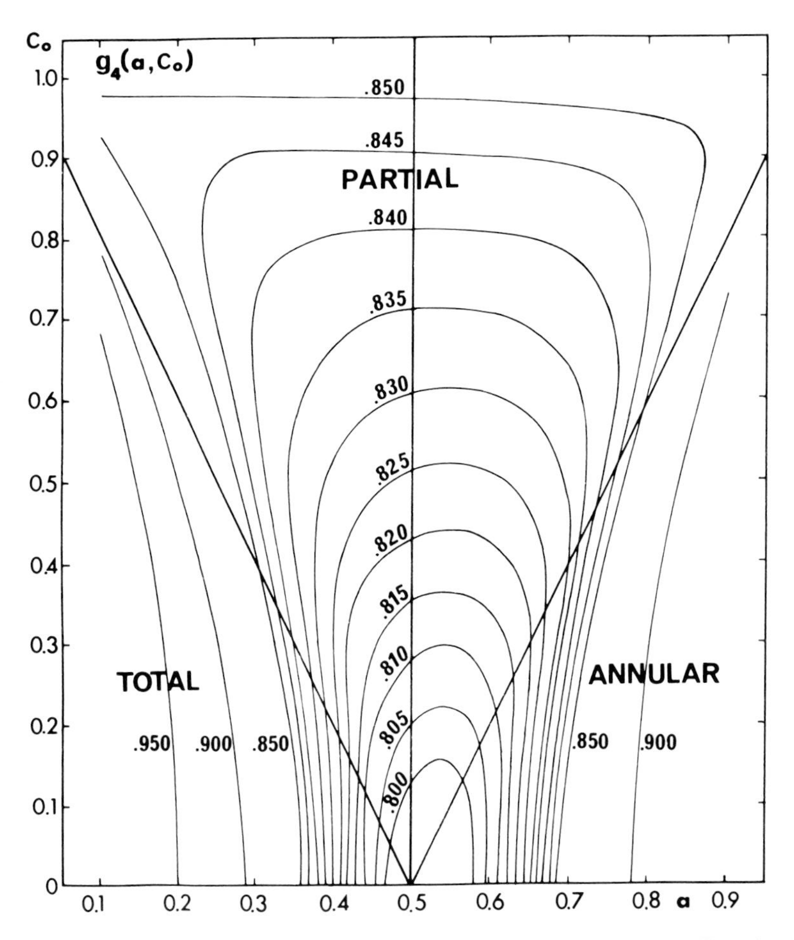

Fig. 5-6. A diagrammatic representation of the function $g_4(a, c_0)$ as defined by Equation (3.6) for every type of eclipse (after Kopal and Demircan, 1978).

where, for total eclipses,

$$f_0(a, c_0) \equiv \alpha(a, c_0) = 1. \tag{3.18}$$

A. OPTIMIZATION OF SOLUTIONS

Once we have done all this the question is, however, bound to be asked: do the g-functions as given, for example, by Equations (3.5) and (3.6) represent optimum means for a determination of the parameters a and c_0 on which everything else depends? This question remains so far unanswered; for the adoption of $m = m' = 2,4$ and $\mu = 0,2$ or $\mu' = 4,6$ underlying Equations (3.5) and (3.6) for $g_{2,4}$ represented an arbitrary act – sufficient, but not necessary, for obtaining a solution. *The best pair of such functions*, leading to an *optimum* solution, *is obviously one for which the square of the Jacobian*

$$J \equiv \frac{\partial(g_{m_1 m'_1;\mu_1\mu'_1}, g_{m_2,m'_2;\mu_2,\mu'_2})}{\partial(a, c_0)} \tag{3.19}$$

becomes a maximum; and *it is this condition which should determine the appropriate choice of the constants* $m_{1,2}$, $m'_{1,2}$; $\mu_{1,2}$ and $\mu'_{1,2}$.

These constants need not necessarily be integers; nor do they have to be separated by any particular interval. They should, however, be *positive*; and such that

$$m_{1,2} + m'_{1,2} = \mu_{1,2} + \mu'_{1,2}. \tag{3.20}$$

Moreover, the values of these constants most suitable for our purpose are the *smallest* ones which maximize the square of the Jacobian (3.19); for since, within eclipses, $\sin\theta < 1$ (and, therefore, the inequality $\sin^{2m}\theta \ll 1$ becomes the stronger, the larger the value of m), it follows the numerical values $A_{2m} > A_{2\mu}$ for $m < \mu$; and if, therefore, $m_{1,2}$ (or $\mu_{1,2}$) in Equation (3.19) becomes too large, it may be difficult to determine the values of the respective moments from the observations with the requisite significance. Conversely, the values of $m_{1,2}$ or $\mu_{1,2}$ for which the Jacobian determinant (3.19) vanishes must be avoided; for a simultaneous solution of the g-functions corresponding to such a case would become indeterminate.

The problem of an *optimization* of a solution – i.e., a determination of the set of m_j's and μ_j's which maximize the square of the Jacobian (3.19) – remains still unsolved. A systematic study of the magnitude of this Jacobian for all permissible values of a and c_0 has recently been undertaken by Edalati (1978). His results confirm that in those sections of the a–c_0 plane corresponding to total or annular eclipses (cf. again Figure 5-2), the Jacobian (3.19) does not vanish; accordingly, any pair of distinct g-functions will furnish a determinate (though not equally accurate) solution for a and c_0. In the domain of partial eclipses this can, unfortunately, no longer be guaranteed in advance; for – especially for shallow partial eclipses – specific quartets of $m_{1,2}$ and $\mu_{1,2}$ exist for which the Jacobian (3.19) of our problem does become zero, rendering its solution indeterminate. Moreover *in certain domains of the a–c_0 plane the value of the Jacobian J*, while strictly non-zero, *may become so small as to render a solution based on observational data of finite accuracy practically indeterminate.* If so, however, this implies that a significant solution for the elements of the eclipse cannot be obtained by an analysis of observations of one minimum alone; to regain its significance, recourse must be had to a *combined evidence furnished by the alternate* minima – a problem to which we now wish to turn our attention.

B. COMBINATION OF ALTERNATE MINIMA

Suppose that *both* alternating eclipses of a system under consideration give rise to observable minima of sufficient amplitude to permit an empirical determination of *two* sets of the moments $(A_{2m})_{p,s}$; where p, s will hereafter refer to properties of the primary (deeper) and secondary (shallower) minima, respectively – regardless of which one happens to be an occultation or a transit. An example of such a system is, e.g., WW Aurigae, the light curve of which can be seen on Figure 5-7. In order to utilize them for a simultaneous determination of geometrical elements of the system, it is sufficient to recall that *the roles of the fractional radii $r_{1,2}$ of the eclipsed and eclipsing component in the alternate minima are interchanged*; and what was r_1 at the time of (say) the primary

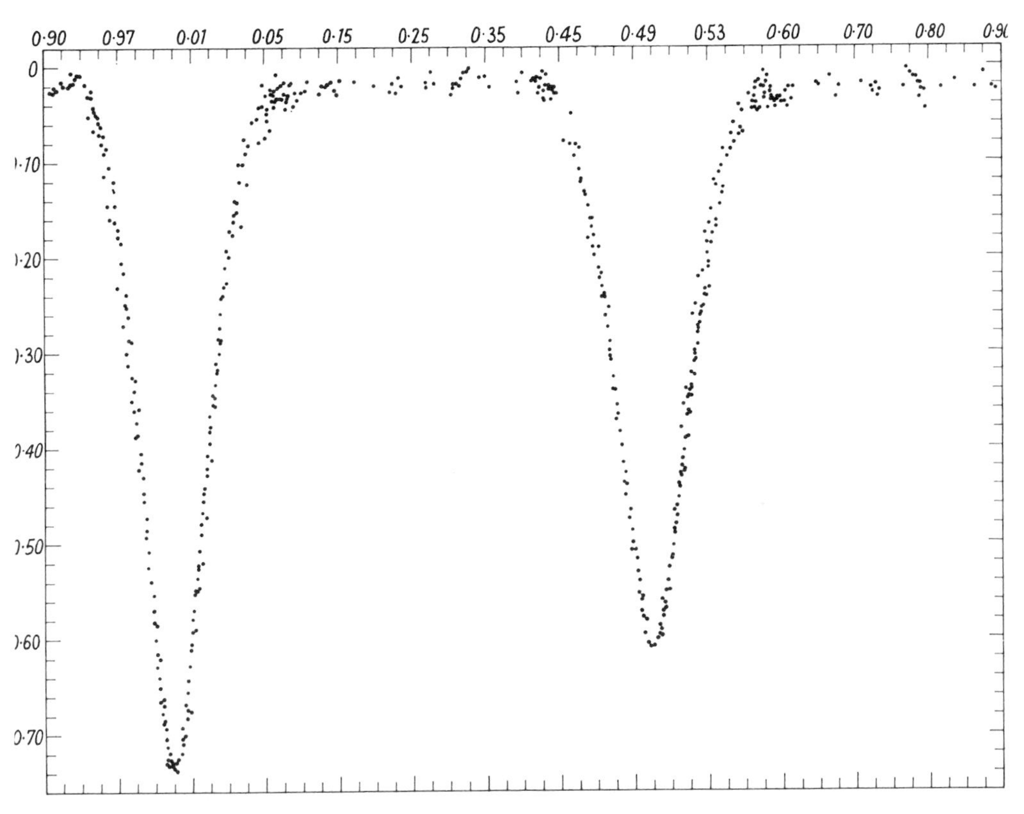

Fig. 5-7. Light curve of the eclipsing system of WW Aurigae – a prototype of those exhibiting an alternation of the primary and secondary minima of comparable depth (in contrast with RZ Cassioppeiae, as shown on Fig. 1-1, where the depths of primary and secondary minima are very unequal), caused by mutual partial eclipses of the two stars, according to photoelectric observations of C. M. Huffer (cf. Huffer and Kopal, 1951).

Abscissae: the relative brightness of the system in stellar magnitudes;
Ordinates: the phase in fractions of the orbital cycle (the time-scale between minima has been contracted for convenience of presentation).

minimum, will become r_2 at the time of the secondary (or vice versa). The orbital inclination i resulting from an analysis of both minima should, of course, remain the same; while (in accordance with our choice of the unit of light) $L_2 = 1 - L_1$.

Let, in agreement with Equation (3.1),

$$(A_{2m})_p = L_1 \sin^{2m} \theta_1 f_{2m}(a, c_0) \tag{3.21}$$

define the moments of the light curve arising from an eclipse of the component of fractional radius r_1 and luminosity L_1; while half a revolution later – at the time of the secondary minimum,

$$(A_{2m})_s = L_2 \sin^{2m} \theta_2 f_{2m}(b, c_0), \tag{3.22}$$

where

$$L_2 = 1 - L_1, \quad b = 1 - a, \tag{3.23}$$

and $\theta_2 = \pi + \theta_1$, so that (for circular orbits)

$$\sin^2 \theta_1 = \sin^2 \theta_2. \tag{3.24}$$

Accordingly, unlike Equation (3.2) the ratio of Equations (3.21) and (3.22)

$$\frac{(A_{2m})_p}{(A_{2m})_s} = \frac{L_1}{L_2} \frac{f_{2m}(a, c_0)}{f_{2m}(1 - a, c_0)} \tag{3.25}$$

becomes independent of the angles $\theta_{1,2}$ of first contact of both eclipses, but contains the ratio L_1/L_2.

The value of the latter can, however, be readily ascertained if we particularize Equation (3.25) for $m = 0$. Since

$$(A_0)_{p,s} = 1 - \lambda_{p,s}, \tag{3.26}$$

a combination of Equations (3.25)–(3.26) with Equation (3.25) discloses that

$$\frac{L_1}{L_2} = \left\{\frac{\alpha(1 - a, c_0)}{\alpha(a, c_0)}\right\} \frac{1 - \lambda_p}{1 - \lambda_s} \tag{3.27}$$

– an equation identical with Equation (3.25) of Chapter IV, in which we abbreviated

$$\frac{\alpha(1 - a, c_0)}{\alpha(a, c_0)} = \left(\frac{r_1}{r_2}\right)^2 Y(a, c_0). \tag{3.28}$$

Let us next return to Equations (3.21)–(3.25), which for $m = 0$ reduce to

$$(A_0)_p = L_1 \alpha(a, c_0), \quad (A_0)_s = L_2 \alpha(b, c_0), \tag{3.29}$$

and combine them with Equations (3.23): an elimination of L_1 and L_2 between them yields

$$\frac{(A_0)_p}{\alpha(a, c_0)} + \frac{(A_0)_s}{\alpha(b, c_0)} = 1; \tag{3.30}$$

or, in view of Equations (3.26) and (3.28) can be rewritten as

$$\alpha(a, c_0) = 1 - \lambda_p + \frac{1 - \lambda_s}{(r_1/r_2)^2 Y(a, c_0)} \tag{3.31}$$

in agreement with Equation (1.19) of Chapter IV. Should, moreover, $a < \frac{1}{2}$ and $c_0 < 1 - 2a$, the eclipse becomes total – in which case $\alpha(a, c_0) = 1$; and if so,

Equation (3.31) can be solved for the ratio r_1/r_2 to yield

$$\left(\frac{r_1}{r_2}\right)^2 = \frac{1-\lambda_s}{\lambda_p Y(a, c_0)}, \tag{3.32}$$

again in agreement with Equation (1.22) of Chapter IV.

For $m > 0$, a division of Equation (3.21) by (3.21) discloses that

$$\frac{(A_{2m})_p}{(A_{2m})_s} = \frac{L_1}{L_2}\frac{f_{2m}(a, c_0)}{f_{2m}(b, c_0)}; \tag{3.33}$$

which on insertion for L_1/L_2 from Equation (3.27) yields

$$\frac{(A_{2m})_p}{(A_{2m})_s}\frac{(A_0)_s}{(A_0)_p} = \frac{\alpha(1-a, c_0)f_{2m}(a, c_0)}{\alpha(a, c_0)f_{2m}(1-a, c_0)}. \tag{3.34}$$

The value of the left-hand side of the preceding equation can be determined from the light curves if at least the depths of both minima are known (as, for example, in the case of Algol or RZ Cassiopeiae whose light curve can be seen on Figure 1-1); while its right-hand side can be expressed in terms of the unknown quantities a and c_0 with the aid of Equations (2.8)–(2.10). With $(A_0)_{p,s} \equiv 1 - \lambda_{p,s}$ and $(A_{2m})_{p,s}$ regarded as known, Equations (3.30) or (3.31) and (3.34) particularized for any value of m (say, 1) can then be used to solve for a and c_0; and the rest of the solution for r_1, r_2, cos i and L_1 from known values of $(A_0)_p$, $(A_0)_s$, $(A_{2m})_p$ and $(A_{2m})_s$ carried out in the same way as before.

Alternatively, one could evaluate the unknown parameters a and c_0 from a pair of equations

$$(A_2^2/A_0A_4)_p = g_2(a, c_0), \tag{3.35}$$

$$(A_2^2/A_0A_4)_s = g_2(1-a, c_0), \tag{3.36}$$

similar to Equation (3.5), but pertaining to each minimum separately. In such a case, however, six (rather than four) distinct moments – e.g., $(A_{0,2,4})_{p,s}$ – would be required to accomplish the task.

A symmetry of the functions (3.35)–(3.36) about the line $a = \frac{1}{2}$ suggests another method by which their simultaneous solution can be carried out. Suppose that the numerical value of g_2 has been established from the empirical moments for one (say, the occultation) eclipse; and an appropriate value of g_2 for the alternate minimum. Each specifies a certain relation between a and c_0 on each half of Figure 5-5. Mark each on a transparent paper, and overlay on each other flopped about the line $a = \frac{1}{2}$; the coordinates of intersection of these two curves will indicate the respective values of a and c_0.

If, lastly, the depth $1 - \lambda_s$ of the secondary minimum becomes negligible – as it would be if the component eclipsed at that time were effectively dark – the fractional luminosity L_1 of the star eclipsed at the time of the primary minimum would be equal to 1 (or very close to it). In such a case, the second independent relation between a and c_0 to be adjoined to Equations (3.35) or (3.36) would follow from (3.31) as

$$a(a, c_0) = (A_0)_p = 1 - \lambda_p, \tag{3.37}$$

which together with Equations (3.35) or (3.36) then specifies the desired solution for the elements of the system $r_{1,2}$ and i consistent with $L_1 = 1$ and $L_2 = 0$.

C. ECCENTRIC ORBITS

In all preceding parts of this chapter we tacitly considered the relative orbit of the stars mutually eclipsing each other to be circular; with both minima of light being symmetrical with respect to the moments of conjunctions. However, in Section 4 of Chapter IV we already saw that if the eccentricity e of the orbit of the two components is non-vanishing, the descending and ascending branches of both minima will cease to be mirror-images of each other; and, consequently, the moments A_{2m} of the light curves defined by Equation (1.1) obtained by a planimetry of each half of the minimum will be different; and the extent of their difference will depend on the longitude ω of periastron of the respective orbit (and vanish only if the apsidal line is parallel with the line of sight). If so, the problem still to be considered consists of a determination of the elements of the eclipse L_1, r_1, r_2 and i – to which we must add now e and ω – from the moments A_{2m} of the light curves that can be extracted from the observations.

The determination of these moments must be considered first; for the following reason. If we continue to define the A_{2m}'s as areas subtended by a plot of the light changes as a function of the time, it follows that the phase-angle θ on Figure 4-2 will be identical with the *mean* anomaly

$$\theta = (2\pi/P)(t - t_0) \equiv M \tag{3.38}$$

of the eclipsing component of the respective system reckoned from the moment t_0 of mid-primary minimum; while θ on the right-hand side of Equation (1.1) signifies (but for its zero point) the *true* anomaly v, related with θ by Equation (4.15) of Chapter IV. In other words, the element $d(\sin^{2m}\theta)$ of integration on the r.h.s. of Equation (1.1) defining the moments of the light curves will then be equal to $d[\cos^{2m}(v + \omega)]$ rather than $d[\sin^{2m} M]$; and, in consequence, the empirical values of A_{2m} cannot be ascertained from the observed data by planimetry (or otherwise) until the times of the individual observations have been converted into the true anomalies $\nu + \omega$ needed to specify the phase angle θ.

This can indeed by accomplished by a resort to the well-known expansion of elliptic motion, asserting that

$$v = \omega + M + (2e - \tfrac{1}{4}e^3)\sin M + \tfrac{5}{4}e^2 \sin 2M + \tfrac{13}{12}e^3 \sin 3M + \cdots, \tag{3.39}$$

correctly to quantities of the order of e^3;* but its use to convert M into v – and, subsequently, θ by Equation (4.15) of Chapter IV – obviously requires *a priori* knowledge of e as well as ω. Suppose that their values have indeed been deduced from the displacement and unequal durations of the alternate minima in eccentric systems with the aid of the formulae already established in Section

* Brown and Shook (1933, p. 79) gave an explicit form of this expansion accurate to terms of the order of e^7.

IV-4 or (in the absence of the requisite photometric information) from the spectroscopic data. If so, it should be possible to re-plot the light curve from $(1-l)\text{-}\sin^{2m} M$ to $(1-l)\text{-}\sin^{2m}\theta$ coordinates for adopted values of e and ω, and to proceed with the solution for $r_{1,2}$, i and L_1 as before in terms of so re-defined moments A_{2m}.

Such a procedure is, however, prone to certain restrictions which should be observed in advance. First, note (cf. Figure 4-2) that unless $\omega = 90°$ or $270°$, the descending the ascending branch of either minimum will be asymmetric; and that, therefore, both minima (primary and secondary) of the same system should furnish *four* distinct moments of the light curve (which for $\omega = 90°$ or $270°$ would reduce to two). To what extent will their definition in terms of the elements of the eclipse differ from that appropriate for circular orbits?

In order to answer this question, let us change over from $\mathrm{d}(\sin^{2m}\theta)$ to $\mathrm{d}(\delta/R)^2$ as the element of integration on the r.h.s. of Equation (1.1) by means of the equation

$$\mathrm{d}(\sin^{2m}\theta) = m \csc^{2m} i\{(\delta/R)^2 - \cos^2 i\}^{m-1}\,\mathrm{d}(\delta/R)^2, \tag{3.40}$$

where $(\delta/R)^2$ continues to be given by Equation (4.14) of Chapter IV; and – in accordance with Equation (1.2) – we continue to have

$$1 - l = L_1\alpha = L_1 \sum_{l=0}^{\infty} C^{(l)}\alpha_l^0(a, c), \tag{3.41}$$

where the functions $\alpha_l^0(a, c)$ for any type of eclipses have been given by Equations (3.91), (3.94) or (3.98)–(3.99) of Chapter I in terms of nondimensional parameters a and c defined by Equations (3.56)–(3.58) of that chapter.

Since both these parameters represent ratios of two lengths, the unit in which these lengths are expressed is immaterial – and can be identified with any particular radius-vector R. Accordingly, the moments A_{2m} should be expressible again in the form

$$A_{2m} = mL_1 \csc^{2m} i \sum_{l=0}^{\Lambda} C^{(l)} \int_{(\delta/R)_0^2}^{(\delta/R)_1^2} \alpha_l^0(a, c)\{(\delta/R)^2 - \cos^2 i\}^{m-1}\,\mathrm{d}(\delta/R)^2, \tag{3.42}$$

an expression which differs from that appropriate for circular orbits, and given by Equation (2.1) only through the limits. In the elliptical case, the upper limit

$$\left(\frac{\delta}{R}\right)_1^2 = \left(\frac{a_1 + a_2}{R_1}\right)^2, \tag{3.43}$$

where R_1 represents the actual radius-vector at the moment of the outer contact of the eclipses and $a_{1,2}$ the actual radii of the components, while, at the lower limit,

$$\left(\frac{\delta}{R}\right)_0^2 = \sin^2\theta_{1,2}\sin^2 i + \cos^2 i, \tag{3.44}$$

where the phase angles $\theta_{1,2}$ of deepest eclipses are given by Equations (4.17) of Chapter IV.

By inserting from Equation (4.17) of that chapter in Equation (3.44) above we find that, correctly to terms of the order of e^2,

$$(\delta/R)_0^2 = (1 + e^2 \cos^2 \omega \cot^2 i) \cos^2 i, \tag{3.45}$$

which differs so little from $\cos^2 i$ (in fact, for $e \sim 0.1$ and $\cos i \sim 0.1$, by less than one part in 10^4) that the difference can as a rule be ignored. If so, however, it follows that *the empirical 'elliptic' moments of the light curves* (obtained by planimetry of a plot of $1 - l$ versus $\cos^{2m}(v + \omega)$) *can furnish the elements of the eclipse exactly as in the 'circular' case; care being only taken to note that the fractional radii resulting from such a solution are expressed in terms of the instantaneous radius-vector R_1 at the moment of outer* (first, or last) *contact of the respective eclipse.* Accordingly, each one of the alternating eclipses (in fact, each separate half of such an eclipse) should yield different values of the fractional radii r_1 and r_2, as these are measured in terms of a different unit of length. Conversely, *an identity of the resulting values of r_1 and r_2 reduced to the same unit of length then provides a sufficient proof of the correctness of the adopted values of e and ω.*

On the other hand, the resulting value of orbital inclination i should be identical for each type of eclipse (or a half thereof); and the same should be likewise true of the fractional luminosity L_1 of the eclipsed star.

Bibliographical Notes

A transfer of the solution of the light curves from the time- to the frequency-domain was first envisaged by Z. Kopal (1958); and its advantages outlined in subsequent publications (Kopal, 1960, 1962). In the 1960's a similar approach was advocated also by M. Kitamura (1965) or H. Mauder (1966); but the 'royal road' to the frequency-domain, as explained in the second part of this book, was not hit upon by the present writer until almost ten years later (see pp. 9ff of the *Introduction*); and its subsequent development will constitute the main subject of the last three chapters of this book.

V-1. The subject matter of this section follows (with some subsequent additions) its presentation in Z. Kopal (1975a); it was in that paper that the general usefulness of the 'moments of the light curves' A_{2m} for an analysis in the frequency-domain was first pointed out; and these moments evaluated – for every type of eclipses – in the case of uniformly bright discs. For a practical evaluation of the A_{2m}'s see also I. Jurkevich, W. W. Willman and A. F. Petty (1976).

V-2. The entire procedure outlined in the preceding section was subsequently generalized to eclipses of the stars exhibiting arbitrary (nonlinear) limb-darkening by Z. Kopal (1975b, c, d; 1976a); but it was not till with the formulation of the fractional loss of light α_l^0 in terms of the Hankel transforms that general expressions were obtained for the moments A_{2m} of the light curves, valid for any type of eclipses, as given in this section. The fundamental series (2.8) and (2.9) for the A_{2m}'s were obtained first by Z. Kopal (1977c).

Alternative expressions for the A_{2m}'s of the form (2.10), exhibiting faster convergence in certain domains of the eclipses, were subsequently obtained by O. Demircan (1978, 1979).

Closed-form expressions (2.25)–(2.27) for the A_{2m}'s, valid for total eclipses, have first been derived by Z. Kopal (1975b) by a more elementary method; but their derivation as given in this section follows that developed in the writer's subsequent paper (Z. Kopal, 1977c).

For the differential properties of the moments A_{2m} cf. Z. Kopal (1976a), where the derivation of the difference-differential equation (2.64) can be found stated for the first time; while the 'generating function' (2.69) for the A_{2m}'s goes back to P. Appell (1880).

An analysis of the atmospheric eclipses (in which the eclipsing components exhibit an arbitrary degree of transparence) follows a method initiated by S. A. H. Smith (1976) for the evaluation of the moments A_{2m} of the eclipses as given by Equations (2.79)–(2.80); though an alternative approach is possible via the Fourier transform $G(u, v)$ as given by Equation (3.21) of Chapter I (cf. Z. Kopal, 1977b). The exponential-atmosphere model considered as an example goes back to Z. Kopal (1945); cf. also Section IV.7 of the writer's treatise on *Close Binary Systems* (1959).

V-3. A solution for the elements of eclipsing binary systems in terms of the moments A_{2m} of the light curves for the case of total eclipses (where a solution can be obtained in a closed form) goes back to Z. Kopal (1975a); and, for partial eclipses (which require a resort to successive approximations) in Z. Kopal (1975a, d). A more general discussion of the problem (including combination of the alternate minima) was given by Z. Kopal and O. Demircan (1978).

For a discussion of the optimization of the solutions by a suitable choice of the frequencies m of the moments A_{2m} cf. T. Edalati (1978, 1979), or O. Demircan (1979); though a large part of the subject matter is new and has not previously been published.

For a treatment of eccentric systems in the frequency-domain cf. Z. Kopal and H. M. Al-Naimiy (1979).

CHAPTER VI

ANALYSIS OF THE LIGHT CHANGES IN THE FREQUENCY-DOMAIN: DISTORTED STARS

In the preceding chapter of this book we outlined the basic principles of the analysis of the light curves of eclipsing binary systems in the frequency-domain; and presented an explicit form of such an analysis appropriate for the case in which both components can be regarded as spheres. With the completion of this task the question is, however, bound to arise as to the extent to which a model consisting of spherical stars can be regarded as a satisfactory representation of eclipsing systems actually encountered in the sky; for axial rotation of their components and their mutual tidal distortion is bound to cause them to depart from spherical form. In many known eclipsing systems, to be sure, the components are separated widely enough – and, consequently, their distortion is sufficiently small – that (within the limits of observational errors) they can be regarded as spheres; and methods expounded in Chapter V applied to the observations as they stand.

However, if the two components are brought closer – and, apart from the greater intrinsic interest of such binaries, observational selection then strongly favours their discovery – a new situation arises which requires careful approach. A significant feature of 'close' eclipsing systems is the fact that observable light changes are no longer confined only to the times of the minima, but extend over the entire cycle. This is due to two reasons. First, since the components of close binaries possess, in general, the form of distorted ellipsoids with longest axes oriented constantly in the direction of the line joining their centres, their apparent cross-sections – and, therefore, the light – exposed to the observer should vary continuously in the course of a revolution ('ellipticity effect'). Secondly, it is also inevitable in close binaries that a fraction of the radiation of each component will be intercepted by its mate, to be absorbed and re-emitted (or scattered) in all directions – including that of the line of sight. The amount of light so 'reflected' by each component towards the observer will again vary with the phase ('reflection effect').

The changes of light arising from the ellipticity and reflection are independent of, and supplementary to, the light changes which arise if our binary system happens to be an eclipsing variable; and the magnitude of all increases with increasing proximity of the components. How to separate them, and disentangle one from the other?

In the time-domain – in which we plot the observed light intensities against the time or the phase angle – the customary procedure practised so far has been known under the name of 'rectification', and consisted of the following strategy. We separate (by inspection, or otherwise) such parts of the light curve around quadratures from those which are (or may be) affected by eclipses, and expand the observed luminosity l in a Fourier cosine series in terms of the multiples of the phase angle ψ. The empirically determined coefficients of such an expansion are then utilized (by addition or division) to 'free' the eclipse parts of the light curve from the proximity effects (ellipticity, reflection) in the hope that so rectified a light curve can be made to represent one arising from eclipses of two spherical stars, to which methods of solution appropriate for eclipses of circular discs could be applied.

Such a procedure – the only one available so far – suffers, however, from grave theoretical as well as practical difficulties as well as obscurities – which are indeed so grave that, at best, it opens the way to serious systematic errors which will propagate through the rest of the solution affecting its outcome to an unspecified extent; while, at worst, it may furnish results that are devoid of any actual meaning. In order to exemplify first the practical difficulties inherent in the process of rectification, consider the task of expanding the observed light intensity l as a series of the form

$$l = \sum_{j=0}^{n} c_j \cos^j \psi, \tag{0.1}$$

where the c_j's are constants to be subsequently used for rectification.

As many equations of condition of the form (0.1) can obviously be set up as there are data observed at different phase angles ψ, and solved for the coefficients c_j of the individual harmonic terms by least squares. Such a strategy – while possible in principle – encounters, however, grave difficulties in practice only too well known to those who attempted to do so for $n > 2$ – especially for close systems where the c_j's are apt to grow large. The cause of such difficulties is not difficult to trace, and goes back to the diminishing fraction of the light curve (i.e., a diminishing range of ψ) which lends itself legitimately for such a purpose. For reasonably well-separated systems the portion of the light curves which are unaffected by eclipses are usually well-marked (cf. Figure 6-1) and can be distinguished by inspection. However, for closer systems this is not necessarily the case (Figure 6-2); and a mere inspection of the observed light

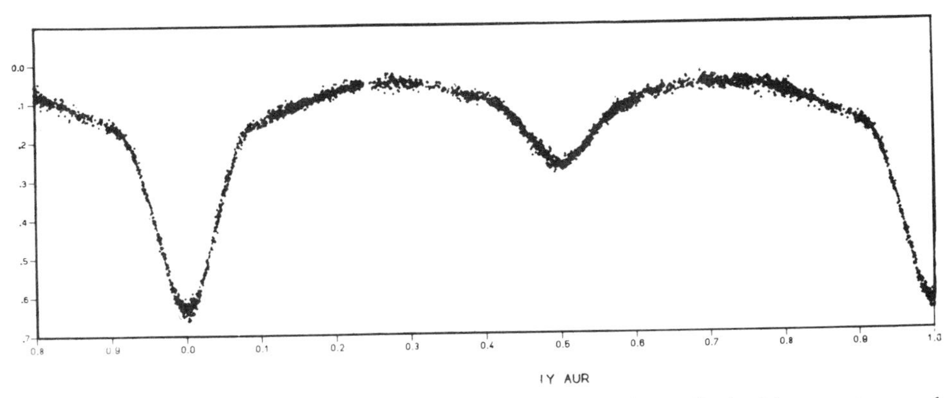

Fig. 6-1. Light changes exhibited by IY Aurigae – a moderately close eclipsing binary system – and reproduced from Fracastoro's *Atlas of the Light Curves of Eclipsing Binaries* (Torino, 1972) based on the observations by P. Tempesti.

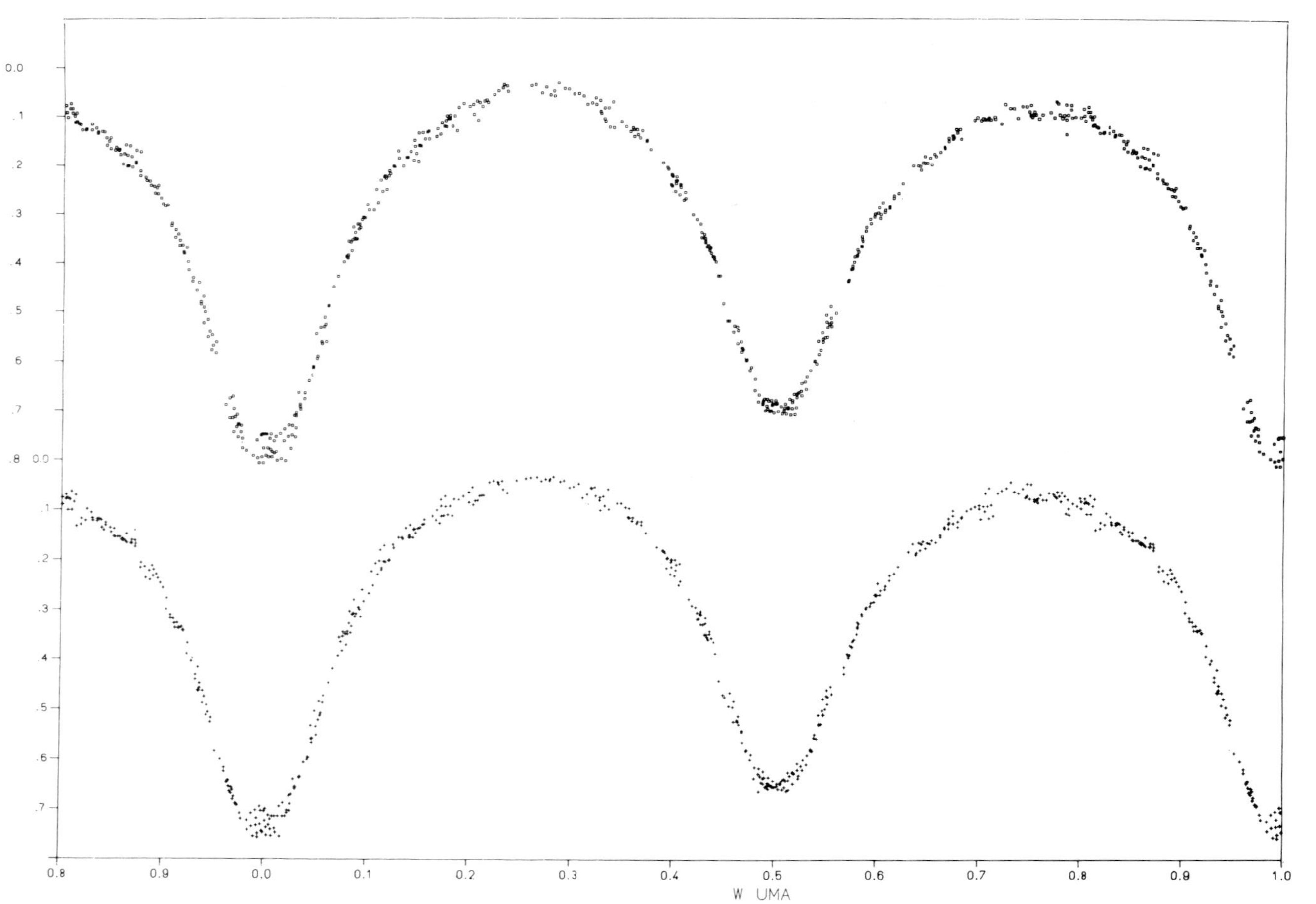

Fig. 6-2. Light changes exhibited by the close eclipsing system W Ursae Maioris, after L. Binnendijk (1966).

changes becomes of little avail for disclosing where the eclipses may begin or end.

It is, to be sure, known from the geometry of contact configurations based on the Roche model (cf. Kopal, 1954, 1972) that (for a very wide range of the mass-ratios exceeding 10:1) the light changes are unaffected by eclipses for all phase angles in excess of $\pm 58°$ – even if the plane of the orbit is perpendicular to the celestial sphere. Therefore, such light changes as are exhibited at phases of $\pm 32°$ around each quadrature must be due solely to the proximity effects, and can be used for their determination without fear of interference from eclipse phenomena.

This is, however, true only in theory, while in practice it becomes a very different story. A restriction of the entire half-cycle of 180° in phase to $\pm 32°$ around each quadrature entails adverse consequences for the determination of the coefficients c_j on the right-hand side of Equation (0.1) from known values of l – by least squares or otherwise – which with increasing number n of terms on the right side, become truly disastrous; the system can become so ill-conditioned that a right combination of the c_j's uncertain with very wide limits can match the observations without any distinguishable difference. The cause of this unpleasant phenomenon is the fact that although $\cos^j \psi$ (or $\cos j\psi$) constitute an orthogonal set of independent functions only over the closed interval ± 1 in $\cos \psi$ (or $0, \pi$ in ψ), they commence to simulate functional dependence as soon as we restrict this interval to a smaller range; and this fact is bound to bring about a loss of weight of the entire solution.

In order to demonstrate this in a practical case, consider a family of Tchebyshev polynomials $T_j(x)$ of the first kind in $x \equiv \cos \psi$, defined by

$$T_j(x) = \cos jt, \quad \cos t = (2x - b - a)/(b - a), \tag{0.2}$$

where a, b represent the limits of their orthogonality. Earlier in this section we mentioned that, for close binaries exhibiting continuous light variation throughout the cycle, it may not be safe to extend these limits beyond $90° \pm 32°$ in phase. If (in order to remain on the safe side) we restrict this range further to $\pm 30°$, it follows that $a, b = \cos(90° \pm 30°) = \pm\frac{1}{2}$, rendering

$$T_j(\cos \psi) = \cos j(\cos^{-1} 2 \cos \psi); \tag{0.3}$$

which for $j = 4, 5, 6, \ldots.$ assume the explicit forms

$$T_4(\cos \psi) = 128 \cos^4 \psi - 32 \cos^2 \psi + 1, \tag{0.4}$$

$$T_5(\cos \psi) = 512 \cos^5 \psi - 160 \cos^3 \psi + 10 \cos \psi, \tag{0.5}$$

$$T_6(\cos \psi) = 2048 \cos^6 \psi - 768 \cos^4 \psi + 72 \cos^2 \psi - 1, \tag{0.6}$$

etc.

Since, however, it follows from Equation (0.2) that

$$T_j(\cos \psi)| \leqslant 1 \tag{0.7}$$

for any argument $\cos \psi$ within the limits of orthogonality – and, therefore, the left-hand sides of Equations (0.4)–(0.6) cannot exceed unity – these equations can be solved to yield, approximately,

$$\cos^4 \psi = \tfrac{1}{4}\cos^2 \psi - \tfrac{1}{128}, \tag{0.8}$$

$$\cos^5 \psi = \tfrac{5}{16}\cos^3 \psi - \tfrac{5}{256}\cos \psi, \tag{0.9}$$

$$\cos^6 \psi = \tfrac{3}{8}\cos^4 \psi - \tfrac{9}{256}\cos^2 \psi + \tfrac{1}{2048}, \tag{0.10}$$

which within the range of $60° \leqslant \psi \leqslant 120°$ are subject to errors not exceeding 2^{-7}, 2^{-9} and 2^{-11}, respectively. Outside this range, the foregoing approximations rapidly deteriorate and eventually become useless, but within the range of interest higher powers of $\cos \psi$ are seen to simulate linear combinations of lower powers with remarkable fidelity; and this causes a system of equations of the form (1.1) to become ill-conditioned with increasing value of n.

That this is so has been experienced by many investigators in the past, who were forced by this fact to terminate the expansion on the right-hand side of Equation (0.1) with $n = 2$. This strategy has indeed been at the basis of 'rectification' of the light curves practised so far. Around the turn of this century – when the process of rectification was first introduced by Myers (1898) and Roberts (1903) – it was considered reasonable to regard the components of close binaries as similar ellipsoids of arbitrary shape which could indeed be described by a single second harmonic. However, when in the 1930's we progressed to a stage at which the shape of each component came to be regarded as a resultant of the rotational and tidal forces prevalent in the system, it became obvious that not one, but several partial tides must be taken into account, corresponding to $n = 2, 3, 4, \ldots$; and that the amplitudes of these tides will diminish with increasing n the more slowly, the closer the components of the systems. Under these conditions, an arbitrary amputation of the series on the right-hand side of Equation (0.1) at $n > 2$ would violate the physics underlying all photometric proximity effects, and cannot be accepted with equanimity in any work claiming serious attention.

But even if the underlying photometric observations were accurate enough to furnish significant values of the constants c_j for $n > 2$ in spite of the ill-conditioned nature of the linear system defining them, their possession would still not permit us to perform any meaningful rectification of the light curve – in the sense that, with their aid, we could convert the observed light curve into one to which methods of solution appropriate for mutual eclipses of circular discs could legitimately be applied. This can be justified only if the two components were similar ellipsoids exhibiting uniformly bright discs. In the case of nonuniform distribution of brightness caused by limb- and gravity-darkening, the legitimacy of 'rectification' proves to be severely restricted by requirements which were investigated only much later by Russell (1942) and – more completely – by Kopal (1945). The necessary condition – a symmetry of the respective isophotes with respect to the centre of the apparent discs at all phases – turned out to be met only for a certain non-realistic combination of the coefficients of limb- and gravity-darkening (cf. Kopal, 1945); and for stars suffering only from second-harmonic rotational or tidal distortion. With the emergence of higher harmonics commencing with the third, the symmetry of isophotes will be irretrievably lost whatever the amount of limb- or gravity-darkening – and with it the justification of any kind of rectification in the usual sense. A 'rectification' carried out under these conditions is really tantamount to a 'falsification' of the observational evidence, which is bound to obscure the true nature of the system rather than to elucidate it.

If, in the face of this situation, the use of rectification persisted throughout the years as a preparatory step for the solutions for the elements of the eclipse, this was due to the fact that nothing better could be offered up to the present to replace it. It was an acute awareness of the need to do so which led the present writer several years ago (Kopal, 1960) to envisage a procedure in which a specification of the proximity effects and a solution for the elements of the eclipse could be performed in parallel rather than consecutively. This task – which necessitated a fundamental re-orientation in the strategy of approach – is impossible to carry out in the time-domain; for no operation using empirical parameters can, in general, convert the observed light curve into one which can be solved directly for the elements of the eclipse. In the frequency-domain this task turns out, however, to be relatively easy: in fact – as we shall show in subsequent sections of this chapter – it can be carried out in a closed form and without the need of any auxiliary tables.

VI-1. Modulation of the Light Curves

In order to demonstrate the method by which this can be done, let us return to Chapter II of the first part of this book, in which we established that the observed variation of the light l of a close eclipsing system with phase θ can be expressed as

$$l = \mathfrak{L}_1 + \mathfrak{L}_2 - L_1 \Delta \mathfrak{L}, \tag{1.1}$$

where $\mathfrak{L}_{1,2}$ denote the luminosities of the two components, distorted by rotation and tides, as given by Equations (1.28) and (1.29) of Chapter II, and $\Delta \mathfrak{L}_1$ stands for the loss of light arising from an eclipse of the component of fractional luminosity L_1, as given by Equations (2.29)–(2.33) of that chapter.

The sum $\mathfrak{L}_1 + \mathfrak{L}_2$ of fractional luminosities of the distorted components of close binary systems can – within the framework of a first-order equilibrium theory of tides used as a basis of our treatment of the subject in Chapter II – be represented by Equations (1.50)–(1.53) of that chapter, and rewritten in the form

$$\mathfrak{L}_1 + \mathfrak{L}_2 = 1 + c_0 - \sum_{j=1}^{4} c_j \cos^j \psi, \tag{1.2}$$

where, by Equation (1.51) of Chapter II,

$$\begin{aligned} c_0 = {} & \tfrac{1}{3} \sum_{i=1}^{2} L_i (X_2^{(k)})_i (1 + \tau_i) v_i^{(2)} P_2(n_0) \\ & + \tfrac{1}{2} \sum_{i=1}^{2} L_i (X_2^{(k)})_i (1 + \tau_i) w_i^{(2)} - \tfrac{3}{8} \sum_{i=1}^{2} L_i (X_4^{(k)})_i (1 + \tfrac{1}{3} \tau_i) w_i^{(4)}, \end{aligned} \tag{1.3}$$

$$c_1 = -\tfrac{3}{2} \sin i \sum_{i=1}^{2} (-1)^{i+1} L_i (X_3^{(k)})_i (1 + \tfrac{1}{2}\tau_i) w_i^{(3)}, \tag{1.4}$$

$$c_2 = \tfrac{3}{2} \sin^2 i \sum_{i=1}^{2} L_i \{(X_2^{(k)})_i (1 + \tau_i) w_i^{(2)} - \tfrac{5}{4} (X_4^{(k)})_i (1 + \tfrac{1}{3}\tau_i) w_i^{(4)}\}, \tag{1.5}$$

$$c_3 = \tfrac{5}{2} \sin^3 i \sum_{i=1}^{2} (-1)^{i+1} L_i (X_3^{(k)})_i (1 + \tfrac{1}{2}\tau_i) w_i^{(3)}, \tag{1.6}$$

$$c_4 = \tfrac{35}{8} \sin^4 i \sum_{i=1}^{2} L_i (X_4^{(k)})_i (1 + \tfrac{1}{3}\tau_i) w_i^{(4)}, \tag{1.7}$$

correctly to terms of first order in rotational and tidal distortion (but terms arising from the reflection effect yet to be added); all notations being strictly consistent with those used in Chapter II.

On the other hand, the loss of light $\Delta \mathfrak{L}$ arising from eclipses can, in accordance with Equations (2.29) and (2.30) of Chapter II, be expressed as

$$\Delta \mathfrak{L} = \sum_{h=1}^{\Lambda+1} C^{(h)} \{\alpha_{h-1}^{0} + f_{*}^{(h)} + f_1^{(h)} + f_2^{(h)}\}, \tag{1.8}$$

where the first term in curly brackets stands for the fractional loss of light due to eclipse of an arbitrarily limb-darkened star of spherical shape; while the f-terms stand – as defined by Equations (2.31)–(2.33) of that chapter – represent the effects of distortion.

Accordingly, a loss of light $l(\frac{1}{2}\pi) - l(\psi)$ observed at a phase ψ should be expressible as

$$l(\tfrac{1}{2}\pi) - l(\psi) = -\sum_{j=1}^{n} c_j \cos^j \psi + L_1 \sum_{h=1}^{k+1} C^{(h)} \alpha_{h-1}^{0} + L_1 \sum_{h=1}^{k+1} C^{(h)} \times \{f_{*}^{(h)} + f_1^{(h)} + f_2^{(h)}\}. \tag{1.9}$$

Should the system under consideration consist of spherical stars, all the c_j's as well as $f^{(h)}$'s vanish identically, and

$$l(\tfrac{1}{2}\pi) - l(\psi) = L_1 \sum_{h=1}^{\Lambda+1} C^{(h)} \alpha_{h-1}^{0} \equiv L_1 \alpha \tag{1.10}$$

in accordance with Equation (1.2) of Chapter V. Should, on the other hand, the inclination i of the orbital plane to the line of sight be such that no eclipses take place, and the observed variation of light is due to axial rotation of tidally-distorted components of a close binary system, it is the α's and $f^{(h)}$'s that remain identically zero, and Equation (1.9) reduces to

$$l(\tfrac{1}{2}\pi) - l(\psi) = -\sum_{j=1}^{n} c_j \cos^j \psi. \tag{1.11}$$

Moreover, within the framework of a first-order theory of tides the summation on the r.h.s. of Equation (1.11) can be terminated with $n = 4$.

In reality, an inspection of such light curves as shown on Figure 6-2 will generally not disclose in advance whether or not the observed variation of light of the respective binary is caused by distortion of its components alone, or is complicated further by eclipses; this can be established only by an appropriate analysis along the following lines. In order to do so, let us integrate both sides of Equation (1.9) with respect to $\sin^{2m}\psi$ between 0 and 90°, and regard

$$\int_0^{\pi/2} \{l(\tfrac{1}{2}\pi) - l(\psi)\}\, d(\sin^{2m}\psi) \equiv \bar{A}_{2m} \tag{1.12}$$

as a 'generalized' moment of the light curve, equal to the area cross-hatched on Figure 6-3 and, as such, obtainable (by planimetry or otherwise) from the observed data. Moreover, by direct integration we easily find that

$$\int_0^{\pi/2} \cos^j\psi\, d(\sin^{2m}\psi) = mB(m, \tfrac{1}{2}j + 1). \tag{1.13}$$

If, lastly we denote by

$$L_1 \sum_{h=1}^{\Lambda+1} C^{(h)} \int_0^{\pi/2} \{f_*^{(h)} + f_1^{(h)} + f_2^{(h)}\}\, d(\sin^{2m}\psi) \equiv \mathfrak{P}_{2m}; \tag{1.14}$$

the 'photometric perturbations' arising from distortion in the course of an eclipse of the star of fractional luminosity L_1, the integral of Equations (1.9) with respect to $\sin^{2m}\psi$ between 0 and 90° can be rewritten as

$$\bar{A}_{2m} = -m \sum_{j=1}^{4} B(m, \tfrac{1}{2}j + 1)c_j + A_{2m} + \mathfrak{P}_{2m}, \tag{1.15}$$

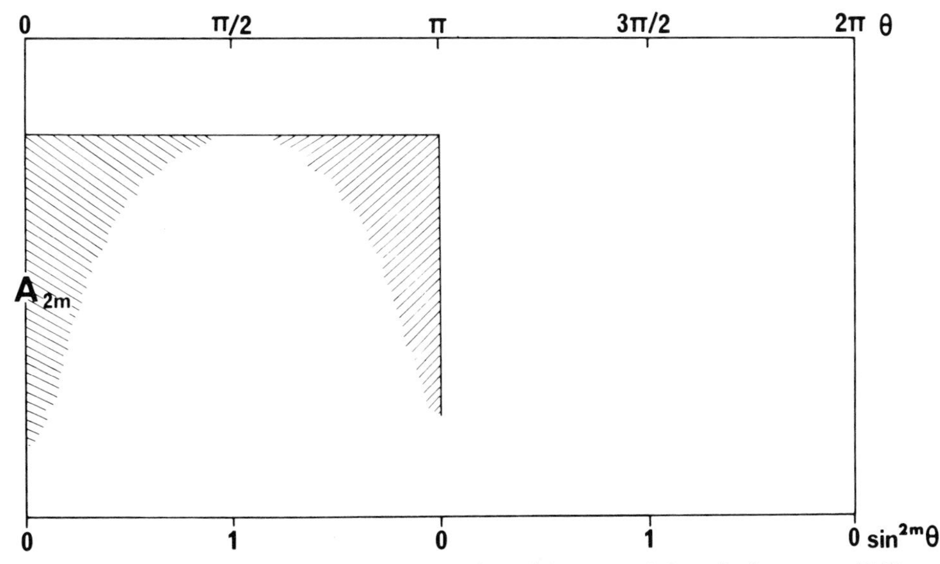

Fig. 6-3. Areas representing the moments A_{2m} of the light curve of the eclipsing system W Ursae Maioris, shown on Figure 6-2. Separate moments can be determined for the primary and secondary minima.

where

$$A_{2m} = \sum_{h=1}^{\Lambda+1} C^{(h)} \int_0^{\pi/2} \alpha_{h-1}^0 \, d(\sin^{2m} \psi) \qquad (1.16)$$

becomes identical with the 'moments of the light curves' A_{2m} due to mutual eclipses of spherical stars, and introduced by Equation (1.1) of Chapter V. This identity becomes indeed obvious if we stop to realize that the upper limit of $\frac{1}{2}\pi$ on the r.h.s. of Equation (1.16) can be reduced to the angle ψ_1 of first contact of the eclipse since, for $\psi > \psi_1$, all associated α-functions in the integrand on the r.h.s. of Equation (1.16) vanish); and the same is, of course, true of the integral on the l.h.s. of Equation (1.14) as well.

In Chapter V we demonstrated that the geometrical elements of the eclipses of spherical stars can be determined uniquely in terms of the moments A_{2m} deducible directly from the observed light curves. If, however, the stars eclipsing each other are distorted, the moments directly obtainable from the light curve are the $\bar{A}_{2m}$'s as defined by Equation (1.12); and related with the A_{2m}'s by Equation (1.15). The task of determining the elements of the eclipses of close (i.e., distorted) binary systems presupposes, therefore, our ability to convert the 'observed' moments $\bar{A}_{2m}$ into their 'rectified' values A_{2m}; for once we have done so, all desired elements can be obtained by methods expounded already in Chapter V.

The only tool available to us for such a conversion is represented by Equation (1.15); and it should enable us to accomplish our end if the observed moments $\bar{A}_{2m}$ can be freed from the effects of photometric perturbations within minima as well as between eclipses. How do these compare in magnitude with the A_{2m}'s The fact (which will be established in the next section) that, numerically, $A_{2m} \gg \mathfrak{P}_{2m}$ should permit an iterative approach to the solution of our problem (in which $\mathfrak{P}_{2m}$, ignored at first, is transposed to the l.h.s. of Equation (1.15) in subsequent approximations); but is the same true also of the first term on the r.h.s. of Equation (1.15) involving the c_j's? This cannot in general be asserted: in point of fact, *for close systems exhibiting shallow eclipses the weighted summation of the* c_j's *may dominate the r.h.s. of Equation* (1.15) *by its numerical magnitude*; and if so, their neglect would render any attempt at an iterative solution of Equation (1.15) for the true values of A_{2m} meaningless. In such a case, *a solution for the elements based on Equation* (1.15) *becomes possible only if the value of the* c_j's *on the r.h.s. of Equation* (1.15) *can be ascertained prior to the actual solution*, and transposed to the left-hand side of that equation as an already known quantity.

How to accomplish this? Fortunately, it is at this critical point that Nature lends us a helpful hand in the following manner. It has been known for many years (cf. Kopal, 1954; or, more recently, 1972 or 1978b) that if the shape of the components of a close binary can be represented by the respective equipotentials of the Roche model (*op. cit.*), it follows from the geometry of such surfaces that no eclipses can occur – even if the components are large enough to be in actual contact – for phase angles ψ in excess of 57°31 for a mass-ratio $q = 1$ even when the line of sight lies in the orbital plane (a circumstance which maximizes duration of the eclipses); and increases but slowly with increasing value of q. According to Chanan *et al.* (1976), the limiting value of ψ does not exceed 59° till for $q = 5$, and 60°.6 for $q = 10$. Therefore, *for virtually all close binaries observed in the sky we can anticipate that their eclipses – if any – will not extent beyond* $\psi = \pm 60°$ *in phase angle, on either side of the conjunctions, even if both components constitute a contact system*; and this range may become considerably smaller if the orbital inclination $i \ll 90°$.

If so, however, it follows that, at least for $\psi = \pm 30°$ of either quadrature and possibly more, Equation (1.9) must reduce to Equation (1.11) containing the c_j's as the only unknowns. In order to determine these from this fraction of the light curve, *let us 'modulate' the light changes* $l(\frac{1}{2}\pi) - l(\psi)$ *by the Legendre polynomials* $P_k(\cos\psi)$ *orthogonalized between limits within which Equation* (1.9) *can be represented by Equation* (1.11). If, moreover, we introduce a new family of 'modulated moments' $B_k^{(a)}$, defined by

$$B_{(k)}^{(a)} = \int_{-a}^{a} \{l(\tfrac{1}{2}\pi) - l(\psi)\} P_k(\cos\psi)\, \mathrm{d}(\cos\psi), \tag{1.17}$$

and set

$$\cos\psi \equiv x, \quad \cos\psi_1 = a, \tag{1.18}$$

it follows that

$$P_k(x) \equiv \frac{1}{(2a)^k (k!)} \frac{\mathrm{d}^k}{\mathrm{d}x^k} (x^2 - a^2)^k \tag{1.19}$$

are identical with the ordinary Legendre polynomials (orthogonalized between ± 1) of the argument x/a.

The empirical values of the moments (1.17) of modulated light changes can be again obtained by planimetry of the respective parts of the light curve around the quadratures. On the other hand, by insertion from Equation (1.11) in (1.17) it follows that, theoretically, the moments $B_k^{(a)}$ should be related with the n unknown constants c_j by the equation

$$B_k^{(a,n)} = -\sum_{j=1}^{n} c_j \int_{-a}^{a} x^j P_k\left(\frac{x}{a}\right) \mathrm{d}x. \tag{1.20}$$

In order to evaluate the integrals on the right-hand side of the foregoing equation, let us remember that the Legendre coefficients $P_k(x/a)$ are polynomials of the kth degree in their arguments. Inverting these, we can obviously find any integral power of x by the identity

$$x^j \equiv \sum_{i=0}^{j} \gamma_i^{(j)}(a) P_i(x/a) \tag{1.21}$$

for $j = 0, 1, 2, \ldots, k$. In view of the symmetry $\pm a$ of the limits of integration, for even values of j the coefficients $\gamma_i^{(j)} = 0$ for odd i's and if j is odd, the $\gamma_i^{(j)}$'s again vanish for all even values of i. Moreover, $\gamma_i^{(j)} = 0$ for $i > j$ because of the orthogonality of Legendre polynomials.

Inserting Equation (1.21) in the right-hand side of Equation (1.20) and taking advantage of the orthogonality conditions satisfied by the P_k's, we readily find from Equation (1.20) that, for $j \geqslant k$,

$$B_k^{(a,n)} = -\frac{2a}{2k+1} \sum_{j=1}^{n} \gamma_k^{(j)} c_j = -\frac{2a}{2k+1} \sum_{j=k}^{n} \gamma_k^{(j)} c_j \tag{1.22}$$

since, for $j > k$, $\gamma(j)k = 0$. The foregoing Equation (1.22) represents a simultaneous system of n linear algebraic equations for a determination of n unknown constants c_j in terms of the empirical 'modulated' moments $B_k^{(a,n)}$ of the light curves between minima, as defined by Equation (1.17). If we solve them, the weighted sum of the respective c_j's can then be likewise ascertained from known values of the B's; and its transposition to the left-hand side of Equation (1.15) completes the first part of the prerequisites for the solution for the elements of the eclipses from the 'rectified' moments A_{2m}.

A. PARTICULAR CASES

In conclusion of this section let us particularize our preceding results to the cases in which the observed light changes are expected to be free of any eclipse effects in the following three ranges of the phase angle:

(1) $60° \leqslant \psi \leqslant 120°$, $a = \frac{1}{2}$;
(2) $45° \leqslant \psi \leqslant 135°$, $a = \sqrt{2}/2$;
(3) $30° \leqslant \psi \leqslant 150°$, $a = \sqrt{3}/2$.

For reasons already stated, the case (1) should always be fulfilled by close binaries regardless of the proximity of their components or the inclination of their orbital planes; while cases (2) and (3) may be fulfilled for more moderately distorted and better-separated pairs – as well as for contact systems whose components revolve in planes more inclined to the line of sight.

Let us consider case (1) first to the accuracy of first-order in surficial distortion (corresponding to $n = 4$). The first four corresponding Legendre polynomials to be used for modulation will be of the form

$$P_1(2x) = 2x, \tag{1.23}$$

$$P_2(2x) = 6x^2 - \tfrac{1}{2}, \tag{1.24}$$

$$P_3(2x) = 20x^3 - 3x, \tag{1.25}$$

$$P_4(2x) = 70x^4 - 15x^2 + \tfrac{3}{8}. \tag{1.26}$$

Inverting these we find that

$$x = \tfrac{1}{2} P_1, \tag{1.27}$$

$$x^2 = \tfrac{1}{6} P_2 + \tfrac{1}{12}, \tag{1.28}$$

$$x^3 = \tfrac{1}{20} P_3 + \tfrac{3}{40} P_1, \tag{1.29}$$

$$x^4 = \tfrac{1}{70}P_4 + \tfrac{1}{28}P_2 + \tfrac{1}{80}, \tag{1.30}$$

as particular cases of Equation (1.21) specifying the coefficients $\gamma_i^{(j)}$.

With the aid of these coefficients, Equation (1.20) assumes, for $k = 1(1)4$, the explicit forms

$$B_1^{(1/2,4)} = -\tfrac{1}{6}c_1 - \tfrac{1}{40}c_3, \tag{1.31}$$

$$B_2^{(1/2,4)} = -\tfrac{1}{30}c_2 - \tfrac{1}{140}c_4, \tag{1.32}$$

$$B_3^{(1/2,4)} = -\tfrac{1}{140}c_3, \tag{1.33}$$

$$B_4^{(1/2,4)} = -\tfrac{1}{630}c_4; \tag{1.34}$$

and their solution yields

$$c_1 = -6B_1 \qquad + 21B_3 \tag{1.35}$$

$$c_2 = \qquad -30B_2 \qquad + 135B_4 \tag{1.36}$$

$$c_3 = \qquad\qquad -140B_3 \tag{1.37}$$

$$c_4 = \qquad\qquad\qquad -630B_4. \tag{1.38}$$

In accordance with Equation (1.17), these four equations can be expressed in the form

$$c_j = \int_{-a}^{a} \{l(\tfrac{1}{2}\pi) - l(\psi)\}\mathscr{P}_j^{(a,n)}(x)\,\mathrm{d}x, \tag{1.39}$$

where, for $j = 1(1)4$, and $a = \frac{1}{2}$,

$$\mathscr{P}_1^{(1/2,4)}(x) = -6P_1(2x) + 21P_3(2x) = 15(28x^3 - 5x), \tag{1.40}$$

$$\mathscr{P}_2^{(1/2,4)}(x) = -30P_2(2x) + 135P_4(2x) = \tfrac{105}{8}(720x^4 - 168x^2 + 5), \tag{1.41}$$

$$\mathscr{P}_3^{(1/2,4)}(x) = -140P_3(2x) = -140(20x^3 - 3x), \tag{1.42}$$

$$\mathscr{P}_4^{(1/2,4)}(x) = -630P_4(2x) = -\tfrac{315}{4}(560x^4 - 120x^2 + 3). \tag{1.43}$$

Therefore, in order to obtain the numerical values of the constants c_j of the expansion on the right-hand side of Equation (1.11), it is not really necessary to evaluate the modulated moments $B_k^{(a,n)}$ separately. All we need to do is *to modulate the observed light curve directly with the polynomials* $\mathscr{P}_k^{(a,n)}(x)$, resulting as linear combinations of the B's, given (in the present case) by the foregoing Equations (1.40)–(1.43). In either case, the procedure requires an equal number of quadratures; but those modulated by $\mathscr{P}_k$ rather than P_k lead more directly to the desired results.

Lastly, should we be willing to forego a knowledge of the individual values of c_j and be interested solely in that of their weighted sum occurring on the r.h.s. of (1.15) we may indeed find it from the equation

$$m\sum_{j=1}^{n} B(m, \tfrac{1}{2}j + 1)c_j = \int_{-a}^{a} \{l(\tfrac{1}{2}\pi) - l(\psi)\}\mathscr{Q}_m^{(a,n)}(x)\,\mathrm{d}x, \tag{1.44}$$

where it follows by use of Equations (1.39)–(1.43) that, for $a = \frac{1}{2}$, $n = 4$ and $m = 1, 2, 3$,

$$\mathcal{Q}_1^{(1/2,4)}(x) = -9975x^4 - 840x^3 + \tfrac{4095}{2}x^2 + 118x - \tfrac{735}{16}, \tag{1.45}$$

$$\mathcal{Q}_2^{(1/2,4)}(x) = -4200x^4 - 416x^3 + 840x^2 + 56x - \tfrac{35}{2}, \tag{1.46}$$

and

$$\mathcal{Q}_3^{(1/2,4)}(x) = -\tfrac{4095}{2}x^4 - \tfrac{704}{3}x^3 + \tfrac{1575}{4}x^2 + \tfrac{208}{7}x - \tfrac{231}{32}. \tag{1.47}$$

A generalization of all foregoing results to quantities of second ($n = 7$) and higher orders constitutes a straightforward algebra, and can be left as an exercise for the interested reader.

Next, let us turn our attention to case (2), applicable if the widths of the eclipses do not exceed 45° – so that $a = \cos 45° = \sqrt{2}/2$. The Legendre polynomials $P_k(x/a)$ on the right-hand side of Equation (1.20) will then be

$$P_1(x\sqrt{2}) = x\sqrt{2}, \tag{1.48}$$

$$P_2(x\sqrt{2}) = 3x^2 - \tfrac{1}{2}, \tag{1.49}$$

$$P_3(x\sqrt{2}) = \sqrt{2}(5x^3 - \tfrac{3}{2}x), \tag{1.50}$$

$$P_4(x\sqrt{2}) = \tfrac{35}{2}x^4 - \tfrac{15}{2}x^2 + \tfrac{3}{8}; \tag{1.51}$$

yielding

$$x = \frac{1}{\sqrt{2}}P_1, \tag{1.52}$$

$$x^2 = \tfrac{1}{3}P_2 + \tfrac{1}{6}, \tag{1.53}$$

$$x^3 = \frac{1}{5\sqrt{2}}P_3 + \frac{3}{10\sqrt{2}}P_1, \tag{1.54}$$

$$x^4 = \tfrac{2}{35}P_4 + \tfrac{1}{7}P_2 + \tfrac{1}{20}, \tag{1.55}$$

In such a case, Equation (1.20) yields

$$B_1^{(\sqrt{2}/2,4)} = -\tfrac{1}{3}c_1 - \tfrac{1}{10}c_3, \tag{1.56}$$

$$B_2^{(\sqrt{2}/2,4)} = -\frac{\sqrt{2}}{15}c_2 - \frac{\sqrt{2}}{35}c_4, \tag{1.57}$$

$$B_3^{(\sqrt{2}/2,4)} = -\tfrac{1}{35}c_3, \tag{1.58}$$

$$B_4^{(\sqrt{2}/2,4)} = -\frac{2\sqrt{2}}{315}c_3; \tag{1.59}$$

the solution of which yields

$$c_1 = -3B_1 \qquad + \frac{21}{2}B_3 \tag{1.60}$$

$$c_2 = \quad -\frac{15}{\sqrt{2}} B_2 \quad + \frac{135}{2\sqrt{2}} B_4 \tag{1.61}$$

$$c_3 = \quad -35 B_3 \tag{1.62}$$

$$c_4 = \quad -\frac{315}{2\sqrt{2}} B_4. \tag{1.63}$$

All these can, furthermore, be expressed again in terms of a quadrature given by Equation (1.39), where now

$$\begin{aligned} \mathcal{P}_1^{(\sqrt{2}/2,4)}(x) &= -3P_1(x\sqrt{2}) + \tfrac{21}{2} P_3(x\sqrt{2}) \\ &= \frac{15}{2\sqrt{2}}(14x^3 - 5x), \end{aligned} \tag{1.64}$$

$$\begin{aligned} \mathcal{P}_2^{(\sqrt{2}/2,4)}(x) &= -\frac{15}{\sqrt{2}} P_2(x\sqrt{2}) + \frac{135}{2\sqrt{2}} P_4(x\sqrt{2}) \\ &= \frac{105}{16\sqrt{2}}(180x^4 - 84x^2 + 5), \end{aligned} \tag{1.65}$$

$$\mathcal{P}_3^{(\sqrt{2}/2,4)}(x) = -35 P_3(x\sqrt{2}) = -\frac{70}{\sqrt{2}}(5x^3 - \tfrac{3}{2}x), \tag{1.66}$$

$$\mathcal{P}_3^{(\sqrt{2}/2,4)}(x) = -\frac{315}{\sqrt{2}} P_4(x\sqrt{2}) = -\frac{315}{16\sqrt{2}}(140x^4 - 60x^2 + 3). \tag{1.67}$$

Eventually, the polynomials $\mathcal{Q}_m^{(a,n)}(x)$ on the right-hand side of Equation (1.44) assume the forms

$$\mathcal{Q}_1^{(\sqrt{2}/2,4)}(x) = \sqrt{2}(-\tfrac{2625}{16}x^4 - 35x^3 + \tfrac{945}{16}x^2 + \tfrac{17}{2}x - \tfrac{105}{64}), \tag{1.68}$$

$$\mathcal{Q}_2^{(\sqrt{2}/2,4)}(x) = \sqrt{2}(-\tfrac{525}{16}x^4 - 12x^3 + \tfrac{105}{16}x^2 + 2x + \tfrac{35}{64}), \tag{1.69}$$

$$\mathcal{Q}_3^{(\sqrt{2}/2,4)}(x) = \sqrt{2}(\tfrac{315}{32}x^4 - \tfrac{8}{3}x^3 - \tfrac{315}{32}x^2 - \tfrac{4}{7}x + \tfrac{147}{128}). \tag{1.70}$$

If, lastly, the widths of the eclipses do not exceed 30° – as is generally the case in moderately distorted systems, for which the duration of the minima can safely be estimated from the observed light curve, and for which $a = \cos 30° = \frac{\sqrt{3}}{2}$ – the relevant Legendre polynomials are of the form

$$P_1\left(\frac{2x}{\sqrt{3}}\right) = \frac{2x}{\sqrt{3}}, \tag{1.71}$$

$$P_2\left(\frac{2x}{\sqrt{3}}\right) = 2x^2 - \tfrac{1}{2}, \tag{1.72}$$

$$P_3\left(\frac{2x}{\sqrt{3}}\right) = \sqrt{3}(\tfrac{20}{9}x^3 - x), \tag{1.73}$$

$$P_4\left(\frac{2x}{\sqrt{3}}\right) = \tfrac{70}{9}x^4 - 5x^2 + \tfrac{3}{8}; \tag{1.74}$$

yielding

$$x = \frac{\sqrt{3}}{2} P_1, \tag{1.75}$$

$$x^2 = \tfrac{1}{2} P_2 + \tfrac{1}{4}, \tag{1.76}$$

$$x^3 = \frac{3\sqrt{3}}{20} P_3 + \frac{9\sqrt{3}}{40} P_1, \tag{1.77}$$

$$x^4 = \tfrac{9}{70} P_4 + \tfrac{9}{28} P_2 + \tfrac{9}{80}. \tag{1.78}$$

In consequence,

$$B_1^{(\sqrt{3}/2,4)} = \tfrac{1}{2} c_1 - \tfrac{9}{40} c_3, \tag{1.79}$$

$$B_2^{(\sqrt{3}/2,4)} = -\frac{\sqrt{3}}{10} c_2 - \frac{9\sqrt{3}}{140} c_4, \tag{1.80}$$

$$B_3^{(\sqrt{3}/2,4)} = -\tfrac{9}{140} c_3, \tag{1.81}$$

$$B_4^{(\sqrt{3}/2,4)} = -\frac{\sqrt{3}}{70} c_4; \tag{1.82}$$

from which it follows that

$$c_1 = -2B_1 \qquad + 7B_3 \tag{1.83}$$

$$c_2 = \qquad -\frac{10}{\sqrt{3}} B_3 \qquad + \frac{45}{\sqrt{3}} B_4 \tag{1.84}$$

$$c_3 = \qquad -\tfrac{140}{9} B_3 \tag{1.85}$$

$$c_4 = \qquad -\frac{70}{\sqrt{3}} B_4, \tag{1.86}$$

all expressible again in the form (1.39) where

$$\begin{aligned} \mathscr{P}_1^{(\sqrt{3}/2,4)}(x) &= -2P_1\left(\frac{2x}{\sqrt{3}}\right) + 7P_3\left(\frac{2x}{\sqrt{3}}\right) \\ &= \frac{5}{3\sqrt{3}}(28x^3 - 15x), \end{aligned} \tag{1.87}$$

$$\begin{aligned} \mathscr{P}_2^{(\sqrt{3}/2,4)}(x) &= -\frac{10}{\sqrt{3}} P_2\left(\frac{2x}{\sqrt{3}}\right) + \frac{45}{\sqrt{3}} P_4\left(\frac{2x}{\sqrt{3}}\right) \\ &= \frac{35}{8\sqrt{3}}(80x^4 - 56x^2 + 5), \end{aligned} \tag{1.88}$$

$$\begin{aligned} \mathscr{P}_3^{(\sqrt{3}/2,4)}(x) &= -\frac{140}{9} P_3\left(\frac{2x}{\sqrt{3}}\right) \\ &= -\frac{140}{27\sqrt{3}}(20x^3 - 9x), \end{aligned} \tag{1.89}$$

$$\mathscr{P}_4^{(\sqrt{3}/2,4)}(x) = -\frac{70}{\sqrt{3}} P_4\left(\frac{2x}{\sqrt{3}}\right)$$
$$= -\frac{35}{36\sqrt{3}}(560x^4 - 360x^2 + 27). \tag{1.90}$$

The weighted sums of the c_j's can then be evaluated again directly from Equation (1.44) with the aid of the polynomials

$$\sqrt{3}\mathscr{Q}_1^{(\sqrt{3}/2,4)}(x) = -\tfrac{175}{27}x^4 - \tfrac{280}{27}x^3 - \tfrac{35}{6}x^2 + 2x + \tfrac{35}{16}, \tag{1.91}$$

$$\sqrt{3}\mathscr{Q}_2^{(\sqrt{3}/2,4)}(x) = \tfrac{700}{27}x^4 + \tfrac{32}{27}x^3 - \tfrac{70}{3}x^2 - \tfrac{72}{27}x + \tfrac{35}{12}, \tag{1.92}$$

and

$$\sqrt{3}\mathscr{Q}_3^{(\sqrt{3}/2,4)}(x) = \tfrac{595}{18}x^4 + \tfrac{448}{81}x^3 - \tfrac{105}{4}x^2 - \tfrac{272}{63}x + \tfrac{91}{32}. \tag{1.93}$$

This completes an evaluation of the constants c_j, the contributions of which dominate the r.h.s. of Equation (1.15), by a 'modulation' of those parts of the light curve which are free from eclipse effects. The actual range $\pm a$ to adopt for this purpose must be ascertained by an inspection of the respective light curve. If the light changes are continuous and exhibit no obvious indication of the onset of an eclipse, we can never go wrong by adopting the range $60° < \psi < 120°$ ($a = 0.5$) calling for the use of Equations (1.35)–(1.44). Should, however, a discontinuity in slope of the light curve indicate the phase angle ψ_1 at which eclipses commence, either one of the ranges $45° < \psi < 135°$ ($a = \sqrt{2}/2$) calling for Equations (1.60)–(1.70) or $30° < \psi < 150°$ ($a = \sqrt{3}/2$) with Equations (1.84)–(1.94) may be chosen provided that ψ_1 lies outside it. The more extended the range, the greater the accuracy with which the values of the c_j's can be determined by a modulation of the light curve. On the other hand, should we allow ψ_2 to lie within the adopted range, a determination of the c_j's may be vitiated by eclipse effects.

It should also be noted that, inasmuch as these constants occur on the r.h.s. of Equation (1.15) only through their sum (1.44), the value of this sum could also be obtained directly by an analogous modulation of the uneclipsed part of the light curve – by use of the polynomials as given by Equations (1.45)–(1.47), (1.68)–(1.70), or (1.91)–(1.93) replacing the $\mathscr{P}_j^{(a,n)}$'s on the r.h.s. of Equation (1.39) – without seeking to specify the c_j's individually. The latter possesses, however, a distinct information content of their own; which can be utilized in due course – for instance, for a specification of the amount of gravity-darkening of the respective stars, or of their mass-ratio, with the aid of Equations (1.4)–(1.7). Therefore, an individual determination of the c_j's and subsequent formation of the sum (1.44) from them should generally be preferable. Should the sum so obtained turn out to be sensibly equal to the corresponding empirical moments A_{2m} (thereby implying the quantities A_{2m} as well as $\mathfrak{P}_{2m}$ to be negligible), this would signify that *the respective system does not eclipse*, and that its observed light changes are due solely to the proximity effects.

Suppose, however, that this is not the case, and that the A_{2m}'s result from Equation (1.15) is significant amounts. Before we can invoke the use of that equation to free the A_{2m}'s from all effects of distortion, it remains still to ascertain the magnitude of the 'photometric perturbations' $\mathfrak{P}_{2m}$, arising from distortion, in terms of the elements of the eclipse; and to this task we shall now turn our attention.

VI-2. Photometric Perturbations

In order to accomplish this task, let us return to Equation (1.14) and rewrite it as

$$\mathfrak{P}_{2m} = L_1 \sum_{h=1}^{\Lambda+1} C^{(h)} \mathfrak{P}_{2m}^{(h)}, \tag{2.1}$$

where

$$\mathfrak{P}_{2m}^{(h)} = \int_0^{\psi_1} \{f_*^{(h)} + f_1^{(h)} + f_2^{(h)}\} \, \mathrm{d}(\sin^{2m} \psi) \tag{2.2}$$

are photometric perturbations appropriate for a h-th term in the adopted law of limb-darkening; and the symbols $f_*^{(h)}, f_{1,2}^{(h)}$ continue to stand for the expressions on the right-hand sides of Equations (2.31)–(2.33) of Chapter II. As all these vanish outside eclipses, the limits of integration on the r.h.s. of Equation (2.2) above can be restricted from $(0, \frac{1}{2}\pi)$ to $(0, \psi_1)$.

The terms in $f_*^{(h)}$ arising from the *tidal* distortion are of the form $l_0^\iota l_2^\kappa \alpha_\lambda^\kappa$, in which the direction cosines l_0, l_2 are given by Equation (2.1) of Chapter II, The letters ι, κ, λ stand for integers (including zero) not exceeding the order j of spherical-harmonic symmetry of a partial tide giving rise to photometric perturbations; and as long as we restrict ourselves to distortion terms of first order, $j = 2$, 3 and 4. Moreover, the structure of $f_*^{(h)}$ is such that $\iota + \kappa = j$ while $\lambda = -1(0)j + \Lambda$ where Λ stands for the degree of the adopted law of limb-darkening. Accordingly, all parts of the perturbations $\mathfrak{P}_{2m}^{(h)}$ arising from terms in $f_*^{(h)}$ of tidal origin should be expressible in terms of integrals of the form

$$A_{\lambda,m}^{\iota,\kappa} = \int_0^{\psi_1} l_0^\iota l_2^\kappa \alpha_\lambda^\kappa \, \mathrm{d}(\sin^{2m} \psi), \tag{2.3}$$

where m is a positive number (not necessarily an integer).

Turning next to the tidal terms in $f_1^{(h)}$ as given by Equation (2.32) of Chapter II, we may note that a resort to Equation (2.17) of Chapter III should permit us to rewrite them also in terms of the products $l_0^\iota l_2^\kappa \alpha_\lambda^\kappa$. Therefore, their contributions to $\mathfrak{P}_{2m}^{(h)}$ should be likewise expressible in terms of the $A_{\lambda,m}^{\iota,\kappa}$'s as given by Equation (2.3). However, the terms in $f_2^{(h)}$ arising from tidal distortion of the shadow cylinder cast by the eclipsing component will (cf. Equation (2.33) of Chapter II) be of the form $l_2 I_{-1,\lambda}^\kappa$; and, therefore, their contributions to $\mathfrak{P}_{2m}^{(h)}$ will be of the form

$$B_{\lambda,m}^{(\kappa)} = \int_0^{\psi_1} l_2^\kappa I_{-1,\lambda}^\kappa \, \mathrm{d}(\sin^{2m} \psi) \tag{2.4}$$

requiring separate treatment. The reader may note, however, that if $\kappa > 0$, the $A^{\iota,\kappa}_{\lambda,m}$'s and the $B^{(\kappa)}_{\lambda,m}$'s are not really independent; for a resort to Equation (2.38) of Chapter III should permit us to express the latter as finite linear combinations of the former; for the explicit forms of such relations cf. Alkan (1979).

Let us return now to Equation (2.3) and set out to evaluate its right-hand side. For $\kappa = 0$ and 1, the $\alpha^{\kappa}_{\lambda}$-functions have already been expressed as series given by Equation (3.77) of Chapter III. If, moreover, we change over from $\sin^{2m}\psi$ to δ^2 on r.h.s. of Equation (2.3) as the variable of integration (by a differentiation of the relation $\delta^2 = 1 - \cos^2\psi\sin^2 i$ and replace δ by $c(r_1 + r_2)$, Equation (2.3) can be rewritten, more explicitly, as

$$A^{\iota,\kappa}_{\gamma,m} = \frac{m[(r_1 + r_2)\cos i]^{2m}}{(\nu)_{\kappa+1}} r_1^{\kappa} \sum_{j=0}^{\infty} (-1)^j \frac{(j+1)_\kappa}{(j+1)!}$$
$$\times (\kappa + \nu + 2j + 2)(\nu + j + 2)_\kappa (\kappa + \nu + 1)_j$$
$$\times \{G_{j+1}(\kappa + \nu, \kappa + \nu + 1, a)\}^2$$
$$\times \int_{c_0^2}^{1} c^{2\kappa}(1 - c^2)^{\nu+1}(c^2 - c_0^2)^{m-1}[1 - (r_1 + r_2)^2 c^2]^{1/2}$$
$$\times G_j(\kappa + \nu + 2, \kappa + 1, c^2)\, \mathrm{d}c^2, \tag{2.5}$$

where now $2\nu = \lambda + 2$, and, as before, $c_0 \equiv \delta_0/(r_1 + r_2)$.

In order to evaluate the integral on the right-hand side of Equation (2.5), let us introduce (as in Equation (2.4) of Chapter V) the substitution

$$u = \frac{c^2 - c_0^2}{1 - c_0^2}, \tag{2.6}$$

in terms of which

$$1 - c^2 = (1 - c_0^2)(1 - u), \tag{2.7}$$

$$l_0^2 = (1 - c_0^2)(1 - u\sin^2\psi_1), \tag{2.8}$$

$$\delta^2 - \delta_0^2 = (r_1 + r_2)^2(1 - c_0^2)u, \tag{2.9}$$

and

$$c^{2\kappa}G_j(\kappa + \nu + 2, \kappa + 1, c^2)$$
$$= j! \sum_{i=0}^{j} \frac{\kappa!}{i!(j-i)!} (\kappa + \nu + j + 2)_i \sum_{k=0}^{i+\kappa} \frac{(-1)^{i+k}}{k!(\kappa + i - k)!} (1 - c_0^2)^k (1 - u)^k \tag{2.10}$$

for $\kappa = 0$ or 1; as a result of which

$$A^{\iota,\kappa}_{\lambda,m} = \frac{m r_1^{\kappa} \sin^{\iota}\kappa}{(\nu)_{\kappa+1}} [(r_1 + r_2)\csc i]^{2m}(1 - c_0^2)^{m+\nu+1}$$
$$\times \sum_{j=0}^{\infty} \frac{(j+1)_\kappa}{j+1} (\kappa + \nu + 2j + 2)(\nu + j + 2)_\kappa (\nu + \kappa + 1)_j$$

$$\times \{G_{j+1}(\kappa+\nu, \kappa+\nu+1, a)\}^2 \sum_{i=0}^{j} \frac{\kappa!}{i!(j-i)!}$$
$$\times (\kappa+\nu+j+2)_i \sum_{k=0}^{\iota+\kappa} (-1)^{i+j+k} \frac{(1-c_0^2)^k}{k!(\kappa+i-k)!}$$
$$\times \int_0^1 u^{m-1}(1-u)^{k+\nu+1}(1-u\sin^2\psi_1)^{\iota/2}\,du, \tag{2.11}$$

where the remaining integral on the r.h.s. can be easily evaluated as

$$\int_0^1 u^{m-1}(1-u)^{k+\nu+1}(1-u\sin^2\psi_2)^{\iota/2}\,du$$
$$= B(m, \nu+k+2)\,{}_2F_1(-\tfrac{1}{2}\iota, m; m+\nu+k+2; \sin^2\psi_1) \tag{2.12}$$

in the form of an ordinary hypergeometric series in ascending powers of $\sin^2\psi_1$; and since this latter quantity is always less than 1 (in fact, less than 0.75 even for central eclipses of contact systems), an absolute and uniform convergence of the series on the r.h.s. of Equation (2.12) is assured.

If $\iota = 0$, the r.h.s. of Equation (2.12) reduces to a constant numerical factor; while if both $\iota = \kappa = 0$, the moments A_{2m} of the light curves also expressed as

$$A_{2m} = L_1 \sum_{h=1}^{\Lambda+1} \frac{2C^{(h)}}{h+1} A^{0,0}_{h-1,m}. \tag{2.13}$$

It should be reiterated that Equations (2.11) or (2.12) hold good only for the particular values of $\kappa = 0$ and 1. When $\kappa > 1$ it is, however, no longer longer necessary to evaluate the respective values of $A^{\iota,\kappa}_{\lambda,m}$ *ab initio*; for they can be expressed in terms of the already known functions $A^{\iota,0}_{\lambda,m}$ and $A^{\iota,1}_{\lambda,m}$ with the aid of the recursion formulae for the α^2_n, α^3_n and α^4_n as given by Equations (2.32)–(2.34) of Chapter III.

If we insert these behind the integral sing on the r.h.s. of Equation (2.3), a term-by-term integration discloses that

$$2(\lambda+4)r_1 A^{\iota,2}_{\lambda,m} = 2r_1\{A^{\iota,0}_{\lambda,m} - A^{\iota+2,0}_{\lambda,m}\}$$
$$+ (\lambda+3)\{(r_1^2 - r_2^2 + 1)A^{\iota,1}_{\lambda,m} - A^{\iota+2,1}_{\lambda,m}\} - (\lambda+5)r_1^2 A^{\iota,1}_{\lambda+2,m}; \tag{2.14}$$

$$12r_1^2 A^{\iota,3}_{\lambda,m} = \{2(2r_1)^2 + (r_1^2 - r_2^2 + 1)^2\}A^{\iota,1}_{\lambda,m}$$
$$- 2\{(2r_1)^2 + r_1^2 - r_2^2 + 1\}A^{\iota+2,1}_{\lambda,m} + A^{\iota+4,1}_{\lambda,m}$$
$$- 2\{r_1^2 - r_2^2 + 5\}r_1^2 A^{\iota,1}_{\lambda+2,m} + 10r_1^2 A^{i+2,1}_{\lambda,m} + r_1^4 A^{\iota,1}_{\lambda+4,m}, \tag{2.15}$$

and

$$32(\lambda+4)(\lambda+6)r_1^3 A^{\iota,4}_{\lambda,m} = 96r_1^3\{A^{\iota,0}_{\lambda,m} - 2A^{\iota+2,0}_{\lambda,m} + A^{\iota+4,0}_{\lambda,m}\}$$
$$+ \{(\lambda+4)(\lambda+6)(r_1^2 - r_2^2 + 1)^2$$
$$+ 12(\lambda^2 + 10\lambda + 20)r_1^2\}$$
$$\times \{(r_1^2 - r_2^2)(A^{\iota,1}_{\lambda,m} - 3A^{\iota+2,1}_{\lambda,m}) + 3A^{\iota+4,1}_{\lambda,m}\}$$
$$- 12(\lambda^2 + 10\lambda + 20)r_1^2 A^{\iota+2,1}_{\lambda,m} - (\lambda+4)(\lambda+6)A^{\iota+6,1}_{\lambda,m}$$

$$
\begin{aligned}
&- 3\{4(\lambda + 6)^2 r_1^4 + 4(\lambda + 4)(\lambda + 5) r_1^2 (r_1^2 - r_2^2 + 1) \\
&+ (\lambda + 4)(\lambda + 6) r_1^2 (r_1^2 - r_2^2 + 1)^2\} A_{\lambda+2,m}^{\iota,1} \\
&+ 6\{2(\lambda + 4)(\lambda + 5) + 2(\lambda + 6)^2 r_1^2 \\
&+ (\lambda + 4)(3\lambda + 16)(r_1^2 - r_2^2 + 1)\} \\
&\times r_1^2 A_{\lambda+2,m}^{\iota+2,1} + 3(\lambda + 4)(5\lambda + 26) r_1^2 A_{\lambda+2,m}^{\iota+4,1} + 3(\lambda + 4) \\
&\times \{4(\lambda + 7) + (\lambda + 6)(r_1^2 - r_2^2)\} r_1^4 A_{\lambda+4,m}^{\iota,1} \\
&- 3(\lambda + 4)(5\lambda + 34) r_1^4 A_{\lambda+4,m}^{\iota+2,1} \\
&- (\lambda + 4)(\lambda + 6) r_1^6 A_{\lambda+6,m}^{\iota,1}.
\end{aligned}
\tag{2.16}
$$

The only terms yet to be evaluated are integrals of the form (2.4), representing photometric perturbations due to the distortion of the shadow cylinder of the eclipsing star (and arising from the function $f_2^{(h)}$ in the integrand on the right-hand side of Equation (2.2). By use of Equation (3.75) of Chapter III, Equation (2.4) can be rewritten for $\kappa = 0, 1$ as

$$
\begin{aligned}
B_{\lambda,m}^{(\kappa)} &= m(r_1 + r_2)^\kappa (r_1/r_2)^{2\nu} \{(r_1 + r_2) \csc i\}^{2m} \frac{b^{2-\kappa}}{\nu \Gamma(\nu + 1)} \\
&\times \sum_{j=0}^{\infty} \frac{(-1)^j}{j!} (\kappa + \nu + 2j + 1) \Gamma(\kappa + \nu + j + 1) \\
&\times \{G_{j+k}(\nu + 1 - \kappa, \nu + 1, a)\}^2 \int_{c_0^2}^{1} (c^2 - c_0^2)^{m-1} (1 - c^2)^\nu \\
&\times c^{2\kappa} G_j(\kappa + v + 1, \kappa + 1, c^2)\, \mathrm{d}c^2,
\end{aligned}
\tag{2.17}
$$

which on changing over from c^2 to u as the variable of integration can be rewritten as

$$
\begin{aligned}
B_{\lambda,m}^{(\kappa)} &= m(r_1 + r_2)^\kappa (r_2/r_1)^{2\nu} \{(r_1 + r_2) \csc i\}^{2m} \frac{(1 - c_0^2)^{m+\nu}}{b^\kappa \nu \Gamma(\nu + 1)} \\
&\times \sum_{j=0}^{\infty} (\kappa + \nu + 2j + 1) \Gamma(\kappa + \nu + j + 1) \\
&\times \{(1 - a) G_{j+\kappa}(\nu + 1 - \kappa, \nu + 1, a)\}^2 \\
&\times \sum_{i=0}^{j} \frac{\kappa!}{i!(j - i)!} (\kappa + \nu + j + 1)_i \sum_{k=0}^{\iota+\kappa} \frac{(-1)^{i+j+k} (1 - c_0^2)^k}{k!(\kappa + i - k)!} \\
&\times \int_0^1 u^{m-1} (1 - u)^{\nu+k}\, \mathrm{d}u;
\end{aligned}
\tag{2.18}
$$

and since

$$
\int_0^1 u^{m-1} (1 - u)^{\nu+k}\, \mathrm{d}u = \frac{\Gamma(m) \Gamma(\nu + k + 1)}{\Gamma(m + \nu + k + 1)} \equiv B(m, \nu + k + 1),
\tag{2.19}
$$

Equation (2.18) can be further simplified to

$$
B_{\lambda,m}^{(\kappa)} = \Gamma(m + 1)(\sin^{2m} \psi_2)(r_1/r_2)^{2\nu} (r_2/b^2)^\kappa
$$

$$
\begin{aligned}
&\times \sum_{j=0}^{\infty} (2j+\kappa+\nu+1)\{(1-a)G_{j+\kappa}(\nu+1-\kappa,\nu+1,a)\}^2\\
&\times \sum_{i=0}^{j} \frac{\kappa!}{i!(j-i)!}\Gamma(i+j+\kappa+\nu+1)\sum_{k=0}^{i+\kappa}\frac{(-1)^{i+j+k}}{k!(\kappa+i-k)!}\\
&\times \frac{(\nu+1)_k(1-c_0^2)^{k+\nu}}{\nu\Gamma(k+m+\nu+1)}\\
&= \frac{\sin^{2m}\psi_1}{\nu(m+1)_\nu}\left\{\frac{r_1^2(1-c_0^2)}{r_2^2}\right\}\left(\frac{r_2}{b^2}\right)^{\kappa}\\
&\times \sum_{j=0}^{\infty}(2j+\kappa+\nu+1)\{bG_{j+\kappa}(\nu+1-\kappa,\nu+1,a)\}^2\\
&\times \sum_{i=0}^{j}\frac{(-1)^{i+j}\kappa!}{i!(j-i)!}\Gamma(i+j+\kappa+\nu+1)\\
&\times G_{j+\kappa}(\nu+1-i-\kappa,\nu+m+1,1-c_0^2),
\end{aligned}
\tag{2.20}
$$

where $\kappa=0$ or 1; but m and λ can be any positive numbers – integral or fractional.

Moreover, when $\kappa>1$ advantage can be taken of Equation (2.38) of Chapter II permitting us to express $I^{\kappa}_{-1,\lambda}$ as linear combinations of associated α-functions of order 1; therefore, the corresponding $B^{(\kappa)}_{\lambda,m}$'s should be expressible in terms of the $A^{\iota,\kappa}_{\lambda,m}$'s whose explicit forms have already been established earlier in this section. In doing so (cf. Alkan, 1979) we find that

$$
\begin{aligned}
B^{(2)}_{\lambda,m} &= \frac{\lambda}{2r_2}\left(\frac{r_1}{r_2}\right)^{\lambda+1}(1-r_1^2+r_2^2)A^{0,1}_{\lambda-2,m} - \frac{\lambda}{4r_2}\left(\frac{r_1}{r_2}\right)^{\lambda+1}A^{1,1}_{\lambda-2,m}\\
&+\frac{\lambda+2}{4}\left(\frac{r_1}{r_2}\right)^{\lambda+2} r_1 A^{0,1}_{\lambda,m},
\end{aligned}
\tag{2.21}
$$

$$
\begin{aligned}
B^{(3)}_{\lambda,m} &= \frac{\lambda}{4r_2^2}\left(\frac{r_1}{r_2}\right)^{\lambda+1}\{(1-r_1^2+r_2^2)^2 A^{0,1}_{\lambda-2,m}\\
&-2(1-r_1^2+r_2^2)A^{2,1}_{\lambda-2,m}+A^{4,1}_{\lambda-2,m}\}\\
&+\frac{\lambda+2}{4}\left(\frac{r_1}{r_2}\right)^{\lambda+3}\{(1-r_1^2+r_2^2)A^{0,1}_{\lambda,m}-A^{2,1}_{\lambda,m}\}\\
&+\frac{\lambda+4}{8}r_1^2\left(\frac{r_1}{r_2}\right)^{\lambda+3}A^{0,1}_{\lambda+2,m},
\end{aligned}
\tag{2.22}
$$

$$
\begin{aligned}
B^{(4)}_{\lambda,m} &= \frac{\lambda}{16r_2^3}\left(\frac{r_1}{r_2}\right)^{\lambda+1}\{(1-r_1^2+r_2^2)^3 A^{0,1}_{\lambda-2,m}\\
&-3(1-r_1^2+r_2^2)^2A^{2,1}_{\lambda-2,m}+3(r_2^2-r_1^2)A^{4,1}_{\lambda-2,m}-A^{6,1}_{\lambda-2,m}\}\\
&+\frac{3(\lambda+2)}{16r_2}\left(\frac{r_1}{r_2}\right)^{\lambda+3}\{(1-r_1^2+r_2^2)^2A^{0,1}_{\lambda,m}-2(1-r_1^2+r_2^2)A^{2,1}_{\lambda,m}+A^{4,1}_{\lambda,m}\}\\
&+\frac{3(\lambda+4)}{16}r_2\left(\frac{r_1}{r_2}\right)^{\lambda+4}\{(1-r_1^2+r_2^2)A^{0,1}_{\lambda+2,m}-A^{2,1}_{\lambda+2,m}\}\\
&+\frac{\lambda+6}{16}\left(\frac{r_1}{r_2}\right)^{\lambda+4}r_1^3A^{0,1}_{\lambda+4,m},
\end{aligned}
\tag{2.23}
$$

etc.

Let us proceed now, with the aid of the foregoing results, to formulate the photometric perturbations $\mathfrak{P}_{2m}^{(h)}$ as defined by Equation (2.2). If we limit ourselves to contributions arising from second-harmonic tides, the corresponding perturbations can be expressed in terms of the $A_{\lambda,m}^{\iota,\kappa}$'s and $B_{\lambda,m}^{(\kappa)}$'s defined by Equations (2.3) and (2.4) as

$$\begin{aligned}\mathfrak{P}_{2m}^{(h)} = &-2\tau\{3(A_{h+1,m}^{2,0} + 2A_{h,m}^{1,1} + A_{h-1,m}^{0,2}) - A_{h-1,m}^{0,0}\}w_1^{(2)} \\ &+ \tfrac{1}{2}(h-1)\{3(A_{h+1,m}^{2,0} + 2A_{h,m}^{1,1} + A_{h-1,m}^{0,2}) \\ &- (3A_{h-1,m}^{2,0} - A_{h-1,m}^{0,0}) - (3A_{h-3,m}^{0,2} - A_{h-3,m}^{0,0})\}w_1^{(2)} \\ &+ (r_2/r_1)^{h+1}\{3B_{h-1,m}^{(2)} - B_{h-1,m}^{(0)}\}w_2^{(2)},\end{aligned} \tag{2.24}$$

where τ denotes the coefficient of gravity-darkening of the star undergoing eclipse, $w_1^{(2)}$ its equatorial ellipticity as given by Equation (1.53) of Chapter II; and the requisite $A_{\lambda,m}^{\iota,\kappa}$'s and $B_{\lambda,m}^{(\kappa)}$'s are already known to us in explicit form. Moreover, contributions to arising from partial tides characterized by higher than second-harmonic symmetry ($j > 2$) can be obtained from Equations (2.31)–(2.33) of Chapter II in terms of the same classes of functions: the requisite algebra is not difficult, but lengthy; and its details can be left as an exercise for the interested reader.

So far in this section we have been concerned with photometric perturbations within eclipses arising from mutual tidal distortion of the constituent stars; but similar effects of *rotational* distortion require separate attention. In discussing these let us regard $n_0^2 \equiv \cos^2 i$ as a small quantity of the same order of magnitude as the equatorial ellipticity $v_{1,2}^{(2)}$ of each star as defined by Equation (1.52) of Chapter II; so that the product $n_0^2 v_{1,2}^{(2)}$ can be considered negligible. Within the scheme of such an approximation we can set

$$n_0 = n_2 = 0 \quad \text{and} \quad n_1 = 1 \tag{2.25}$$

on the right-hand sides of Equations (2.31)–(2.33) of Chapter II; in which case the rotational terms in the latter equations will reduce to

$$f_*^{(h)} = \{(2\tau - h - 1)(\alpha_{h+1}^0 + \alpha_{h-1}^2 - \tfrac{2}{3}\alpha_{h-1}^0) + \tfrac{2}{3}\alpha_{h-1}^0\}v_1^{(2)}, \tag{2.26}$$

$$f_1^{(h)} = \{\mathfrak{I}_{1,h-1}^0 - \tfrac{1}{3}\mathfrak{I}_{-1,h-1}^0\}v_1^{(2)}, \tag{2.27}$$

$$f_2^{(h)} = -(r_2/r_1)^{h+1}\{3I_{1,h-1}^0 - I_{-1,h-1}^0\}v_2^{(2)}. \tag{2.28}$$

All terms on the right-hand sides of the foregoing equations are already familiar to us from our preceding discussion of photometric effects of tidal distortion; but two are new – namely, $I_{1,h-1}^0$ and $\mathfrak{I}_{1,h-1}^0$ – and these require special treatment within the scheme of our approximation represented by Equations (2.25), $\mathrm{d}(\sin^{2m}\psi) = m\delta^{2(m-1)}\,\mathrm{d}\delta^2$; in which case Equations (2.22) and (3.16) of Chapter III disclose that

$$\begin{aligned}\int_0^{\psi_1} I_{1,h-1}^0\,\mathrm{d}(\sin^{2m}\psi) &= \frac{2mr_2}{h+1}\int_0^{\delta_1} \delta^{2(m-1)} I_{-1,h+1}^1\,\mathrm{d}\delta \\ &= -mr_1^2(r_1/r_2)^{h+1}\int_0^{\delta_1^2} \delta^{2(m-1)}\frac{\partial\alpha_{h+1}^0}{\partial\delta^2}\,\mathrm{d}\delta^2;\end{aligned} \tag{2.29}$$

while by use of Equations (2.22), (2.43) and (3.16) of Chapter III we similarly find that

$$\int_0^{\psi_1} \mathfrak{I}^0_{1,h-1}\, \mathrm{d}(\sin^{2m}\psi) = m \int_0^{\delta_1} \delta^{2(m-1)} \times \left\{ \frac{\delta^2 - r_1^2 - r_2^2}{2(h+1)} \frac{\partial \alpha^0_{h+1}}{\partial \delta^2} + \frac{r_1^2}{2(h+3)} \frac{\partial \alpha^0_{h+3}}{\partial \delta^2} \right\} \mathrm{d}\delta^2. \qquad (2.30)$$

An evaluation of the integrals on the right-hand sides of the foregoing equations (2.29) and (2.30) is, however, rather tedious and for this reason we shall not reproduce it in this place; for fuller details the reader is referred to Kopal (1975) or Livaniou (1977).

In contemplating a structure of different components constituting the photometric perturbations $\mathfrak{P}^{(h)}_{2m}$ investigated in this section, a more casual reader may perhaps be deterred by their apparent complexity, and possess some doubts as to their usefulness in practice. It is certainly true that if the technical means at our disposal were limited to slide-rules (which represented the 'state of the art' in 1912 when the Russell–Shapley methods were conceived), or even to desk-type computers available to astronomers in the 1940's (when the Kopal–Piotrowski iterative methods were developed), the expansions for photometric perturbations established in the present section would have been of only theoretical, and not practical, interest; for technical means for their implementation were still some decades in the future.

However, with the advent of present-day electronic computers the speed of arithmetical operations has increased 10^6–10^7 times; and computations which would have taken years or decades only a generation ago can now be performed within seconds or minutes – provided that the problem in question can be spelled out in a language intelligible to automatic computers. It is the outstanding feature of our problem that – transcendental as it is in the time-domain – becomes algebraic in the frequency-domain; and all its steps can be written down as a sequence of arithmetic operations which a computer can perform automatically at electronic speed. It is – we repeat – the ability to program the task in algebraic terms which gives the frequency-domain methods their decisive advantage over any alternative avenue of approach; in view of the speed with which arithmetic operations can be performed inside the computer, the number of instructions called for by the programme becomes of secondary importance. In point of fact, an automatic computer may often complete the entire solution in less time than it would take us to copy the results by hand if we knew them.

A. PARTICULAR CASE: TOTAL ECLIPSES

All expressions developed in the preceding parts of this section apply equally well for all types of eclipses – be these occulations or transits; total, partial or annular. We may, however, recall also that although the same was true of Equations (2.8)–(2.10) or Chapter V for the moments A_{2m} of the light curves, for *total* eclipses these infinite expansions reduced to *closed* forms represented by

Equations (2.25)–(2.27) of Chapter V whenever m happens to be a positive integer. Will the same prove to be true of the photometric perturbations as well?

It is easy to show that this is indeed true for perturbations arising from the distortion of the shadow cylinder cast by the secondary component; for in such a case (cf. Kopal, 1975)

$$\int_{\delta_2^2}^{\delta_1^2} \delta^{\kappa+2\mu} I^{\kappa}_{-1,\lambda}\, d\delta^2 = \frac{2}{\lambda+2} r_2^{2(\mu+1)+\kappa} k^{\lambda+2} \times {}_2F_1(-\mu, \kappa-\mu; 2+\tfrac{1}{2}\lambda; k^2), \tag{2.31}$$

where $\kappa = 0, 1$, $\mu = 0, 1, 2, \ldots, m$, $\delta_{1,2} = r_2 \pm r_1$ and $k \equiv r_1/r_2 < 1$.

The integrals on the left-hand side of Equation (2.31) are constituents of the $B^{(\kappa)}_{\lambda,m}$'s as defined by Equation (2.4); and for $\mu \geqslant \kappa$ the hypergeometric series ${}_2F_1$ on the right-hand side reduce to Jacobi polynomials in k^2. It is true that Equation (2.31) holds good as it stands only for $\kappa = 0$ and 1; however, for $\kappa = 1$ the respective integrals can always be expressed in terms of those for $\kappa = 1$ by use of the equation

$$I^{\kappa}_{1,\lambda} = \sum_{j=1}^{\kappa} \binom{\kappa-1}{j-1} \left(\frac{\delta - s}{r_2}\right)^{\kappa-j} \left(\frac{r_2}{2\delta}\right)^{j-1} I^{1}_{-1,\lambda+2j-2}, \tag{2.32}$$

obtained by a combination of Equations (2.22) and (2.38) of Chapter III.

The actual evaluation of the terms in $\mathfrak{P}^{(h)}_{2m}$ which are factored by $v_2^{(2)}$ and $w_2^{(j)}$ $(j = 2, 3, 4)$ is tedious, but not difficult; and the outcome discloses that the contributions to $\mathfrak{P}_{2m}$ arising from the distortion of the secondary's shadow cylinder can be expressed (for an arbitrarily accurate law of limb-darkening as

$$\mathfrak{P}_2 = L_1C_1\{(r_2/r_1)^2[\tfrac{1}{3}v_2^{(2)} - w_2^{(2)} + \tfrac{3}{4}w_2^{(4)}] - \tfrac{3}{4}X_1[\tfrac{1}{3}v_2^{(2)} - w_2^{(2)}]\}, \tag{2.33}$$

$$\mathfrak{P}_4 = 2(r_2/r_1)C_2(1-q^2)\mathfrak{P}_2 + L_1C_1^2\{(r_2/r_1)^2X_1[\tfrac{1}{3}v_2^{(2)} - w_2^{(2)} + \tfrac{15}{4}w_2^{(4)}] + X_2w_2^{(4)}\}, \tag{2.34}$$

and

$$\mathfrak{P}_6 = 3\{(r_2/r_1)^2C_2^2(1-q^4)\mathfrak{P}_2 - \mathfrak{P}_4\cos^2 i\} + 3L_1C_1^3\{(r_2/r_1)^2X_1[\tfrac{1}{2}v_2^{(2)} - 2w_2^{(2)} + \tfrac{9}{2}w_2^{(4)}] + X_2[\tfrac{1}{18}v_2^{(2)} + \tfrac{1}{3}w_2^{(2)} - \tfrac{19}{4}w_2^{(4)}] + \tfrac{3}{8}X_3w_2^{(4)}\}, \tag{2.35}$$

where we have abbreviated

$$C_1 = r_1^2\csc^2 i, \quad C_2 = r_1r_2\csc^2 i, \quad q = r_2^{-1}\cos i, \tag{2.36}$$

and

$$X_j = \sum_{h=1}^{\Lambda+1} \frac{(j+1)!C^{(h)}}{\nu(\nu+1)(\nu+2)\ldots(\nu+j)}, \quad \nu = \frac{h+1}{2} \tag{2.37}$$

are coefficients depending on the limb-darkening of the eclipsed star (of degree Λ).

On the other hand, the contributions to $\mathfrak{P}_{2m}$ which arise from the distortion of the eclipsed star can, for total eclipses, be similarly expressed as

$$\mathfrak{P}_2 = -L_1(r_2/r_1)C_2(1-q^2)\{(1+\tau)Y_1[\tfrac{1}{6}v_1^{(2)} + w_1^{(2)} - \tfrac{3}{4}(1+q^2)r_2^2 w_1^{(2)}] + (1+\tfrac{1}{2}\tau)Y_2 w_1^{(3)} + (1+\tfrac{1}{3}\tau)Y_3 w_1^{(4)}\} + \tfrac{3}{5}L_1(r_2/r_1)C_2(1+4\tau)Y_2 w_1^{(3)}, \tag{2.38}$$

$$\mathfrak{P}_4 = -2\mathfrak{P}_2 \cot^2 i - L_1(r_2/r_1)^2 C_2^2\{(1+\tau)Y_2 \times [(1-q^4)(\tfrac{1}{6}v_1^{(2)} + w_1^{(2)}) - (1-q^6)r_2^2 w_1^{(2)}] + (1-q^4)[(1+\tfrac{1}{2}\tau)Y_2 w_1^{(3)} + (1+\tfrac{1}{3}\tau)Y_3 w_1^{(4)}]\} + \tfrac{6}{5}L_1(r_2/r_1)^2 C_2^2(1-4\tau)Y_2 w_1^{(3)}, \tag{2.39}$$

$$\mathfrak{P}_6 = -3\mathfrak{P}_2 \cot^4 i - 3\mathfrak{P}_4 \cot^2 i - L_1(r_2/r_1)^3 C_2^3\{(1+\tau)Y_1 \times [(1-q^6)(\tfrac{1}{6}v_1^{(2)} + w_1^{(2)}) - \tfrac{9}{8}(1+q^8)r_2^2 w_1^{(2)}] + (1-q^6)[(1+\tfrac{1}{2}\tau)Y_2 w_1^{(3)} + (1+\tfrac{1}{3}\tau)Y_3 w_1^{(4)}]\} + \tfrac{9}{5}L_1(r_2/r_1)^3 C_2^3(1-4\tau)Y_2 w_1^{(3)}, \tag{2.40}$$

where τ denotes the coefficient of gravity darkening of the star undergoing eclipse, and Y_j is related with its limb-darkening by

$$Y_j = j(j+3)\sum_{h=1}^{\Lambda+1} \frac{(\nu - \tfrac{1}{2}j)\left(\nu - \dfrac{j-2}{2}\right)\cdots\left(\nu + \dfrac{j}{2}\right)}{\nu(\nu+\tfrac{1}{2})(\nu+1)\cdots\left(\nu + \dfrac{j+1}{2}\right)} C^{(h)}. \tag{2.41}$$

The reader may note that the corresponding expressions $\mathfrak{P}_{2m}$ for $m = 1, 2, 3$ as given by Equations (2.33)–(2.35) and (2.38)–(2.40) are of opposite algebraic sign and, therefore, tend to counteract each other. It should also be added that Equations (2.38)–(2.40) are incomplete insofar as the cross-products $r_1^2 r_2^2 v_1^{(2)}$ as well as $r_1^2 r_2^2 w_1^{(j)}$ have been omitted from their right-hand sides. For total eclipses $(r_1 < r_2)$ these should be small enough within the scheme of our approximation; but in case of need can be recovered from the work of Livaniou (1977).

Our last analytic property of the expression for the photometric perturbations established in this section may be noted: namely, that – by analogy with Equation (2.64) of Chapter V satisfied by the moments A_{2m} of the light curves due to mutual eclipses of spherical stars – the photometric perturbations $\mathfrak{P}_{2m}$ established in this section and produced by distortion of the components satisfy likewise the difference-differential equation

$$\frac{\partial \mathfrak{P}_{2m}}{\partial C_3} = m\mathfrak{P}_{2(m-1)}, \tag{2.42}$$

valid for any degree of limb-darkening, in which (as before) $C_3 = r_2^2 \csc^2 i - \cot^2 i$.

B. SUMMARY OF PROCEDURE

The actual method of computation of the elements of distorted eclipsing systems can now be summarized by the following scheme:

(1) First, determine the requisite number of the empirical moments A_{2m} of the observed light curves, as defined by Equations (3.1) of Chapter V or Equation (1.12) of Chapter VI.

(2) Next, evaluate the requisite number n of the constants c_j by appropriate modulation of the 'uneclipsed' parts of the light curve, by use of Equation (1.39), and form their weighted sum (1.44).

(3) Transpose the sum (1.44) on the r.h.s. of Equation (1.15) to the left, ignore $\mathfrak{P}_{2m}$; and evaluate the A_{2m}'s.

(4) With the aid of so 'rectified' moments A_{2m} of the light curve, evaluate the elements r_1, r_2, i and $L_{1,2}$ of the system by the methods of Chapter V.

(5) By use of the elements so obtained evaluate the corresponding 'photometric perturbations' $\mathfrak{P}_{2m}$, transpose them together with the sum (1.44) to the l.h.s. of Equation (1.15) to obtain an improved set of the A_{2m}'s; and from these improved elements.

(6) Should the improved elements differ significantly from their first version, repeat the steps 3–5 until both sides of Equation (1.15) can be satisfied by the same set of the values of $r_{1,2}$, $L_{1,2}$, i; and these constitute the final solution of our problem.

It may be noted that, first, unlike in the case of light changes exhibited by mutual eclipses of *spherical* stars treated in the preceding section of this paper – which could be solved *directly* – a solution for the elements of *distorted* systems can be obtained only by *iteration*. The need to iterate arises solely from a presence of the photometric perturbations $\mathfrak{P}_{2m}$ on the r.h.s. of Equation (1.15) relating $\bar{A}_{2m}$ with A_{2m}. Since, however, the numerical magnitudes of the $\mathfrak{P}_{2m}$'s will generally be small, iterative solutions should converge with sufficient rapidity to make more than one repetition of the steps 3–5 of the foregoing cycle unnecessary. Should, however, this cycle fail to converge, our solution – in fact, *any* solution consistent with a physically sound model of the system – would then become indeterminate from the photometric evidence alone; and additional (e.g., spectroscopic) evidence may be required to alter this situation.

Secondly, it should be stressed that – unlike in the previous treatment of the subject in the time-domain by more conventional methods – each step of our present analysis can be expressed in algebraic form (as closed formulae, or convergent series of satisfactory asymptotic properties) which is amenable to automation; and the entire solution can be obtained at high speed with the aid of electronic computers. However, once the empirical values of the moments $\bar{A}_{2m}$ or A_{2m} have been determined, automatic computers can be programmed to perform the rest of the solution internally, and print out only the results, to any desired degree of accuracy. It is the amenability, and ease of programming for automatic computers, that gives our frequency-domain approach its great advantage over the time-domain approach followed by earlier investigators.

Bibliographical Notes

A realization of the fact that the essential part of the light changes exhibited by close eclipsing systems between minima is due to their ellipticity of figure goes back to the closing decades of the last century, and is due to J. Plassmann (1888); and G. W. Myers (1898), who together with A. W. Roberts (1903) were the first to consider a determination of the elements of distorted eclipsing systems on the assumption that their components could be regarded as similar prolate spheroids appearing in projection as uniformly bright discs. This latter assumption was relaxed by H. N. Russell and H. Shapley (1912a) who allowed (approximately) for the effects of (linear) limb-darkening; while K. Walter (1931) was the first to point out that also the geometrical form of the two components must ge generally dissimilar. The modern epoch of physically sound treatment of the photometric proximity effects commenced with S. Takeda (1934); and further relevant references can be found in the Bibliographical Notes for Chapter II.

The process of 'rectification' of the light curves of close eclipsing systems appears to have been introduced in the astronomical literature by H. N. Russell and H. Shapley (1912a, b); but its basic inadquacy was not realized till many years later (cf. H. N. Russell, 1942; or, more fully, Z. Kopal, 1945). No way out of its shortcomings could, however, be found in the time-domain; and a physically sound separation of the photometric proximity effects from those due to eclipses did not become possible till with the translation of the underlying problem from the time- to the frequency-domain.

VI-**1**. An analytical treatment of the photometric proximity effects exhibited by close eclipsing systems – between minima as well as within eclipses – commenced with the writer's investigation (Kopal, 1975d) which laid down the principles by which this can be done, and indicated the ways by which these can be put in practice. The methods of 'modulation' of the light curves, essential for this purpose, was gradually developed in greater detail (cf. Z. Kopal, V. Markellos and P. Niarchos, 1976; Z. Kopal, 1976b); and the presentation of the subject in this section follows largely that of Kopal's (1976b) paper.

VI-**2**. Of the 'photometric perturbations' $\mathfrak{P}_{2m}$ described in this section, only the most important ones can be found in Sections 6 and 7 of Kopal's (1975d) paper. A systematic investigation of these perturbations was undertaken by H. Livaniou (1977) and published subsequently in two papers: H. Livaniou (1977) containing explicit results (complete to first-order distortion terms of rotational as well as tidal origin) for total and annular eclipses; and H. Rovithis–Livaniou (1978) containing the corresponding result for partial eclipses (cf. also Rovithis–Livaniou, 1979).

Unlike the perturbations $\mathfrak{P}_{2m}$ accompanying total eclipses – which can all be expressed in a closed form – those accompanying partial (or annular) eclipses could be formulated only in terms of expansions whose asymptotic properties left something to be desired for practical applications (cf. P. Niarchos, 1978). This led the present writer to investigate the problem from another angle (cf. Z. Kopal, 1978a); and this treatment has been largely followed in the present section; in particular, closed-form $\mathfrak{P}_{2m}$'s appropriate for total eclipses have been taken from this latter reference. Of subsequent investigations of the subject cf., e.g., H. Alkan and T. Edalati (1979).

As regards the numerical magnitude of combined photometric perturbation in the frequency-domain, caused by the rotational as well as tidal distortion of the components of close eclipsing systems within minima, and computed on the basis of the results presented in this section cf. T. Edalati and E. Budding (1978).

CHAPTER VII

ERROR ANALYSIS

Having evaluated the most probable values of the elements of an eclipsing binary system from an analysis of its light curve in the time- (Chapter IV) or frequency- (Chapters V–VI) domain, we have not yet arrived at the end of their investigation; for it still remains for us to determine the *uncertainty* within which such elements can be specified from the available observational data. A quest of this uncertainty constitutes a point whose importance cannot be overemphasized. Indeed, to say that the most probable value of any such element (say, r_1) is equal to $r_1 = 0.123$ represents a statement whose actual information contents is still quite limited: for it obviously would make a great difference to the physical significance of the results if it were to signify that, in fact, $r_1 = 0.123\,000\ldots$ or again that its uncertainty may amount to several units of the third decimal place. The answer is, of course, implied in the photometric observations at the basis of our study – in their quality as well as number – but appropriate procedures for extracting it remain still to be described.

Since such observations are never infinitely numerous or precise, all results based upon them are bound to be inaccurate within certain limits; and their uncertainty should be described quantitatively in terms of the mean (or probable) *errors* of the respective elements. *Such errors* – it cannot be repeated too often – *must be regarded as an inseparable part of the solution*, and should be supplied by each investigator if his results are to merit serious attention. This becomes indeed the more important, the fewer (or less accurate) are the observations at his disposal; for the danger of overestimating the significance of results based upon limited observational evidence becomes then greatest. How often in the past did those who failed to pay due heed to error analysis become victims of their belief in the precision of their results beyond the limits to which the underlying observations could lend them unequivocal support – a mischance whose consequences were compounded if they used them as a basis for generation of further hypothetical constructions!

VII-1. Error Analysis in the Time-Domain

Direct methods for the solution of the elements of eclipsing binary systems from an analysis of their light curves, as outlined in Section IV-1 of this book and associated with the names of Russell and Shapley (or, more recently, Merrill), are ill-suited to serve as a basis of any systematic error analysis of their outcome. Any process relying on the use of 'fixed points' of a light curve drawn by free hand to follow the course of individual observations, and aiming primarily at a mere graphical representation of the observed data assigns arbitrary weights to different part of the light curve (in fact, infinite weight to the adopted fixed points); and cannot, therefore, help us to ascertain the genuine errors of the elements; or the stability of the solution. A mere graphical agreement between theory and observations is indeed necessary, but *not* sufficient, to ensure a correctness of the solution for the elements of the system. For such elements are many (four or more), and must be obtained simultaneously. Any estimates of their reliability from the quality of graphical representation can, therefore, be at best of only heuristic value. In point of fact, the less determinate the solution, the easier it should be to obtain an (almost) equally good representation of the observations by a right combination of wrong elements – as only too many investigators learned the hard way to their sorrow!

In contrast, the *iterative* methods developed by Kopal and Piotrowski (as presented in Sections 2–3 of Chapter IV) have been the first to lend themselves to systematic error analysis; in fact, they were developed with this aim in view. They enable us to treat the ensemble of given data impartially; and while the (O–C)-residuals of the least-squares solutions of the equations of condition of the form (2.25) or (3.15)–(3.16) can specify the error ϵ of any observed point of unit weight; while the elements of the inverse matrix of the respective normal equations then specify the weights with which all unknowns (or their combinations) can be simultaneously obtained. The purely algebraic aspects of the processes by which this can be accomplished have been detailed before (cf. Kopal, 1950; or 1959, Appendix to Chapter VI) and need not be repeated in this place; in what follows we shall confine our attention to a discussion of different sources of errors which contribute to the final outcome.

In Sections 2 and 3 of Chapter IV we found it expedient to express the geometrical elements r_a, r_b and i of our eclipsing system in terms of three auxiliary constants C_1, C_2 and C_3 or p_0 depending on whether the eclipses are total (annular) or partial. The uncertainty of these elements, caused by the errors inherent in a determination of the respective auxiliary constants, can be expressed as

$$\Delta r_a = \frac{\partial r_a}{\partial C_1}\Delta C_1 + \frac{\partial r_a}{\partial C_2}\Delta C_2 + \frac{\partial r_a}{\partial C_3}\Delta C_3, \tag{1.1}$$

$$\Delta r_b = \frac{\partial r_b}{\partial C_1}\Delta C_1 + \frac{\partial r_b}{\partial C_2}\Delta C_2 + \frac{\partial r_b}{\partial C_3}\Delta C_3, \tag{1.2}$$

$$\Delta i = \frac{\partial i}{\partial C_1}\Delta C_1 + \frac{\partial i}{\partial C_2}\Delta C_2 + \frac{\partial i}{\partial C_3}\Delta C_3, \tag{1.3}$$

if the eclipses are total or annular, or from

$$\Delta r_a = \frac{\partial r_a}{\partial C_1}\Delta C_1 + \frac{\partial r_a}{\partial C_2}\Delta C_2 + \frac{\partial r_a}{\partial p_0}\Delta p_0, \tag{1.4}$$

$$\Delta r_b = \frac{\partial r_b}{\partial C_1}\Delta C_1 + \frac{\partial r_b}{\partial C_2}\Delta C_2 + \frac{\partial r_b}{\partial p_0}\Delta p_0, \tag{1.5}$$

$$\Delta i = \frac{\partial i}{\partial C_1}\Delta C_1 + \frac{\partial i}{\partial C_2}\Delta C_2 + \frac{\partial i}{\partial p_0}\Delta p_0, \tag{1.6}$$

if they are partial. The coefficient of the individual corrections on the right-hand sides of the preceding relations can, in turn, be obtained by an appropriate differentiation of Equations (2.35)–(2.37) or (3.19)–(3.21) of Chapter IV, which yield

$$\left.\begin{aligned} \partial r_a/\partial C_1 &= \tfrac{1}{2}(1 + r_b^2 - r_a^2)r_a^{-1}\sin^2 i,\\ \partial r_a/\partial C_2 &= -(r_b - r_a)\sin^2 i,\\ \partial r_a/\partial C_3 &= \tfrac{1}{2}r_a \sin^2 i; \end{aligned}\right\} \quad (1.7)$$

$$\left.\begin{aligned} \partial r_b/\partial C_1 &= -\tfrac{1}{2}(1 - r_b^2 + r_a^2)r_a^{-2}r_b\sin^2 i,\\ \partial r_b/\partial C_2 &= (1 - r_b^2 + r_a r_b)r_a^{-1}\sin^2 i,\\ \partial r_b/\partial C_3 &= \tfrac{1}{2}r_b \sin^2 i; \end{aligned}\right\} \quad (1.8)$$

$$\left.\begin{aligned} \partial i/\partial C_1 &= \tfrac{1}{2}(r_b^2 - r_a^2)r_a^{-2}\sin^2 i\tan i,\\ \partial i/\partial C_2 &= -(r_b - r_a)r_a^{-1}\sin^2 i\tan i,\\ \partial i/\partial C_3 &= \tfrac{1}{2}\sin^2 i\tan i; \end{aligned}\right\} \quad (1.9)$$

if the eclipses are total or annular; and

$$\left.\begin{aligned} \partial r_a/\partial C_1 &= \tfrac{1}{2}r_a^{-1}\sin^4 i + r_a^{-1}r_b\sin^2 i\cos i,\\ \partial r_a/\partial C_2 &= -\sin^2 i\cos i,\\ \partial r_a/\partial p_0 &= -r_a^2\cos i; \end{aligned}\right\} \quad (1.10)$$

$$\left.\begin{aligned} \partial r_b/\partial C_1 &= -\tfrac{1}{2}(1 - 2r_b\cos i + \cos^2 i)r_a^{-2}r_b\sin^2 i,\\ \partial r_b/\partial C_2 &= (1 - r_b\cos i)r_a^{-1}\sin^2 i,\\ \partial r_b/\partial p_0 &= -r_a r_b\cos i; \end{aligned}\right\} \quad (1.11)$$

$$\left.\begin{aligned} \partial i/\partial C_1 &= \tfrac{1}{2}(2r_b - \cos i)r_a^{-2}\sin^3 i,\\ \partial i/\partial C_2 &= -r_a^{-1}\sin^3 i,\\ \partial i/\partial p_0 &= -r_a\sin i; \end{aligned}\right\} \quad (1.12)$$

if they are partial.

In either case, the geometrical elements proved to be non-linear functions of the auxiliary constants C_1, C_2 and C_3 or p_0 resulting from a least-squares solution; and, hence, the errors of such elements must be identified with those of *linear functions* Δr_a, Δr_b and Δi represented by Equations (1.1)–(1.6), the coefficients of which as given by Equations (1.7)–(1.11) depend on the elements whose uncertainty we wish to establish. Moreover, the very fact that the values of the constant $C_{1,2,3}$ result from a *simultaneous* solution of the same set of the equations of condition renders the computed uncertainty of the C_j's, not independent, but *correlated* with each other. Therefore, the reader should beware of identifying the resultant uncertainty of Δr_a, Δr_b or Δi with a square-root of a sum of the squares of the individual terms on the r.h.s. of Equations (1.1)–(1.6). Such a procedure would be legitimate only if the contributing sources of error would be independent of each other – which they are *not* in our case; and, for this reason, the uncertainty of our elements r_a, r_b and i must be identified with that of the right-hand sides of Equations (1.1)–(1.6) considered as linear functions of ΔC_1, ΔC_2 and ΔC_3 or Δp_0.

Furthermore, before we apply Equations (1.1)–(1.6) to establish the actual uncertainty of the respective elements we should keep in mind that ΔC_1, ΔC_2 and ΔC_3 or Δp_0 represent the *total* errors of the respective quantities – arising from the dispersion of individual observations as well as from the rate of the iterative process by which the final values of our auxiliary constants have eventually been established. For a single least-squares solution of the equations of condition (representing each step of our iterative process) knows still nothing about the convergence of our iterations; and an uncertainty of C_1, C_2 and C_3 (or p_0) resulting from a single solution carried out with a fixed value of K would, therefore, reflect solely the dispersion of the individual observations.

A. ERRORS OF AUXILIARY CONSTANTS

The errors of the auxiliary constants C_1, C_2 and C_3 or p_0 stem, therefore, from two distinct sources:

(a) the dispersion of the observed normal points and the uncertainty of the adopted depths of the minima, and

(b) the finite speed of convergence of our iterative process.

The first source of error is obvious enough, but the second is more subtle. It should, however, be remembered that (quite apart from the influence of any observational errors) ***the elements computed from our set of auxiliary constants would be exact only if the assumed value K of the ratio of the radii were identical with the resulting value of k*** (i.e., if the speed of convergence of our iterative process were *infinite*). If, as will always be the case in practice, the equality $K = k$ can be established, in the final iteration, only within a certain mean error, this error will constitute an *additional* source of uncertainty of all computed elements which is independent of, and supplementary to, that arising from the observational errors alone. This is another aspect of the problem, of basic importance, whose analysis we owe to Piotrowski (1948a); and his reasoning can be outlined as follows.

Let us consider, quite generally,

$$\left.\begin{aligned} C_1 &\equiv F_1(K; L_1, L_2, \ldots, L_i), \\ C_2 &\equiv F_2(K; L_1, L_2, \ldots, L_i), \end{aligned}\right\} \tag{1.13}$$

where $L_i \equiv \sqrt{w}\sin^2\theta_i$ stands for the absolute term of the i-th equation of condition relating C_1 and C_2; our aim being to determine them so that

$$K = k = C_1/C_2. \tag{1.14}$$

Insert Equation (1.14) in the preceding relations and differentiate them with respect to L_i: we obtain

$$\frac{\partial C_1}{\partial L_i} = \frac{1}{C_2^2}\frac{\mathrm{d}C_1}{\mathrm{d}K}\left(\frac{\partial C_1}{\partial L_i}C_2 - \frac{\partial C_2}{\partial L_i}C_1\right) + \frac{\partial F_1}{\partial L_i} \tag{1.15}$$

and

$$\frac{\partial C_2}{\partial L_i} = \frac{1}{C_2^2}\frac{\mathrm{d}C_2}{\mathrm{d}K}\left(\frac{\partial C_1}{\partial L_i}C_2 - \frac{\partial C_2}{\partial L_i}C_1\right) + \frac{\partial F_2}{\partial L_i}, \tag{1.16}$$

respectively. If we remember that, in view of Equation (1.14)

$$\frac{1}{C_2}\frac{\mathrm{d}C_1}{\mathrm{d}K} - \frac{k}{C_2}\frac{\mathrm{d}C_2}{\mathrm{d}K} = \frac{\mathrm{d}k}{\mathrm{d}K}, \tag{1.17}$$

a solution of Equations (1.15)–(1.16) for $\partial C_1/\partial L_i$ and $\partial C_2/\partial L_i$ yields

$$\left.\begin{aligned} C_2\left(1-\frac{\mathrm{d}k}{\mathrm{d}K}\right)\frac{\partial C_1}{\partial L_i} &= \left(C_2 + k\frac{\mathrm{d}C_2}{\mathrm{d}K}\right)\frac{\partial F_1}{\partial L_i} - k\frac{\mathrm{d}C_1}{\mathrm{d}K}\frac{\partial F_2}{\partial L_i}, \\ C_2\left(1-\frac{\mathrm{d}k}{\mathrm{d}K}\right)\frac{\partial C_2}{\partial L_i} &= \left(C_2 - \frac{\mathrm{d}C_1}{\mathrm{d}K}\right)\frac{\partial F_2}{\partial L_i} + \frac{\mathrm{d}C_2}{\mathrm{d}K}\frac{\partial F_1}{\partial L_i}. \end{aligned}\right\} \tag{1.18}$$

Now the *total* errors of C_1 and C_2 are, by definition,

$$\Delta C_1 = \sum_i \frac{\partial C_1}{\partial L_i}\,\delta L_i \quad \text{and} \quad \Delta C_2 = \sum_i \frac{\partial C_2}{\partial L_i}\,\delta L_i, \tag{1.19}$$

where δL_i signifies the O–C residual of the i-th equation of condition and the summation is to be extended over all available normal points; while the *partial* errors, arising from the dispersion of the observations alone (with the assumed value of K considered as *fixed*), are

$$(\delta C_1)_K = \sum_i \frac{\partial F_1}{\partial L_i}\,\delta L_I \quad \text{and} \quad (\delta C_2)_K = \sum_i \frac{\partial F_2}{\partial L_i}\,\delta L_i, \tag{1.20}$$

respectively. *The latter are the errors whose root-mean-square values can be found from our least-squares solutions.* Our aim should, therefore, be to express the ΔC's in terms of the $(\delta C)_K$'s; and this can be done if we multiply both sides of Equations (1.18) by δL_i and perform the summation over all i's: we obtain

$$\Delta C_1 = (\delta C_1)_K + X_1\{(\delta C_1)_K - k(\delta C_2)_K\}, \tag{1.21}$$

$$\Delta C_2 = (\delta C_2)_K + X_2\{(\delta C_1)_K - k(\delta C_2)_K\}, \tag{1.22}$$

and, if the eclipses are total,

$$\Delta C_3 = (\delta C_3)_K + X_3\{(\delta C_1)_K - k(\delta C_2)_K\}, \tag{1.23}$$

where we have abbreviated

$$X_j = \frac{\dfrac{1}{C_2}\dfrac{\mathrm{d}C_j}{\mathrm{d}K}}{1-\dfrac{\mathrm{d}k}{\mathrm{d}K}}, \quad j = 1, 2, 3. \tag{1.24}$$

The contributions to the total uncertainty of C_1, C_2 and C_3 and, therefore, of the geometrical elements of the system, arising from the sources (a) and (b) mentioned above can now be distinctly localized. A dispersion of the individual normal points, in Equations (1.21)–(1.22) or (1.23), while the finite speed of convergence of our iterative process will invoke additional terms multiplied by X_1, X_2 or X_3. If we pause to consider the relative magnitudes of (δC_1) and (δC_2) for partially eclipsing systems, we shall notice that their ratio will depend largely on the circumstances of the eclipse. In point of fact, in the simplest case – when C_1 and C_2 satisfy Equation (3.1) of Chapter IV this ratio should be

$$\left(\frac{\partial C_2}{\partial C_1}\right)_K^2 = \frac{1}{4}\frac{[w(p^2 - p_0^2)^2]}{[w(p - p_0)^2]}, \tag{1.25}$$

where the square brackets on the right-hand side denote the customary sums taken over all equations of condition. If the eclipse were grazing or nearly so, this ratio should be in the neighbourhood of unity; the weights with which both C_1 and C_2 are specified by our solution can be substantial (unless the observations are of poor quality) and very nearly the same. As p_0 increases, however, the uncertainty of both C_1 and C_2 will gradually increase, but in such a way that *the ratio* $(\delta C_2)_K \div (\delta C_1)_K$ *diminishes* – which means that *the weight of* C_1 *diminishes, with increasing* p_0 *much more rapidly than that of* C_2. This situation persists until, for extremely shallow eclipses ($p_0 \sim 1$), the weights of C_1 and C_2 may become once more nearly the same – though this time both are effectively nil. These facts lead us to expect that, considering the observational errors alone, *the product* $r_a r_b$ *of the fractional radii* (i.e., C_2) *should be defined by the observations of partially eclipsing systems with considerably greater relative accuracy than their ratio* r_a/r_b (i.e., C_1/C_2).*

Irrespective of the value of the ratio (1.25), the numerical magnitude of the $(\delta C_j)_K$'s can be obviously diminished arbitrarily by increasing the number (or precision) of the underlying observational data. The magnitudes of the constants X_1, X_2 or X_3 in Equations (1.21)–(1.22) or (1.23) depend, however, solely on the geometrical circumstances of the eclipse and cannot, therefore, be controlled by the observer. As long as the speed of convergence of the iterative process is high – which will, in general, be true if the eclipses under investigation are total or annular – the terms in Equations (1.21)–(1.22) or (1.23) multiplied by X_1, X_2 or X_3 are likely to be unimportant (though not negligible). If, however, the eclipses happen to be partial, the *convergence of our process becomes in general the slower, the shallower the eclipses* and, as a result, the terms factored by X_1 or X_2 may become a large, or even dominant, part of the total error.† Hence, *the slower the convergence of the iterative process, the greater the inherent uncertainty with which the elements of the eclipse can be extracted by our method* – or, for that matter, by *any* method – from an analysis of the light changes.

A succinct measure of the rate of convergence of our iterative process is the quantity $\mathrm{d}k/\mathrm{d}K$ related with the derivatives of C_1 and C_2 with respect to K by means of

$$\frac{\mathrm{d}k}{\mathrm{d}K} = \frac{1}{C_2}\frac{\mathrm{d}C_1}{\mathrm{d}K} - \frac{k}{C_2}\frac{\mathrm{d}C_2}{\mathrm{d}K}. \tag{1.26}$$

* It may be observed that C_2 specifies, in effect, the geometric mean of the two fractional radii r_a and r_b. Unless the disparity in radii is very large (i.e., unless k is very small), the geometric mean of the two radii is known to be very nearly equal to their arithmetic mean. Within the scheme of this approximation we are thus entitled to assert that, for shallow partial eclipses, the sum $r_a + r_b$ of the two fractional radii can be inferred from the observations with a much greater relative accuracy than their ratio.

† In particular, it is in the nature of our problem that, in such cases, C_2 is very much less sensitive to a variation in the adopted value K then C_1 and that, in consequence, $X \gg Y$.

If the speed of convergence of our iterations were infinite this quantity would, by definition, be equal to zero; while if it were equal to (or greater than) unity, the iterations would manifestly fail. Therefore, for any eclipsing system whose elements can be deduced from its light cufve,

$$0 \leq \frac{\mathrm{d}k}{\mathrm{d}K} < 1. \tag{1.27}$$

If the derivative is close to its lower limit, the denominator in the expressions for X and Y will be close to unity and the magnitude of these quantities will be controlled by $\mathrm{d}C_1/\mathrm{d}K$ and $\mathrm{d}C_2/\mathrm{d}K$. If, on the other hand, $\mathrm{d}k/\mathrm{d}K$ is close to unity, the denominator in the expressions (1.24) for X_1 and X_2 will tend to zero and, consequently, both ΔC_1 and ΔC_2 (and therefore also ΔC_2) will increase without limit; and this will make the errors of C_1, C_2 or C_3 grow beyond any limit regardless of the smallness of (δC_j)'s. This is an important revelation; for it shows that *if the eclipses under consideration possess such special features that* $\mathrm{d}k/\mathrm{d}K$ *is close to unity,* (*shallow partial eclipses!*) *the elements of such a system are inherently indeterminate – irrespective of the amount or precision of the underlying observational data.**

These considerations alone render $\mathrm{d}k/\mathrm{d}K$ one of the most important auxiliary quantities of any solution for the elements, and a determination of its numerical value – or, what amounts to the same, or numerical values of the derivatives

$$\frac{\mathrm{d}C_1}{\mathrm{d}K}, \frac{\mathrm{d}C_2}{\mathrm{d}K}, \frac{\mathrm{d}C_3}{\mathrm{d}K}$$

is indispensable for establishing the genuine uncertainty of the geometrical elements of an eclipsing system as can be deduced from an analysis of its light changes. In order to ascertain these quantities, Piotrowski suggested carrying out two independent solutions based on two assumed values of K_1 and K_2, and determining the derivatives of C_1, C_2 or C_3 with respect to K from the differences of the respective C's computed under the two assumptions divided by $K_1 - K_2$. Such a process would, however, not only just about double the computer's work, but the values of the derivatives approximated in this way would also pertain to the ratio of the radii half way between K_1 and K_2, which need not necessarily envelop k. Moreover, the approximate values of the derivatives based solely on the quotients of the first differences may be appreciably in error in regions where the variation of the C_j's with K is rapid. As it sometimes happens, however, a determination of actual derivatives $\mathrm{d}C_j/\mathrm{d}K$ $(j = 1, 2, 3)$ turns out to be easier than that of their finite-differences approximation and can be performed in the course of our intermediary solution with a minimum of repetitive work.

In order to do so, let us return again to our fundamental equation $H = \sin^2 \theta$

* It should, however, be clearly understood that this statement holds good whenever we are solving for the elements from an overdeterminate system of equations based on the actual observations, and not on the basis of fixed points equal in number to that of the unknowns. In this latter case, the uncertainty $(\delta C_j)_K$ remains, by definition, zero irrespective of a possible proximity of $\mathrm{d}k/\mathrm{d}K$ to unity.

and consider what happens when we vary the value of K. It is obvious that the values of C_1 and C_2 or C_3 must be altered so as to counter the change of p with K because, if the light curve is to pass through the same points, the right-hand side of the fundamental equation must remain the same. In consequence, the function H on its left-hand side must be invariant with respect to an infinitesimal change in K, and this can be mathematically expressed as

$$\frac{dH}{dK} = 0, \tag{1.28}$$

which represents an implicit equation for determining dC_1/dK, dC_2/dK, or dC_3/dK. On performing the actual differentiation of Equations (2.2) and (3.1) of Chapter IV we find that, if the eclipse is total or annular,

$$(p^2 - 1)\frac{dC_1}{dK} + 2(p+1)\frac{dC_2}{dK} + \frac{dC_3}{dK} = -2C_2(1 + kp)\frac{dp}{dK}; \tag{1.29t}$$

while if it is partial,

$$(p^2 - p_0^2)\frac{dC_1}{dK} + 2(p - p_0)\frac{dC_2}{dK} = 2C_2\left\{(1 + kp_0)\frac{dp_0}{dK} - (1 - kp)\frac{dp}{dK}\right\}. \tag{1.29p}$$

The computation of the ordinary derivatives of p with respect to K requires some care. In general,

$$\frac{dp}{dK} = \left(\frac{\partial p}{\partial K}\right)_\alpha + \left(\frac{\partial p}{\partial \alpha}\right)_K \frac{d\alpha}{dK}. \tag{1.30}$$

The reader may remember that, if the eclipse is *total* (or annular), $\alpha \equiv n$ is a quantity supplied directly by the observations and, therefore,

$$\frac{d\alpha}{dK} = 0. \tag{1.31t}$$

If, on the other hand, the eclipse happens to be *partial*, $\alpha = n\alpha_0$ and α_0 is, in turn, given by Equations (1.19) or (1.20) of Chapter IV – depending on whether the eclipse under investigation is an occultation or a transit – which both involve K. Differentiating them and ignoring, for simplicity, the minor variation of $Y(K, p_0)$ with K we find that

$$\frac{d\alpha}{dK} = -\frac{2n(1 - \lambda_b)}{K^3 Y(k, p_0)} \tag{1.31p'}$$

if the eclipse is an occultation, and

$$\frac{d\alpha}{dK} = -\frac{2n(1 - \lambda_b)}{K^3 Y(k, -1)} \tag{1.31p''}$$

if it is a transit. Moreover, differentiating partially the identity $p \equiv p[K, \alpha(K, p)]$ we have

$$\left(\frac{\partial p}{\partial K}\right)_\alpha = -\left(\frac{\partial p}{\partial \alpha}\right)_K \left(\frac{\partial \alpha}{\partial K}\right)_p. \tag{1.32}$$

Inserting Equations (1.30), (1.31) and (1.32) in (1.33) and multiplying the latter by $\sqrt{w}$, we find that the explicit forms of Equations (1.29) for determining the required derivatives of C_1, C_2 and C_3 with respect to K ultimately become

$$\sqrt{w}(p^2-1)\frac{dC_1}{dK}+2\sqrt{w}(p+1)\frac{dC_2}{dK}+\sqrt{w}\frac{dC_3}{dK}=-(1-\lambda)\frac{\partial\alpha}{\partial K} \tag{1.33t}$$

if the eclipse is total (or annular), and

$$\sqrt{w}(p^2-p_0^2)\frac{dC_1}{dK}+2\sqrt{w}(p-p_0)\frac{dC_2}{dK}=\frac{1-\lambda}{\alpha_0}\left\{\frac{2(1-\lambda_b)}{K^3Y(k,p_0)}\left[\sqrt{\frac{w}{w_0}}-n\right]+\left[\frac{\partial\alpha}{\partial K}\sqrt{\frac{w}{w_0}}-\frac{\partial\alpha}{\partial K}\right]\right\} \tag{1.33p}$$

if it is partial. It should be reiterated that, in this and other equations, α_0 and λ stand for either α_0' and λ_a or α_0'' and λ_b – depending on the type of the eclipse we are considering.

The reader should notice that $\sqrt{w}$ as defined by Equation (2.14) (total eclipses) or (3.6) (partial eclipses) of Chapter IV involves C_2^{-1}; hence, if we exclude this factor from $\sqrt{w}$ the actual unknowns in Equations (1.33) are the ratios $C_2^{-1}(dC_j/dK)$, $j=1,2,3$. Apart from this fact, however, the coefficients of the unknowns in Equations (1.33) are identical with those of the fundamental Equations (2.25) or (3.15)–(3.16) of Chapter IV for the determination of C_1, C_2 or C_3, so that only the absolute terms on the right-hand sides of Equations (1.33) need to be evaluated additionally before a solution for the derivatives can be made. Since, moreover, the corresponding numerical values of $\sqrt{w}$ as well as n are also available at this stage, the only new quantities needed are the derivatives $\partial\alpha/\partial K$ and their values can be extracted with the greatest ease from the tabular differences of the appropriate Tsesevich tables.* A determination of the derivatives dC_1/dK by means of Equations (1.33) is much more accurate and less time-consuming than the one based on a comparison of solutions started with two independent values of K. Moreover, if we set up Equation (1.33t) for more than three values of p, or (1.33p) for more than two phases, we can ascertain not only the most probable value of the derivatives, but also the uncertainty with which these derivatives are defined by the available observational data. A computation of this uncertainty would, however, serve no specific purpose and may be dispensed with if necessary.

Once the values of dC_1/dK and dC_2/dK or dC_3/dK have thus been established, we shall insert them in Equation (1.26) to evaluate dk/dK which, in turn, will permit us to specify the quantities X_1, X_2 and X_3 as defined by Equation (1.24). When this has been done ΔC_1, ΔC_2 or ΔC_3 from Equations (1.21)–(1.23) and, consequently, the expressions for Δr_1, Δr_2 and Δi as defined by (1.1)–(1.6) will become linear functions of $(\delta C_1)_k$, $(\delta C_2)_k$ or $(\delta C_3)_k$ with known

* This is due to the fact that the absolute terms of Equations (1.33) involve only partial derivatives, as compared with ordinary derivatives on the right-hand sides of Equations (1.29). The transformation of Equations (1.29) into (1.33) was effected precisely for this purpose; for whereas partial derivatives $\partial\alpha/\partial k$ can be approximated from tabular differences in columns $k = constant$, to evaluate ordinary derivatives would call for bi-variate differentiation.

numerical coefficients, and their uncertainty can be found from the elements of the inverse matrix of our least-squares solution for C_1, C_2 or C_3.

It should be pointed out, in this connection, that if the first derivative dk/dK can be evaluated by the method of this section while the second derivative d^2k/dK^2 may be considered negligible, our iterative process for a determination of the auxiliary constants characterizing an eclipsing binary system with known types of the minima can be *by-passed* by the following procedure:

1. Assume a plausible value of K and carry through a customary solution for C_1, C_2 (and C_3) leading to $C_2/C_2 = k$.

2. Solve Equations (1.33t) or (1.33p) for the dC_j/dK's and evaluate dk/dK by means of Equation (1.26).

3. The true of k (i.e., the one for which the assumed K and the resulting k would be identical) – let us denote it by $\bar{k}$ – follows from

$$\bar{k} = K - \frac{K - k}{1 - (dk/dK)}, \tag{1.34}$$

while the true (barred) values of the C_j's (such that $\bar{C}_1/\bar{C}_2 = \bar{k}$ are then given by

$$\bar{C}_j = C_j + (dC_j/dK)(\bar{k} - K) = C_j + C_2 X_j(k - K), \quad j = 1, 2, 3; \tag{1.35}$$

and these values should be used to compute the true geometrical elements of our eclipsing system.

B. OTHER COMBINATIONS OF ELEMENTS; LIMB-DARKENING

Having established explicit expressions for the errors of the individual geometrical elements, let us outline the way in which the errors of non-linear combinations of such elements, or of other quantities not so far considered, can be evaluated.

As an illustrative example of the former, let us consider the ratio of the radii $k = C_1/C_2$. Differentiating the latter we find that

$$\Delta k = \frac{\Delta C_1}{C_2} - k\frac{\Delta C_2}{C_2} = \frac{(\delta C_1)_K - k(\delta C_2)_K}{C_2[1 - (dk/dK)]} \tag{1.36}$$

by use of Equations (1.21) and (1.22); so that the error Δk obtains as that of a linear combination of $(\delta C_{1,2})_K$ on the r.h.s. The reader should note that this error may be very much larger than those of $(\delta C_{1,2})_K$ if the quantity dk/dK happens to be close to 1 (as a result of slow convergence of our iterations). Should $dk/dK = 1$, k becomes indeterminate – implying that the observed light curve can be equally represented by *any* value of this parameter.

On the other hand, the uncertainty of the fractional luminosities L_a and L_b can be shown (cf. Kopal, 1959; pp.462–463) to be equal to that of a linear function

$$\Delta L_{a,b} = L_{a,b}\left\{\frac{2(\delta C_1)_K + 2k(\delta C_2)_K}{C_2[1 - (dk/dK)]} \mp \frac{\Delta\lambda_a}{1 - \lambda_a} \pm \frac{\Delta\lambda_b}{1 - \lambda_b} + \frac{L_{a,b}\Delta U}{L_a L_b} \pm \frac{(\lambda_a - \lambda_b\Delta U}{(1 - \lambda_a)(1 - \lambda_b)}\right\}, \tag{1.37}$$

where ΔU denotes the uncertainty of the adopted value for our unit of light, and

$\lambda_{a,b}$ stand for the fractional intensity of light at the time of the component of fractional luminosity $L_{a,b}$ (the upper algebraic sign on the r.h.s. of Equation (1.37) pertains to ΔL_a and the lower one to ΔL_b). Should the eclipse happen to be total, the foregoing Equation (1.37) admits of considerable simplification: in fact, Equations (2.18) and (2.38) of Chapter IV then disclose at once that

$$\Delta L_a = \Delta U - \Delta\lambda_a \tag{1.38}$$

and

$$\Delta L_b = \qquad \Delta\lambda_a. \tag{1.39}$$

Lastly, we may wish to consider the uncertainty of another parameter underlying our solution: namely, the coefficient(s) u_l of limb-darkening, for which so far fixed values have been adopted. Since, however, the uncertainty Δu_l of these adopted quantities may often be large (and the quantities themselves subject to a later revision), it may be desirable to investigate the extent to which the values of the geometrical elements r_a, r_b and i of the system, obtained (say) on the basis of an adopted value u of linear limb-darkening, may be affected (or made more uncertain) if its possible error Δu is included as an additional unknown, to be determined simultaneously with all other elements sought after.

In order to do so, consider again (cf. Piotrowski, 1948a) Equations (2.15) or (3.8) of Chapter IV and remember that the function H depends also (through p) on u. If an error of Δu has been committed in estimating u at the outset, we could in principle hope to determine it from an equation of the form

$$\sqrt{w}H_0 + \sqrt{w}\left(\frac{\partial H}{\partial U}\right)_0 \Delta U + \sqrt{w}\left(\frac{\partial H}{\partial \lambda}\right)_0 \Delta\lambda + \cdots + \sqrt{w}\left(\frac{\partial H}{\partial u}\right)\Delta u = \sqrt{w}\sin^2\theta, \tag{1.40}$$

where

$$\frac{\partial H}{\partial u}\Delta u \equiv \frac{\partial H}{\partial n}\frac{\partial n}{\partial u}\Delta u \tag{1.41}$$

and (with $n \equiv \alpha/\alpha_0$)

$$\frac{\partial n}{\partial u} \equiv \frac{\partial}{\partial u}\left(\frac{\alpha}{\alpha_0}\right) = \frac{1}{\alpha_0}\left\{\frac{\partial\alpha}{\partial u} - n\frac{\partial\alpha_0}{\partial u}\right\}; \tag{1.42}$$

for total eclipses, $\alpha_0 = 1$. On the other hand,

$$\frac{\partial H}{\partial n} = 2\alpha_0 C_2(1 + kp)\frac{\partial p}{\partial n} = -\frac{1-\lambda}{\sqrt{w}}; \tag{1.43}$$

so that

$$\sqrt{w}\frac{\partial H}{\partial u}\Delta u = -\frac{1-\lambda}{\alpha_0}\left\{\frac{\partial\alpha}{\partial u} - n\frac{\partial\alpha_0}{\partial u}\right\}\Delta u \tag{1.44p}$$

if the eclipse is partial, and

$$\sqrt{w}\,\frac{\partial H}{\partial u}\,\Delta u = -(1-\lambda)\,\frac{\partial \alpha}{\partial u}\,\Delta u \tag{1.44t}$$

if it is total (since $\alpha_0 = 1$ and the derivative $\partial\alpha_0/\partial u$ then vanishes).

The coefficient of Δu on the r.h.s. of the preceding equation can be easily evaluated. However, it turns out to be so small (of the order of 0.01–0.1) and – which is worse – is so closely correlated with that of C_1, that *the inclusion of Δu among the unknowns to be determined simultaneously with C_1, C_2, etc. may entail so drastic a loss of weight of the entire solution as to render it* often (especially for partial eclipses!) *indeterminate within wide limits.* But – and this is what Piotrowski suggested – *we may transpose the term* (1.44) *in our equations of condition of the form* (1.40) *to the right-hand side as a part of its absolute term, and evaluate C_1, C_2 etc. in terms of Δu* retained as an arbitrary quantity. If so, their solution can assume the form $C_j = a_j + b_j\Delta u$, where a_j stands for the value of C_j corresponding to the adopted degree of darkening, and $b_j \equiv \mathrm{d}C_j/\mathrm{d}u$.

Moreover – as has been shown by Kopal (1948) – this latter task can be accomplished more simply in the following manner. Let us return again to Equations (2.15) or (3.8) of Chapter IV and consider the consequences of a small change of u on its left-hand side. All p's will change as a consequence; but so must C_1 and C_2 or C_3 if the resultant light curve is to remain the same. This is equivalent to an assertion that

$$\frac{\mathrm{d}H}{\mathrm{d}u} = 0 \tag{1.45}$$

or, explicitly,

$$(p^2-1)\frac{\mathrm{d}C_1}{\mathrm{d}u} + 2(p+1)\frac{\mathrm{d}C_2}{\mathrm{d}u} + \frac{\mathrm{d}C_3}{\mathrm{d}u} = -2C_2(1+kp)\frac{\mathrm{d}p}{\mathrm{d}u} \tag{1.46t}$$

if the eclipse is total, and

$$(p^2-p_0^2)\frac{\mathrm{d}C_1}{\mathrm{d}u} + 2(p-p_0)\frac{\mathrm{d}C_2}{\mathrm{d}u} = 2C_2\left\{(1+kp_0)\frac{\mathrm{d}p_0}{\mathrm{d}u} - (1+kp)\frac{\mathrm{d}p}{\mathrm{d}u}\right\} \tag{1.46p}$$

if it is partial. Differentiating the function $\alpha(k,p)$ we find that, in general,

$$\mathrm{d}\alpha = \frac{\partial\alpha}{\partial k}\,\mathrm{d}k + \frac{\partial\alpha}{\partial p}\,\mathrm{d}p + \frac{\partial\alpha}{\partial u}\,\mathrm{d}u. \tag{1.47}$$

Now if α is supplied to us by the observations (i.e., if $\mathrm{d}\alpha = 0$) and if we adopt a fixed value of K (thus rendering $\mathrm{d}k = 0$), the foregoing equation discloses that

$$\frac{\mathrm{d}p}{\mathrm{d}u} = -\frac{\partial p}{\partial\alpha}\frac{\partial\alpha}{\partial u}. \tag{1.48}$$

This equation specifies the rate at which p would change, for given K and α, if we vary the adopted degree of limb-darkening. If we insert this result in the preceding Equations (1.46) and multiply them by the square-root of the intrinsic weight $\sqrt{w}$ we eventually find that

$$\sqrt{w}(p^2-1)\frac{\mathrm{d}C_1}{\mathrm{d}u}+2\sqrt{w}(p+1)\frac{\mathrm{d}C_2}{\mathrm{d}u}+\sqrt{w}\frac{\mathrm{d}C_3}{\mathrm{d}u}=-(1-\lambda)\frac{\partial\alpha}{\partial u} \tag{1.49t}$$

if the eclipse is total, and

$$\sqrt{w}(p^2-p_0^2)\frac{\mathrm{d}C_1}{\mathrm{d}u}+2\sqrt{w}(p-p_0)\frac{\mathrm{d}C_2}{\mathrm{d}u}=(1-\lambda)\left\{\frac{\partial\alpha_0}{\partial u}\sqrt{\frac{w}{w_0}}-\frac{\partial\alpha}{\partial u}\right\} \tag{1.49p}$$

if it is partial. It should again be remembered that, inasmuch as the expressions (2.14) or (3.8) of Chapter IV for $\sqrt{w}$ involves C_2, the actual unknowns in these quantities are the ratios $C_2^{-1}(\mathrm{d}C_j/\mathrm{d}u)$, $j=1,2,3$; furthermore, since $\sqrt{w}$ as defined by Equations (2.14) or (3.6) is itself multiplied by $1-\lambda$, this latter fact can be cancelled on both sides of Equations (1.49). If these equations are set up for more than two or three points of the light curve, we can again ascertain not only the most probable values of the requisite derivatives, but also their probable errors – if one is interested to know them.

Once equations of the form (1.49t) or (1.49p) have been solved and the requisite derivatives $\mathrm{d}C_1/\mathrm{d}u$, $\mathrm{d}C_2/\mathrm{d}u$ or $\mathrm{d}C_3/\mathrm{d}u$ obtained, the derivatives of r_a, r_b and $\sin i$ with respect to u follow readily. The relations between the auxiliary constants and the geometrical elements are provided by Equations (2.35)–(2.37) of Chapter IV if the eclipses are total (or annular), or Equations (3.19)–(3.21) if they are partial. Differentiating them we obtain

$$\frac{1}{C_2}\frac{\mathrm{d}C_1}{\mathrm{d}u}=\frac{2}{r_b}\frac{\mathrm{d}r_\alpha}{\mathrm{d}u}-\frac{2k}{\sin i}\frac{\mathrm{d}\sin i}{\mathrm{d}u} \tag{1.50}$$

and

$$\frac{1}{C_2}\frac{\mathrm{d}C_2}{\mathrm{d}u}=\frac{1}{r_a}\frac{\mathrm{d}r_a}{\mathrm{d}u}+\frac{1}{r_b}\frac{\mathrm{d}r_b}{\mathrm{d}u}-\frac{2}{\sin i}\frac{\mathrm{d}\sin i}{\mathrm{d}u}; \tag{1.51}$$

moreover, if the eclipses are total

$$\frac{1}{C_2}\frac{\mathrm{d}C_3}{\mathrm{d}u}=-2\frac{r_b-r_a}{r_ar_b}\frac{\mathrm{d}r_a}{\mathrm{d}u}+2\frac{r_b-r_a}{r_ar_b}\frac{\mathrm{d}r_b}{\mathrm{d}u}+2\frac{\cos^2\theta''\sin i}{r_ar_b}\frac{\mathrm{d}\sin i}{\mathrm{d}u}; \tag{1.52}$$

while if they are partial,

$$\frac{\partial p_0}{\partial\alpha}\frac{\partial\alpha_0}{\partial u}=\frac{p_0}{r_a}\frac{\mathrm{d}r_a}{\mathrm{d}u}+\frac{1}{r_a}\frac{\mathrm{d}r_b}{\mathrm{d}u}+\frac{\tan i}{r_a}\frac{\mathrm{d}\sin i}{\mathrm{d}u}. \tag{1.53}$$

The left-hand sides of these equations being known, the rates of change of r_a, r_b and $\sin i$ corresponding to a given small change u in the adopted value of the coefficient of limb-darkening can be readily evaluated from Equations (1.50)–(1.51) and (1.52)) if the eclipses are total (or annular), or from Equations (1.50)–(1.51) and (1.53) if they are partial; and the geometrical elements themselves expressed as

$$r_a=(r_a)_u+\left(\frac{\mathrm{d}r_a}{\mathrm{d}u}\right)\Delta u+\cdots, \tag{1.54}$$

$$r_b=(r_b)_u+\left(\frac{\mathrm{d}r_b}{\mathrm{d}u}\right)\Delta u+\cdots, \tag{1.55}$$

$$\sin i = (\sin i)_u + \left(\frac{\mathrm{d} \sin i}{\mathrm{d}u}\right) \Delta u + \cdots, \tag{1.56}$$

where $(r_a)_u$, $(r_b)_u$ and $(\sin i)_u$ stand for the values of the respective elements based on the initially assumed value of u.

VII-2. Differential Corrections

In discussions of photometric measures of eclipsing variables, the ultimate aim of the investigator should be to derive a set of the most probable elements of the eclipsing system under investigation, and to demonstrate that the entire observational evidence can be represented by such elements within the limits of observational errors. The final residuals between theory and observations should enable him to ascertain the uncertainty within which his elements are defined by the underlying observational data. In the preceding section, systematic methods were developed for a determination, not of the elements themselves, but of certain auxiliary constants which specify the elements of our eclipsing system; and once the most probable values of these constants have been established, the rest of the solution follows unambiguously. Of these auxiliary constants, C_1, C_2 and C_3 or p_0 are arrived at by iteration, combined with the process of differential corrections for establishing the most probable values of λ_a, λ_b and U at the same time.

The light residual Δl between our theory of analysis in the time-domain and the observations are likewise implied already in the solution outlined in Chapter IV, since (by Equations (2.14) or (3.6) of that chapter) we defined the weight $\sqrt{w}$ of each equation of condition in such a way that

$$\Delta l \equiv l_{\text{obs}} - l_{\text{comp}} \equiv O - C = -\Delta(\sqrt{w} \sin^2 \theta). \tag{2.1}$$

Hence, the *(O–C)-residuals of least-squares solution of properly weighted equations of condition* of the form (2.25) and (3.15) or (3.16) of Chapter IV for $C_1, C_2, \ldots,$ etc. *are, by definition, identical with the light residuals* $(-\Delta l)$. If, moreover, the average error of a single photometric observation l_{obs} be ϵ, the theory of least-squares processes discloses (cf. Otrebski, 1948) that the average error of $l_{\text{comp}} \equiv l_{\text{obs}} - \Delta l$ will be equal to $\epsilon\sqrt{m/n}$, where m denote the number of the unknown quantities to be simultaneously determined and n, the number of available equations of condition.

The reader should also keep in mind that the foregoing identity (2.1) holds good only *provided that our equations of condition have been multiplied by the complete expression for* $\sqrt{w}$ *as given by Equations* (2.14) *or* (3.6) *of Chapter IV*; which will, however, be feasible but seldom in practice. In particular, the value of C_2 which occurs in the denominator of these equations will rarely be known beforehand with a precision warranting its use in the formation of the weights; and if only one minimum has been observed (or if both minima are being treated separately), the term $1 - \lambda$ in the numerator on the right-hand sides of Equations (2.14) or (3.6) can likewise be ignored as an irrelevant constant. If both C_2 and $1 - \lambda$ were disregarded in the formation of our weights, we should expect our light residuals to follow from

$$\Delta l = -[(1-\lambda)/C_2](O-C) \tag{2.2}$$

where the value of the conversion factor $(1-\lambda)/C_2$ can be determined accurately from the outcome of our solution. Furthermore, if (in treating the visual or photographic observations) we satisfied ourselves that the errors of observation were constant on the magnitude rather than intensity scale – i.e., if we multiplied our equations of condition by $\sqrt{w}$ as defined by Equation (2.11b) rather than by Equation (2.11a) of Chapter IV – it would follow that

$$\Delta l = -l(O-C). \tag{2.3}$$

In general, if we multiplied the weight $\sqrt{w}$, as defined by Equations (2.11), of any given equation of condition by any number N to account for unequal observational weight of the respective normal point, or simply in order to keep its numerical value within reasonable limits, the corresponding light residual should be specified by

$$\Delta l = -N^{-1}(O-C). \tag{2.4}$$

Once the conversion factor between Δl and $(O-C)$ has been appropriately determined, the task of converting the $(O-C)$-residuals of the intermediary least-squares solution into the corresponding light residuals is completely straightforward and can be performed without difficulty.

The light residuals Δl thus determined should now be scrutinized with care for their possible origin. If the timing of our observations was sufficiently accurate, and the correctness of the underlying photometric scale beyond any doubt, the remaining residuals should arise from

(a) accidental errors of observation,

(b) systematic errors in the determination of any (or all) intermediary elements of the eclipse.

The accidental errors are characterized by a random distribution of signs and should be equal in magnitude to those exhibited by the observations of full light. If, however, the Δl's within minima are found to be noticeably larger and (or) their variation turns out to be systematic, we are led to suspect that our intermediary elements do not represent the observed light changes within the limits of observation errors and are in need of further refinement. The aim of the present section will, therefore, be to outline a process by which our elements can be further improved by differential corrections.

A. EQUATIONS OF THE PROBLEM

Let – as in previous parts – L_1 denote the fractional luminosity of the component undergoing eclipse, of fractional radius r_1; and let r_2 denote the fractional radius of the eclipsing star. The fractional light of the system at any phase should then follow from

$$l = 1 - fL_1, \tag{2.5}$$

where the fractional loss of light $f \equiv \Delta\Omega$ is given in terms of the elements of the eclipse by Equations (2.29) and (2.30) of Chapter II. If the two components could be regarded as spheres,

$$\Delta\Omega = \sum_{h=1}^{\infty} C^{(h)}\alpha_{h-1}^{0}; \tag{2.6}$$

while if the effects of distortion cannot be ignored, the r.h.s. of the preceding equation is to be augmented by

$$\Delta\Omega = \sum_{h=1}^{\infty} C^{(h)}\{f_{*}^{(h)} + f_{1}^{(h)} + f_{2}^{(h)}\}, \tag{2.7}$$

where the photometric perturbations $f_{*}^{(h)}$, $f_{1}^{(h)}$ and $f_{2}^{(h)}$ are given by Equations (2.31)–(2.33) of Chapter II.

Let, furthermore, Δl denote the differences between the light observed at any particular phase, and that computed by any method developed in Chapters IV–VI. If these $(O-C)$-differences prove to be systematic, this indicates that one (or more) of the elements at the basis of the computed model is in need of an improvement; but even the errors Δl are accidental, their magnitude can disclose the uncertainty of the respective elements. In order to carry out these tasks, let us differentiate Equation (2.5) partially with respect to all elements involved on its right-hand side: in doing so we find that each observed point should furnish an equation of the form

$$\Delta l = -f\Delta L_1 - L_1\Delta f, \tag{2.8}$$

where

$$\Delta f \equiv \frac{\partial f}{\partial r_1}\Delta r_1 + \frac{\partial f}{\partial r_2}\Delta r_2 + \frac{\partial f}{\partial i}\Delta i + \cdots, \tag{2.9}$$

and where the quantities ΔL_1, $\Delta r_{1,2}$ or Δi stand for the values of differential corrections to be added to the respective elements to improve the fit. As many equations of condition of the form (2.8)–(2.9) can be set up as there are observed $(O-C)$-differences; and provided that their number sufficiently exceeds that of the unknowns, a least-squares solution of such a system of equations should then yield the most probable values of the requisite corrections, as well as their errors.

In order to set up such equations, the explicit form of partial derivatives of the function f with respect to all elements involved must be established, and numerically evaluated in terms of these elements before their solution can be carried out. This preparatory work was, however, already accomplished in Section 3 of Chapter II; and the results obtained there enable us to assert that, for *spherical* stars,

$$\frac{\partial f}{\partial r_1} = \sum_{h=1}^{\infty} C^{(h)}\frac{\partial \alpha_{h-1}^{0}}{\partial r_1} = \frac{2}{r_1}\sum_{h=1}^{\infty} C^{(h)}\left(\frac{r_2}{r_1}\right)^{h+1}\left\{\frac{\delta}{r_2}I^{1}_{-1,h-1} - I^{0}_{-1,h-1}\right\} \tag{2.10}$$

and

$$\frac{\partial f}{\partial r_2} = \sum_{h=1}^{\infty} C^{(h)}\frac{\partial \alpha_{h-1}^{0}}{\partial r_2} = \frac{2}{r_2}\sum_{h=1}^{\infty} C^{(h)}\left(\frac{r_2}{r_1}\right)^{h+1} I^{0}_{-1,h-1} \tag{2.11}$$

by Equations (3.14) and (3.15) of Chapter III; while Equation (3.16) permits us to assert that

$$\frac{\partial f}{\partial i} = \frac{\partial f}{\partial \delta}\frac{\partial \delta}{\partial i} = -\frac{1}{\delta}\frac{\partial f}{\partial \delta}\sin i\cos i\cos^2\theta$$
$$= \frac{\cos^2\theta}{\delta r_2}\left\{\sum_{h=1}^{\infty} C^{(h)}\left(\frac{r_2}{r_1}\right)^{h+1} I^{1}_{-1,h-1}\right\}\sin 2i, \tag{2.12}$$

where the integrals of the form $I^{m}_{-1,n}$, defined by Equations (2.28) of Chapter II, have been adequately investigated in Chapter III; (cf. also Appendix IV) and their numerical values tabulated by Kopal (1947) so that, in what follows, they can be regarded as known.

If, moreover, one (or both) components of the system under investigation are *distorted*, and Equation (2.7) must be added to Equation (2.6) to obtain the complete expression for f, its partial derivatives can be similarly obtained if, in the derivatives of $f^{(h)}_{*}$, Equations (3.10)–(3.12) of Chapter III replace (3.14)–(3.16) used above; and those of $f^{(h)}_{1,2}$ are deduced in terms of the relevant recursion formulae established in Section 3 of that chapter. Care should only be exercised to note that the direction cosines l_0 and l_2 or n_0 and n_2 on the right-hand sides of Equations (2.31)–(2.33) of Chapter II depend also on the angle i, and must be differentiated with respect to it to obtain the complete coefficients of $\Delta r_{1,2}$ or Δi.

An interesting point arises in connection with the partial derivatives of f with respect to the coefficients u_1 of limb-darkening involved in the $C^{(l)}$'s. A resort to Equations (2.7)–(2.9) of Chapter I defining the latter discloses that, correctly to quadratic terms in the law of limb-darkening (1.2) of that chapter with the coefficients $u_{1,2}$,

$$\frac{\partial f}{\partial u_1} = -\frac{6}{(6-2u_1-3u_2)^2}\{4(\alpha_0^0 - \tfrac{3}{2}\alpha_1^0) - u_2(\alpha_0^0 - 3\alpha_1^0 + 2\alpha_2^0)\}$$
$$= -\frac{12(2-u_2)}{(6-2u_1-3u_2)^2}(\alpha_0^0 - \tfrac{3}{2}\alpha_1^0) + \frac{6u_2}{(6-2u_1-3u_2)^2}\mathfrak{I}^0_{-1,2} \tag{2.13}$$

and

$$\frac{\partial f}{\partial u_2} = -\frac{6}{(6-2u_1-3u_2)^2}\{3(\alpha_0^0 - 2\alpha_2^0) + u_1(\alpha_0^0 - 3\alpha_1^0 + 2\alpha_2^0)\}$$
$$= -\frac{12u_1}{(6-2u_1-3u_2)^2}(\alpha_0^0 - \tfrac{3}{2}\alpha_1^0) + \frac{6(3-u_1)}{(6-2u_1-3u_2)^2}\mathfrak{I}^0_{-1,2}; \tag{2.14}$$

so that

$$\frac{\partial f}{\partial u_1}\Delta u_1 + \frac{\partial f}{\partial u_2}\Delta u_2 = -\frac{12(2\Delta u_1 - u_2\Delta u_1 + u_1\Delta u_2)}{(6-2u_1-3u_2)^2}(\alpha_0^0 - \tfrac{3}{2}\alpha_1^0)$$
$$+\frac{6(3\Delta u_2 - u_1\Delta u_2 + u_2\Delta u_1)}{(6-2u_1-3u_2)^2}\mathfrak{I}^0_{-1,2}, \tag{2.15}$$

which, if the darkening were linear (i.e., if $u_2 = 0$) would reduce to

$$\frac{\partial f}{\partial u_1}\Delta u_1 = -\frac{6}{(3-u_1)^2}(\alpha_0^0 - \tfrac{3}{2}\alpha_1^0)\Delta u_1. \tag{2.16}$$

Moreover, for *any* arbitrary degree Λ of the law of limb-darkening higher than the second, the partial derivatives $\partial f/\partial u_l$ can likewise be built up from the difference $\alpha_0^0 - \frac{3}{2}\alpha_1^0$ and of different $\mathfrak{I}^0_{-1,l}$-functions (for $l = 2, 3, \ldots, \Lambda$); with coefficients depending on $u_1, u_2, \ldots, u_\Lambda$.

The explicit forms of these functions of the phase can be obtained by the methods already established in Chapter III. All of them vanish at the first as well as second contacts of the eclipse, and remain numerically of the order of 10^{-3} to 10^{-2} during partial phases – in contrast with the coefficients of other differential corrections on the r.h.s. of Equation (2.9) which are generally of zero order. This fact exposes at once the difficulty to be encountered in any attempt to determine the degree of darkening of the star undergoing eclipse simultaneously with other elements of the system; but, unfortunately, this is not yet the worst. It appears that Nature really conspired to thwart any efforts at an empirical determination of the limb-darkening of the components of eclipsing binary systems from an analysis of their light changes in a very effective manner: for *not only the coefficients $\partial f/\partial u_l$ are numerically so small, but their variation with the phase turns out to simulate also that of $\partial f/\partial r_1$* (cf. Equation (3.14) of Chapter III) *so closely as to render a separation of Δr_1 and Δu_l from a simultaneous solution a very difficult task*, which can be successfully accomplished only on the basis of observational evidence of very high quality. And this makes it also evident that a determination of u (by *any* method) which does not provide for a simultaneous adjustment of all other elements would possess but little meaning; for an arbitrary change of u can be countered to a large extent by a suitable change in r_1 to restore an almost identical fit.

B. COMBINATION OF BOTH MINIMA

The actual techniques by which the least-squares solutions of the equations defining our differential corrections should be performed will not detain us, since a full account of them can be found in other sources (cf. Kopal, 1950; or 1959) and need not be repeated in this place. Suffice it only to point out a few safeguards which should be observed in combining observational evidence based on alternate eclipses of the same system into a single set of normal equations.

It goes without saying that whenever two minima have been observed, all elements of the system should be adjusted so as to ensure the best possible fit to the light changes observed during *both* alternate eclipses. If the secondary minimum is shallow, the weight of the observational evidence based upon it may be too small to make its neglect of any consequence; but whenever this is not true, the equations of condition based upon both minima should be combined in a single set of normal equations before proceeding to the final solution. As regards the weights of the individual equations of condition of the form (2.8)–(2.9), it should be borne in mind that *their intrinsic weights are already included in the numerical values of the coefficients of the respective differential corrections*, so that it is only appropriate observational weights which should be applied to Equations (2.8)–(2.9) if the underlying normal points themselves are of unequal quality.

In combining Equations (2.8)–(2.9) into a single set of normal equations we should, furthermore, observe that – since the roles of eclipsing and eclipsed component in the alternate minima are interchanged – the corrections Δr_1 and ΔL_1 for one minimum become identical with Δr_2 and ΔL_2 in the other; just as the corrections Δu_l to the assumed degree of limb-darkening pertain always to the star undergoing eclipse (which is different in each minimum).

As regards the differential corrections to L_1 and L_2, the reader may recall that their sum $L_1 + L_2$ has been adopted as our unit of light; and if the value of this unit were known exactly, we could obviously put $\Delta L_1 = -\Delta L_2$. In actual practice, the value of our light unit can again be inferred from the observations with only a finite degree of accuracy, in which case the foregoing relation between ΔL_1 and ΔL_2 can no longer be regarded as exact and should be replaced by

$$\Delta L_1 = \Delta U - \Delta L_2, \tag{2.17}$$

where ΔU represents, as before, a possible correction to a preliminary determination of our light unit. The foregoing equation makes it evident that if we retain both ΔL_1 and ΔL_2 as independent unknowns of our least-squares solution, their sum will furnish ΔU and the uncertainty of the linear function $\Delta L_1 + \Delta L_2$ will be equal to that of ΔU. If so, Equation (2.8) should be replaced by

$$\Delta l = \Delta L_1 + \Delta L_2 - f\Delta L_1 - L_1 \Delta f. \tag{2.18}$$

Moreover, the observations between minima provide us with an additional equation of condition of the form

$$\Delta L_1 + \Delta L_2 \equiv \Delta U = 0, \tag{2.19}$$

which is fulfilled within a certain error. If, on the other hand, we decide to eliminate ΔL_2 from our solution by putting it equal to $-\Delta L_1$ we should automatically include ΔU among the unknowns of our solution and replace Equation (2.18) by

$$\Delta l = \Delta U - f\Delta L_1 - L_1 \Delta f, \tag{2.20}$$

while the observations between minima assert that

$$\Delta U = 0 \tag{2.21}$$

within the uncertainty of available observations.

The square-root $\sqrt{w}$ of the relative weight with which the foregoing equations (2.19) or (2.21) should join the rest of our overdeterminate system is – as in Equations (2.28) of Chapter IV – equal to the ratio ϵ/σ_U – where σ_U denotes again the probable error of our unit of light as determined from the observations between minima (if the light remains sensibly constant out of eclipses, σ_U becomes simply the probable error of the approximate mean of all observations secured during full light); and ϵ, the corresponding error of a single normal, within minima, of unit weight.

The foregoing considerations will permit us to bring the whole photometric evidence, between minima as well as within eclipses, to bear on the determination of differential corrections to all elements of our eclipsing system, which are required to establish the best possible fit. If the elements which we thus set out to improve represented a close approximation to reality, all corrections resulting from such a solution may be on the limit of significance. Should this *not* be the case, and should the magnitude of resultant differential correction exceed their uncertainty by an appreciable margin, the solution should be *repeated* – with the coefficients of all differential corrections recomputed with the aid of the improved set of the elements – and this procedure repeated until the resulting corrections are no longer significant.

Suppose, however, that an acceptable set of differential corrections has eventually been established, and their uncertainty found in the usual manner. The most probable values of all elements of our eclipsing systems are then clearly obtained b6 an addition of the respective differential correction to the previously adopted value of the element; but *will the uncertainty of this element be identical with that of its differential correction*? The answer is known to be unequivocally in the affirmative only if the equations of condition defining the differential corrections possess constant coefficients. In our present case, however, the coefficients in the equations (2.6)–(2.7) are functions of the variables we seek to determine; and if so, the problem raised by our inquiry turns out to be by no means simple.

This subtle point from the theory of linear equations has been investigated by Piotrowski (1948b); and his answer turned out to be a conditional 'yes'. The uncertainty of our differential corrections, as defined by our least-squares solution, remains sensibly (though not exactly) equal to that of the respective element, *provided* that the sums of products of all second partial derivatives of f multiplied by the respective $(O-C)$ differences are negligible in comparison with the sums of cross-products of the first partial derivatives of f. If, for example, any one of the second derivatives of f could become numerically large at any phase of the eclipse, Piotrowski's condition might be violated regardless of the smallness of the $(O-C)$'s. The same situation could possibly occur also if the signs of the $(O-C)$ differences were not distributed at random, but were correlated with the sign of any one of the second derivatives of f. Fortunately, none of the second partial derivatives of f become very large for any type of the eclipse. If, in addition, the algebraic signs of the $(O-C)$'s are found to oscillate approximately at random, we can derive confidence from Piotrowski's work to conclude that the probable errors Δx of x will, for all practical purposes, be indeed the same – whatever element x may stand for.

VII-3. Error Analysis in the Frequency-Domain

When we set out to ascertain the errors of the elements of an eclipsing system, obtained by an analysis of the requisite evidence in the frequency-domain, the reader can rightfully raise the question: why not utilize the elements so obtained to establish the shape of the theoretical light curve in the time-domain by the methods outlined in the first part of this book (Chapters I–III), from the respective $O-C$ residuals, and proceed to minimize their ensemble by least-squares method as developed in Section VII-2? This is indeed a very legitimate question; and the answer to it should be largely in the affirmative. Yes, this is possible; for the elements obtained by the methods of Chapters V–VI should furnish a set of $O-C$ differences not inferior to those obtained by the iterative methods of Chapter IV; and definitely superior to them for close systems exhibiting conspicuous proximity effects between eclipses – provided that the solution has been *optimized* by an adoption of the *minimum* values of the indices m requisite for this purpose (cf. Section V-3).

The reason goes back to the fact that these m's represent, in effect, *frequencies* in terms of which the light changes are being analyzed; and that (for the supporting data cf. Kopal, 1977a) *virtually all information contained in the light changes during mutual eclipses of the stars is confined to the low-frequency end of Fourier spectra of the respective data*; while *higher frequencies are primarily generated by the proximity effects* (ellipticity, reflection) *at discrete* (rather than continuous) *frequencies.*

The Fourier transforms of the light changes arising from mutual eclipses of spherical stars have already been investigated in Section I-4 of this book. Between eclipses – when, for close binary systems,

$$f(\psi) \equiv 1 - l = -\sum_{j=1}^{\infty} c_j \cos^j \psi, \tag{3.1}$$

in accordance with Equation (1.11) of Chapter VI – the Fourier cosine transform $F_1(\nu)$ of $f(\psi)$ becomes (in accordance with Equations (4.8) of Chapter I) equal to

$$\begin{aligned} F_1(\nu) &= -2\sum_{j=1}^{\infty} c_j \int_0^{\pi} \cos^j \psi \cos h\psi \, \mathrm{d}\psi \\ &= 0 \quad \text{if } j < h, \\ &= -\binom{j}{j-h} \frac{\pi c_j}{2^{j-1}} \quad \text{if } j \geqslant h. \end{aligned} \tag{3.2}$$

Accordingly, a Fourier transform of the light changes of the type (3.1) arising from the proximity effects consists of a discrete set of values, given by

$$F_1(2\pi) = \pi(c_1 + c_2 + \tfrac{3}{4}c_3 + \tfrac{1}{2}c_4 + \cdots), \tag{3.3}$$

$$F_1(4\pi) = \pi(\tfrac{1}{2}c_2 + \tfrac{3}{4}c_3 + \tfrac{3}{4}c_4 + \cdots), \tag{3.4}$$

$$F_1(6\pi) = \pi(\quad + \tfrac{1}{4}c_3 + \tfrac{1}{2}c_4 + \cdots), \tag{3.5}$$

$$F_1(8\pi) = \pi(\qquad + \tfrac{1}{8}c_4 + \cdots), \tag{3.6}$$

where the constants c_j may be positive or negative (for effects produced by equilibrium tides, their values were given by Equations (1.4)–(1.7) of Chapter VI).

In other words, a cosine Fourier transform acts on the light changes represented by Equation (3.1) like a 'comb-filter', transmitting only discrete frequencies $\nu \equiv 2\pi h$ for $h = 1, 2, 3, \ldots$; and the 'intensity' of the individual 'spectral lines' is specified by the magnitude of the constants c_j; the 'continuous spectrum' of $F_1(\nu)$ for non-integral values of h is identically zero. Or – to express ourselves still in different words – while the light changes arising from the eclipses can be decomposed in a *continuous* frequency-spectrum, those arising from the proximity effects (ellipticity, reflection) are *multiplexed* by Nature at their source; and their frequency-spectrum is, therefore, *discrete.*

A. ERROR ANALYSIS BY FOURIER SERIES

After these preliminaries, let us return to Chapter V in which the principal tools for a determination of the elements of eclipsing binary systems in the frequency-domain turned out to be the moments A_{2m} of the light curves, defined by Equation (1.1) of that chapter for arbitrary constants m; for once their numerical values have been ascertained from the observations (by planimetry or otherwise), all elements of the respective system can be determined from them by a sequence of purely algebraic operations; and their uncertainty deduced from that of the underlying moments A_{2m}.

What are the sources of uncertainty besetting the A_{2m}'s? In order to localize these, let us return to Figure 5-1 of Chapter V and recall that the times t (or phase angles θ) of individual photometric observations can be established by appropriate measurements without appreciable error, the ordinates on Figure 5-1 can be regarded as exact. It is only the abscissae of the individual points representing the observed light changes which are subject to finite observational errors; and their implications alone will concern us in the present discussion.

The uncertainty of the empirical moments A_{2m} goes back to two sources: namely, to the *accidental* errors of individual observations which define the shape of the light curve (representing one boundary of the area A_{2m} on Figure 5-1); and the *systematic* errors arising from an inaccurate knowledge of our unit of light (representing the upper boundary of the respective area). If the light changes between eclipses are sufficiently well covered by the observations, the latter errors may be small enough to be negligible; but if the latter are not available in sufficient numbers (or quality), this need not necessarily be the case.

Let us, therefore, keep in mind both types of errors as potential contributors to the uncertainty of the empirical moments A_{2m}; what will be the cumulative magnitude of their effects? In order to determine this, let us expand the observed changes of light $1-l$ in a Fourier series in terms of the phase angle θ. As long as the light changes due to eclipses are symmetrical with respect to the conjunctions, only the cosine terms of the respective series will possess nonvanishing coefficients (cf. Section 4 of Chapter I); and, in accordance with Equation (4.11) of that chapter

$$1-l=\tfrac{1}{2}a_0+\sum_{n=1}^{\infty} a_n\cos(n\pi\theta/\theta_1), \tag{3.7}$$

where θ_1 denotes the phase angle of the beginning (or end) of the eclipse – or as close an estimate of it as can be made from the observations.

As many equations of the form (3.7) can obviously be set up as there are values of $l(\theta)$ observed for $0<\theta<\theta_1$. Let their number be denoted by M. Moreover, the series on the r.h.s. of Equation (3.7) converges as a rule rapidly enough to make their terms beyond (say) the N-th insignificant. By breaking off in Equation (3.7) the sum after $n=N$, its right-hand side will contain $N+1$

coefficients $a_0, a_1, a_2, \ldots, a_N$, the values of which are originally unknown. If, however, the number M of the observed values of $l(\theta)$ exceeds $N+1$ by a sufficient margin, Equation (3.7) can be treated as one of condition for the determination of the most probable values of a discrete set of the constants $a_0, a_1, a_2, \ldots, a_N$ by the method of least squares; and such a solution should also furnish (by standard methods) the *uncertainty* of the individual a_n's arising from *accidental* errors affecting the observed values of $l(\theta)$.

Moreover, the *systematic* errors arising from an inexact knowledge of our unit of light (and contributing likewise to the total uncertainty of the outcome) can be taken into account in the following manner. Let the adopted unit of light be identified with the arithmetic (straight, or weighted) mean of m observations of the brightness of the star between eclipses (when the light of the system is supposed to be constant). Let, moreover, ϵ denote the standard deviation (mean error) of a single observation of unit weight; and σ, the standard deviation from the mean. Lastly, if ΔU denotes the error of the estimated unit of light (expected to be less than σ), Equation (3.7) should be augmented to

$$1 + \Delta U - l = \tfrac{1}{2}a_0 + \sum_{n=1}^{N} a_n \cos(n\pi\theta/\theta_1). \tag{3.8}$$

No matter how many observations of the brightness $l(\theta)$ are available to us within eclipse, the equations of condition of the form (3.8) based upon them will not enable us to separate ΔU from a_0. However, the m observations made between eclipses will provide an additional equation of condition of the form

$$(\epsilon/\sigma)\Delta U = 0 \tag{3.9}$$

equivalent to Equation (2.16) of Chapter IV. If, moreover, $\Sigma(O-C)^2$ denotes a sum of the squares of the errors of m individual observations made between minima then, as is well known,

$$\epsilon = \left\{\frac{\Sigma\,(O-C)^2}{m-1}\right\}^{1/2}, \tag{3.10}$$

while

$$\sigma = \left\{\frac{\Sigma\,(O-C)^2}{m(m-1)}\right\}^{1/2}. \tag{3.11}$$

Accordingly, $\epsilon/\sigma = \sqrt{m}$, and Equation (3.9) of condition can be rewritten as

$$\sqrt{m}\,\Delta U = 0. \tag{3.12}$$

For large values of m the weight of this equation will be very much larger than the weight of the equations of condition of the form (3.8); and a simultaneous least-squares solution of Equation (3.8) *with* (3.12) should furnish the most probable values of $a_0, a_1, \ldots, a_N$ *and* ΔU, together with their probable errors.

A similar differential improvement of the adopted value of θ_1 for the phase angle of first contact involved in the time-dependent functions $\cos(n\pi\theta/\theta_1)$ on the r.h.s. of Equation (3.7) is, unfortunately, impossible at this stage, because of non-uniform convergence on the Fourier series (3.7) at $\theta = \theta_1$, where the time-derivative of the light changes becomes discontinuous (the Gibbs phenomenon!). Since, however, the constants specifying θ_1 – see Equation (2.5) of Chapter IV – are also implied in the a_n's, we shall return to an improvement of θ_1 as a by-product of our entire solution at a later stage of our analysis.

B. RELATION BETWEEN THE FOURIER TRANSFORM AND MOMENTS OF THE LIGHT CURVES

Suppose, then, that the available observational data permit us to determine the most probable values of the Fourier coefficients a_n and of ΔU by a least-squares solution of the equations of condition of the form (3.8)–(3.9), together with their probable errors. In what relation are the latter with the uncertainty of the moments A_{2m} of the light curves? A clue to the answer is provided by Equations (4.12) of Chapter I, which discloses that

$$a_n = \frac{1}{\theta_1} F_1\left(\frac{n}{2\theta_1}\right); \tag{3.13}$$

i.e., that the Fourier coefficients a_n on the r.h.s. of Equations (3.7) or (3.8) are equal to the local values of the Fourier cosine transform

$$F_1(\nu) = 2 \int_0^{\theta_1} (1 - l) \cos h\theta \, \mathrm{d}\theta \tag{3.14}$$

for discrete values $\nu = h/2\pi$ of the frequency.

Explicit expressions of $F_1(\nu)$ for any type of eclipse have already been established in Chapter I, Equations (4.26)–(4.28). On the other hand, from the identity

$$\cos h\theta \, \mathrm{d}\theta \equiv h^{-1} \, \mathrm{d}(\sin h\theta) \tag{3.15}$$

where

$$\sin h\theta = h \sin \theta \; {}_2F_1\left(\frac{1+h}{2}, \frac{1-h}{2}; \frac{3}{2}; \sin^2\theta\right) \tag{3.16}$$

for any value of h (integral or fractional), it follows that Equation (3.15) can be rewritten as

$$\cos h\theta \, \mathrm{d}\theta = \mathrm{d}\left\{\sin\theta \; {}_2F_1\left(\frac{1+h}{2}, \frac{1-h}{2}; \frac{3}{2}; \sin^2\theta\right)\right\}. \tag{3.17}$$

If we insert this latter equation in the definition (3.14) of the Fourier cosine transform and remember that the moments A_n of the light curves are defined by

$$A_n = \int_0^{\theta_1} (1 - l) \, \mathrm{d}(\sin^n \theta), \tag{3.18}$$

a combination of Equations (3.14), (3.17) and (3.18) discloses that

$$F_1(\nu) = 2\sum_{j=0}^{\infty} \frac{\left(\frac{1+h}{2}\right)_j \left(\frac{1-h}{2}\right)_j}{j!\left(\frac{3}{2}\right)_j} A_{2j+1}$$

$$= 2\left\{A_1 + \frac{1^2-h^2}{3!}A_3 + \frac{(1^2-h^2)(3^2-h^2)}{5!}A_5 + \frac{(1^2-h^2)(3^2-h^2)(5^2-h^2)}{7!}A_7 + \frac{(1^2-h^2)(3^2-h^2)(5^2-h^2)(7^2-h^2)}{9!}A_9 + \cdots\right\}, \tag{3.19}$$

i.e., that the *Fourier cosine transform* $F_1(\nu)$ *can be readily expressed as a series in odd moments* A_{2j+1} *of the light curves of ascending orders*; which, if $h \equiv 2\pi\nu$ happens to be an odd integer, reduces to polynomials.

In particular,

$$F_1\left(\frac{1}{2\pi}\right) = 2A_1, \tag{3.20}$$

$$F_1\left(\frac{3}{2\pi}\right) = 2A_1 - \tfrac{8}{3}A_3, \tag{3.21}$$

$$F_1\left(\frac{5}{2\pi}\right) = 2A_1 - 12A_3 + \tfrac{32}{5}A_5, \tag{3.22}$$

$$F_1\left(\frac{7}{2\pi}\right) = 2A_1 - 16A_3 + 32A_5 - \tfrac{120}{7}A_7, \tag{3.23}$$

etc.; by inversion of which it follows that

$$A_1 = \tfrac{1}{2}F_1\left(\frac{1}{2\pi}\right), \tag{3.24}$$

$$A_3 = \tfrac{3}{8}F_1\left(\frac{1}{2\pi}\right) - \tfrac{3}{8}F_1\left(\frac{3}{2\pi}\right), \tag{3.25}$$

$$A_5 = \tfrac{35}{64}F_1\left(\frac{1}{2\pi}\right) - \tfrac{45}{64}F_1\left(\frac{3}{2\pi}\right) + \tfrac{5}{32}F_1\left(\frac{5}{2\pi}\right), \tag{3.26}$$

$$A_7 = \tfrac{35}{48}F_1\left(\frac{1}{2\pi}\right) - \tfrac{77}{80}F_1\left(\frac{3}{2\pi}\right) + \tfrac{7}{24}F_1\left(\frac{5}{2\pi}\right) - \tfrac{7}{120}F_1\left(\frac{7}{2\pi}\right), \tag{3.27}$$

etc.

In setting up to express the right-hand sides of the foregoing equations in terms of the empirical Fourier coefficients a_n from Equations (3.7) or (3.8) with the aid of their definition (3.13) in terms of the Fourier transform $F_1(\nu)$, we must remember that the arguments ν of $F_1(\nu)$ in Equations (3.13) and (3.20)–(3.23) are *not* the same: in Equation (3.13), $\nu = n/2\theta_1$; while in Equations (3.20)–(3.23), $\nu = (2j+1)/2\pi$. It is, however, possible to *interpolate* for the latter from the former in the following way. On insertion of Equation (3.7) in (3.14) we obtain

$$F_1(\nu) = a_0 \int_0^{\theta_1} \cos h\theta \, d\theta + 2 \sum_{n=1}^{\infty} a_n \int_0^{\theta_1} \cos h\theta \cos k\theta \, d\theta, \tag{3.28}$$

where $h = 2\pi\nu$ and $k = n\pi/\theta_1$.

An evaluation of the integrals on the r.h.s. of the preceding equation presents no difficulty: indeed,

$$\int_0^{\theta_1} \cos h\theta \, d\theta = \frac{\sin h\theta_1}{h}, \tag{3.29}$$

while

$$\int_0^{\theta_1} \cos h\theta \cos k\theta \, d\theta = (-1)^n \left\{ \frac{h\theta_1^2}{(h\theta_1)^2 - (n\pi)^2} \right\} \sin h\theta_1. \tag{3.30}$$

Inserting Equations (3.29) and (3.30) in (3.28) we readily find that, for an arbitrary frequency ν, the value of the transform $F_1(\nu)$ can be obtained in terms of its local values $F_1(n/2\theta) \equiv a_n\theta_1$ with the aid of the interpolation formula

$$F_1(\nu) = \frac{\sin h\theta_1}{h} \sum_{n=0}^{\infty} \frac{(-1)^n a_n}{1 - (n\pi/h\theta_1)^2}, \tag{3.31}$$

a formula of the same general form as Equations (4.26)–(4.28) of Chapter I. Moreover, for the arguments $(2j+1)/2\pi$ of the F_1's on the right-hand sides of Equations (3.24)–(3.27) the foregoing equation particularizes to

$$F_1\left(\frac{2j+1}{2\pi}\right) = \frac{\sin(2j+1)\theta_1}{2j+1} \sum_{n=0}^{\infty} \frac{(-1)^{n+1} a_n}{\left[\dfrac{n\pi}{(2j+1)\theta_1}\right]^2 - 1}, \tag{3.32}$$

where $j = 0, 1, 2, \ldots$; and the a_n's are coefficients of the Fourier cosine series (3.7). If, lastly, we insert for the $F_1(2j+1/2\pi)$'s on the right-hand sides of Equations (3.24)–(3.27) from (3.32), the values of the odd moments A_{2j+1} can be expressed in terms of the empirically-determined a_n's.

By this our task is, however, still far from being completed; for it is not these *odd* moments that are particularly needed. In Section V-1 we have seen that, for total eclipses, the entire solution can be expressed in a closed form in terms of *even* moments A_{2m}; and in Section V-3 ('Optimization of Solutions') it transpired that, for other types of eclipses, the moments A_μ of unrestricted (not necessarily integral) order may be required to obtain the best solution. To be sure, expansions have been set up in Section V-2 which define such moments in terms of the elements of the eclipses (of any type) for unrestricted (non-negative) values of μ; and their empirical values can always be obtained by numerical quadratures of planimetry in the $(1-l)$-sin θ coordinates. However, if we were to restrict ourselves to planimetry, we could not determine – only, at best, estimate – the *uncertainty* with which moments A_μ are defined by the observed data. Naturally we do not wish to remain indefinitely at this stage. The avowed aim of this chapter has, therefore, been to develop systematic procedures – in the time – as well as frequency-domains – for determining the uncertainty of our solutions for the elements by an impersonal process which is free of bias and immune to subjective judgments.

Earlier in this section we pointed out already the way in which this task can be approached: namely, by use of the Fourier coefficients a_n on the r.h.s. of Equation (3.7), the empirical values of which – together with their uncertainty – can be obtained by a least-squares fit of Equation (3.7) to the observed data. Moreover, equations of the type (3.24)–(3.27) combined with (3.32) should enable us to express all odd moments A_{2j+1} of integral orders as a weighted mean of the a_n's and to evaluate their uncertainty as that of the appropriate linear function of the a_n's from the same least-squares solution.

In order to extend such a process to moments $A\mu$ of the unrestricted index μ, let us depart from a well-known expansion (cf. Oberhettinger, 1973; p. 31) asserting that

$$\sin^{\mu}\theta = \frac{\Gamma(\mu+1)}{2^{\mu-1}} \sum_{j=0}^{\infty} \frac{(-1)^j \sin(2j+1)\theta}{\Gamma\left(\frac{\mu+3}{2}+j\right)\Gamma\left(\frac{\mu+1}{2}-j\right)} \tag{3.33}$$

for $0 \leqslant \theta \leqslant \pi$. Let us take a differential of both sides of this equation, multiply by $(1-l)$ and integrate with respect to $\mathrm{d}(\sin^{2m}\theta)$ between $\theta = 0$ and θ_1; the result discloses that

$$\begin{aligned}\int_0^{\theta_1} (1-l)\,\mathrm{d}(\sin^{2m}\theta) &\equiv A_{2m} \\ &= \frac{\Gamma(2m+1)}{2^{2m-1}} \sum_{j=0}^{\infty} \frac{(-1)^j}{\Gamma(m+j+\frac{3}{2})\Gamma(m-j+\frac{1}{2})} \int_0^{\theta_1} (1-l)\,\mathrm{d}\{\sin(2j+1)\theta\}.\end{aligned} \tag{3.34}$$

Since, however,

$$\mathrm{d}\{\sin(2j+1)\theta\} = (2j+1)\cos(2j+1)\theta\,\mathrm{d}\theta \tag{3.35}$$

while – in accordance with Equation (3.32),

$$\begin{aligned}\int_0^{\theta_1} (1-l)\cos(2j+1)\theta\,\mathrm{d}\theta &= \tfrac{1}{2}F_1\left(\frac{2j+1}{2\pi}\right) \\ &= \frac{\sin(2j+1)\theta_1}{2(2j+1)} \sum_{n=0}^{\infty} \frac{(-1)^n[(2j+1)\theta_1]^2}{[(2j+1)\theta_1]^2-[n\pi]^2}\,a_n\end{aligned} \tag{3.36}$$

by Equation (3.32), a combination of Equations (3.34)–(3.36) yields

$$\begin{aligned}A_{2m} = \frac{\Gamma(2m+1)}{4^m} &\sum_{j=0}^{\infty} \frac{(-1)^j \sin(2j+1)\theta_1}{\Gamma(m+j+\frac{3}{2})\Gamma(m-j+\frac{1}{2})} \\ &\times \sum_{n=0}^{\infty} \frac{(-1)^n[(2j+1)\theta_1]^2}{[(2j+1)\theta_1]^2-[n\pi]^2}\,a_n,\end{aligned} \tag{3.37}$$

where the second summation may have to be restricted to $n = 0, 1, 2, \ldots, N$ if no more than $N + 1$ coefficients a_n of the Fourier series on the r.h.s. of Equation (3.7) can be determined significantly from the observed data (a retention of more terms would require extrapolation). For $2m = 1, 3, 5, 7, \ldots,$ Equations (3.34) – (3.37) reduce to Equations (3.24)–(3.27) already given; but the present Equation (3.37) holds good for any value of $m \geqslant 0$ – irrespective of whether it is integral (even as well as odd) or fractional.

Equation (3.37) should permit us to evaluate the moments A_{2m} of the light curve as a linear combination of the Fourier coefficients a_n on the r.h.s. of Equation (3.7), the values of which can be determined by a least-squares fit of Equation (3.7) to available observations. Once such a solution has been performed, any planimetry to obtain the A_{2m}'s becomes superfluous; and can be resorted to only for the sake of an independent check of numerical work. The reader can also convince himself of the speed with which the expansions on the r.h.s. of Equation (3.37) converge to the desired result.

This speed can, moreover, be increased further if we restrict the range of validity of the Fourier expansion (3.33) underlying (3.37) from $(0, \pi)$ to $(0, \theta_1)$. The former range is really superfluous for our purposes; for the limits of integration on both sides of Equation (3.34) exclude any angles $\theta > \theta_1$; and within a narrower range the convergence of the expansion for $\sin^\mu \theta$ can be speeded up by the following device.

C. RECURSION RELATIONS BETWEEN DIFFERENT MOMENTS OF THE LIGHT CURVES

Let us consider the expressions

$$y = \sin^\mu \theta \equiv x^\mu, \tag{3.38}$$

defining y as a function of x. A single differentiation of y with respect to x discloses at once they y represents a solution of the first-order differential equation

$$x \frac{\mathrm{d}y}{\mathrm{d}x} - \mu y = 0, \tag{3.39}$$

whose integration constant is specified by a single boundary condition which we shall locate at $\theta = \theta_1$, requiring that

$$y(x_1) = x_1^\mu = \sin^\mu \theta_1. \tag{3.40}$$

Our aim will be to develop a method by which the desired solution of the foregoing boundary-value problem (3.39)–(3.40) can be approximated by polynomials in *integral* powers of x.

This can indeed be done if we replace Equation (3.39) by a substitute of the form

$$x \frac{\mathrm{d}y_n}{\mathrm{d}x} - \mu y_n = \tau_n P_n(x), \tag{3.41}$$

defining an auxiliary function y_n which satisfies the same boundary condition

(3.40) as y, but obeys Equation (3.41) where $P_n(x)$ on the right-hand side stands for an (as yet arbitrary) polynomial of n-th degree in x (n being a positive integer), and τ_n is a constant. Equations (3.39) for y and Equation (3.41) for y_n differ only in the expressions on their right-hand sides; and as $\tau_n \to 0$, obviously $y_n \to y$.

In order to construct the solutions of Equation (3.41) for $y_n(x)$, consider an associated boundary-value problem represented by the equations

$$\left\{x \frac{\mathrm{d}}{\mathrm{d}x} - \mu\right\} Q_n(x) = x^n, \quad Q_n(0) = 0; \tag{3.42}$$

the solution of which is

$$Q_n(x) = \frac{x^n}{n - \mu}. \tag{3.43}$$

Let, moreover, the polynomials $P_n(x)$ on the r.h.s. of Equation (3.41) be expressible as

$$P_n(x) = \sum_{j=0}^{n} c_j^{(n)} x^j, \tag{3.44}$$

where the $c_j^{(n)}$'s represent numerical coefficients.

If so, however, the solution of Equation (3.41) can obviously be expressed as a combination of the $Q_n(x)$'s defined by Equation (3.43), of the form

$$y_n(x) = \tau_n \sum_{j=0}^{n} \frac{c_j^{(n)} x^j}{j - \mu}; \tag{3.45}$$

and the value of τ_n can, moreover, be obtained from the boundary condition (3.40), which assumes now the form

$$y_n(x_1) = \tau_n \sum_{j=0}^{n} \frac{c_j^{(n)} x_1^j}{j - \mu} = \sin^\mu \theta_1. \tag{3.46}$$

Inserting from τ_n from Equation (3.46) in (3.45) we find that our n-th approximation to the function $y = \sin^\mu \theta$ will be a polynomial of n-th degree in $\sin \theta$, and of the form

$$\sin^\mu \theta = \frac{\sin^\mu \theta_1 \sum_{j=0}^{n} \frac{c_j^{(n)} \sin^j \theta}{j - \mu}}{\sum_{j=0}^{n} \frac{c_j^{(n)} \sin^j \theta_1}{j - \mu}}, \tag{3.47}$$

which for $j = \mu$ reduces to an identity.

In choosing the family of polynomials to adopt for $P_n(x)$ on the r.h.s. of Equation (3.41) expressed by Equation (3.44), let us keep in mind that our aim is not to minimize a difference $|y_n(x) - y(x)|$ for any particular value of the argument, but rather over the entire range $0 < x < x_1$ of integration on both sides of Equation (3.34). Now, as is well known (cf., e.g., Szegö, 1939), a square of this difference will be minimized between $\pm x_1$ if we identify the $P_n(x)$'s with the Legendre polynomials orthogonalized within that range.

Suppose, for the sake of illustration, that the first and last contact of the eclipse occurs at $\theta_1 = \pm 30°$, corresponding to a range of $\pm\frac{1}{2}$ in x. If so, the Legendre polynomials orthogonalized within this range will be of the form

$$P_n(x) = \frac{1}{4^n n!} \frac{\mathrm{d}^n}{\mathrm{d}x^n} (4x^2 - 1); \tag{3.48}$$

yielding, for

$$n = 0: P_0 = 1,$$

$$n = 1: P_1 = 2x,$$

$$n = 2: P_2 = 6x^2 - \tfrac{1}{2},$$

$$n = 3: P_3 = 20x^3 - 3,$$

$$n = 4: P_4 = 70x^4 - 15x^2 + \tfrac{3}{8},$$

etc.

The values of $c_j^{(n)}$ for the orthogonality limits $\pm\frac{1}{2}$ are implied in the foregoing equations: e.g., for $n = 4$ the non-zero coefficients are

$$c_0^{(4)} = \tfrac{3}{8}, \quad c_2^{(4)} = -15, \quad c_4^{(4)} = 70, \tag{3.49}$$

which on insertion in Equation (3.47) yield a three-term approximation to

$$\sin^\mu \theta = \left\{\frac{3(\mu - 2)(\mu - 4)}{8(\mu + 1)(\mu + 3)}\right\} \frac{1}{2^\mu} - \left\{\frac{15\mu(\mu - 4)}{(\mu + 1)(\mu + 3)}\right\} \frac{\sin^2 \theta}{2^\mu} + \left\{\frac{70\mu(\mu - 2)}{(\mu + 1)(\mu + 3)}\right\} \frac{\sin^4 \theta}{2^\mu}, \tag{3.50}$$

which for $\mu = 0, 1, 2$ reduces to identities. Moreover, for $n = 6, 8, 10, \ldots$ higher approximations to $\sin \theta$ can be obtained in terms of a greater number of even powers of $\sin \theta$; and for odd values of $n = 1, 3, 5, \ldots$ similar approximations follow in odd powers of $\sin \theta$. Inserting these in Equation (3.34) we can translate, e.g., Equation (3.50) into

$$A_\mu = \left\{\frac{3(\mu - 2)(\mu - 4)}{8(\mu + 1)(\mu + 3)}\right\} \frac{A_0}{2^\mu} - - \left\{\frac{15\mu(\mu - 4)}{(\mu + 1)(\mu + 3)}\right\} \frac{A_2}{2^\mu} + \left\{\frac{70\mu(\mu - 2)}{(\mu + 1)(\mu + 3)}\right\} \frac{A_4}{2^\mu}, \tag{3.51}$$

valid for any positive value of μ; and similar expressions (based on the use of polynomials $P_n(x)$ of odd degree) for $A\mu$ can be obtained in terms of odd moments A_{2j+1} ($j = 0, 1, 2, \ldots$). For fuller details of such approximations, based also on other choice of the $P_n(x)$'s than those adopted in this section, cf. Kopal (1976c).

D. SOLUTION FOR THE MOMENTS AND THEIR UNCERTAINTY

Let us return now to Equation (3.37) which should permit us to express the moments A_{2m} of the light curve for arbitrary value of $m \geqslant 0$ in terms of the coefficients a_n of the Fourier cosine series on the r.h.s. of Equation (3.7). This way of establishing the numerical values of the requisite moments A_{2m} eliminates planimetry, and can be performed by the method of least-squares, which lends itself for complete automation.

The method itself is well known to astronomers; so that only its salient features need to be recalled in this place. Let Equation (3.7) with its Fourier expansion truncated to $N+1$ terms be rewritten symbolically as

$$\sum_{n=0}^{N} f_n a_n = g, \tag{3.52}$$

where

$$f_n = \begin{cases} \frac{1}{2} & \text{for } n = 0, \\ \cos(n\pi\theta/\theta_1) & \text{for } n > 0, \end{cases} \tag{3.53}$$

$$g = 1 - l(\theta), \tag{3.54}$$

and the coefficients a_n are unknowns whose most probable values are to be determined by our analysis.

Suppose that, to this end, we are in possession of M observations of $l(\theta_m)$, from which $M \gg N+1$ *equations of condition* of the form (3.52) can be constructed. Let, moreover, square brackets define the cross-product sums

$$[f_i f_j] \equiv \sum_{m=1}^{M} \cos\frac{i\pi\theta_m}{\theta_1} \cos\frac{j\pi\theta_m}{\theta_1} \tag{3.55}$$

and

$$[f_i g] \equiv \sum_{m=1}^{M} \{1 - l(\theta_m)\} \cos\frac{i\pi\theta_m}{\theta_1} \tag{3.56}$$

for $i, j = 0, 1, 2, \ldots, N$; such that, for $i = 0$,

$$[f_0 f_0] \equiv \tfrac{1}{4}M; \tag{3.57}$$

while, for $j > 0$,

$$[f_0 f_j] \equiv \frac{1}{2} \sum_{m=1}^{M} \cos\frac{j\pi\theta_m}{\theta_1} \tag{3.58}$$

and

$$[f_0 g] = \frac{1}{2} \sum_{m=1}^{M} \{1 - l(\theta_m)\}. \tag{3.59}$$

If so, the *most probable* values of the coefficients a_n satisfying our overdeterminate system of the equations of condition of the form (3.52) will result from the solution of the determinate system of *normal equations*

$$\sum_{i=0}^{N} \sum_{n=0}^{N} [f_i f_n] a_n = \sum_{i=0}^{N} [f_i g], \tag{3.60}$$

the coefficients of which are diagonally-symmetrical.

Let a_n $(n = 0, 1, 2, \ldots, N)$ be the constants satisfying the system of $N+1$ normal equations (3.60). These will not, in general, satisfy all equations of condition of the form (3.52) identically; but their individual insertion in Equation (3.52) will leave residuals which we shall denote by $O-C$. Suppose that we form a sum $\Sigma\,(O-C)^2$ of the squares of all such residuals – which can also be found from the expression

$$\sum_{m=1}^{M} (O-C)^2 = [gg] - [f_0g]a_0 - [f_1g]a_1 - \cdots - [f_Ng]a_N. \tag{3.61}$$

The mean error* of a single equation of unit weight will then be given by

$$\epsilon = \left(\frac{\Sigma\,(O-C)^2}{M-N-1}\right)^{1/2}; \tag{3.62}$$

while the mean error of any one of the coefficients a_n resulting from our solution will be given by

$$\epsilon_n = \epsilon(D_{nn}/D_0)^{1/2}, \tag{3.63}$$

where D_0 stands for the determinant of the coefficients on the l.h.s. of the system of normal equations (3.60); and the D_{nn}'s, for the minors of D_0 obtained by omitting its n-th column and row.

Having obtained the values of $a_n \pm \epsilon_n$ $(n = 0, 1, 2, \ldots, N)$ of each coefficient of the Fourier series on the r.h.s. of Equation (3.7) truncated to $N+1$ terms, the reader may be tempted to ask: why proceed to the A_{2m}'s and not solve the a_n's for the elements of the system directly? This should indeed be possible in theory; but *not*, unfortunately, in practice; for the following reason. As we have seen in Section V-2, the moments A_{2m} can be theoretically factorized as $L_1 \sin^{2m}\theta_1 f(a, c_0)$ for any type of eclipse; while the Fourier coefficients a_n (see Section I-4) can be written only as $L_1 f(\sin\theta_1, a, c_0)$. In other words, the g-functions formulate in Section V-3 from the A_{2m}'s to aid the solution for the elements depend on only *two* unknowns (i.e., a and c_0); while similar expressions formed from the a_n's would have to be simultaneously solved for three unknowns (a, c_0 and $\sin\theta_1$). To do so numerically (or otherwise) for three rather than two unknowns constitutes a much more difficult task; and its difficulty fully justifies a transition from a_n to A_{2m}, as an auxiliary intermediate quantity, with the aid of Equation (3.37) of this chapter.

If we insert the most probable values of the a_n's found from the solution of normal equations (3.60) on the r.h.s. of Equation (3.37) the most probable values of the corresponding moments A_{2m} should be obtained. Care is, however, required in specification of the mean errors of the A_{2m}'s in terms of those of the a_n's; which can be performed as follows. Let Equation (3.37) be symbolically rewritten as

$$A_{2m} = \phi_0 a_0 + \phi_1 a_1 + \cdots + \phi_N a_N, \tag{3.64}$$

* The 'probable error' (defined so as to equalize the chances that the absolute magnitude of actual deviation exceeds its limits, or remains within them) is equal to 0.674 49 (or, approximately, two-thirds) of the mean error ϵ.

where the ϕ_n's are coefficients of the respective a_n's on the r.h.s. of Equation (3.37). If so, the mean error ϵ_A of the moments A_{2m} will be identical with that of the linear function (3.64) of the a_n's; and given by

$$\epsilon_A = \epsilon/\sqrt{w_A} \tag{3.65}$$

where ϵ continues to be given by Equation (3.62), and the weight $\sqrt{w_A}$ of the respective moment A_{2m} will be given by the quadratic form

$$\begin{aligned}\frac{D_0}{w_A} &= \sum_{n=0}^{N} \phi_n^2 D_{nn} + \sum_{n=0}^{N} \sum_{r=n+1}^{N} \phi_n \phi_r D_{nr} \\ &= \phi_0^2 D_{00} + \phi_1^2 D_{11} + \phi_2^2 D_{22} + \cdots + 2(\phi_0\phi_1 D_{01} + \phi_0\phi_2 D_{02} + \cdots \\ &\quad + 2(\phi_1\phi_2 D_{12} + \cdots)\end{aligned} \tag{3.66}$$

where D_{nr} denote the minors of the determinant D_0 of the system of equations (3.60), in which the n-th column and r-th row has been omitted.

Should all but one of the ϕ_n's be zero (and the remaining one be equal to 1), the error ϵ_A above would become identical with ϵ_n as given by Equation (3.63). Should, however, A_{2m} be a linear function of more than one a_n obtained by a simultaneous solution of Equations (3.60), the expression (3.65)–(3.66) for the total error of A_{2m} will contain not only a sum of the squares of the individual ϵ_n's under the square-root (represented by the first sum on the r.h.s. of Equation (3.66)), but also their cross-products (see the second sum on the r.h.s. of Equation (3.66)). The latter may increase or diminish the uncertainty which would affect A_{2m} if those of the a_n's were independent. Since, however, cases may arise (especially for shallow partial eclipses!) when the cross-products can dominate numerically the entire right-hand side of Equation (3.66), their existence should not be overlooked!

And once this all has been accomplished, we can translate the uncertainty of the A_{2m}'s into those of the g's from Section V-3, and proceed with the solution as outlined in that section. In doing so we should only keep in mind that, while the observational errors affecting a determination of the A_{2m}'s should be essentially independent of m (i.e., the 'noise' affecting their determination is essentially 'white'), the absolute values of the A_{2m}'s will diminish very rapidly with increasing value of m – the more rapidly, the wider the binary system. Therefore, the proportional errors $(\delta A_{2m})/A_{2m}$ are to be expected to increase rapidly with m; and since the quality of the solution based upon the moments A_{2m} will depend on their proportional – rather than absolute – errors, we recommended in Section V-3 to 'optimize' the solution by use of lowest moments furnishing a sufficiently large value of the Jacobian as given by Equation (3.23) of Chapter V.

In the case of total eclipses, it is sufficient for our ends to convert any three $A\mu$'s ($\mu > 0$) into three even moments A_{2j} by use of Equation (3.37), to express the elements r_1, r_2 and i in terms of the requisite A_{2j}'s; and then to ascertain their uncertainty in terms of those of the A_{2j}'s by means of Equations (3.65)–(3.66).

And by this we have come to the end of our narrative, and of this book. That its last chapter has been devoted to an error analysis of the processes developed in Chapters IV–VI, and to an outline of methods by which the uncertainty of the results obtained by an analysis of the light changes in the time- or frequency-domain, is only natural. For no results obtained on the basis of the data of finite accuracy can ever be exact; and to investigate the degree of uncertainty affecting results based upon them should – we expect – be regarded as an integral part of any solution, without which the elements resulting from such a solution do not merit serious consideration (nor do any results based upon them).

In more general terms, what really matters is the total *probability* p with which the respective elements are defined to lie within a given range. To summarize the subject matter of the last chapter in language of formalism customary in the theory of probability, let x stand for any one element (or combination thereof), deduced on the basis of a 'hypothesis' (or model) y. Let, moreover, $p(x)$ stand for the probability of the element in question, and $p_x(y)$ the probability of the hypothesis ('model') on which the analysis has been based. If so, the total probability $p_y(x)$ of the element x based upon the model y is given by Bayes's equation of the form

$$p_y(x) = p(x)p_x(y). \tag{3.67}$$

Direct methods of Section IV-1 based upon the use of fixed points, or numerical construction of 'synthetic' light curves (cf. pp. 10–11 of this book), are indeed equivalent to a 'pattern recognition' process, in which the probability $p_x(y)$ is arbitrarily set to 0 or 1 – a process which (in the latter case) is bound spuriously to increase the degree of confidence of the results. It is well to keep in mind, in this connection, that *any* process of reduction can only *lose* information contained in the observations (i.e., increase its 'entropy'), but never generate new information. This entropy becomes indeed a minimum if we set (arbitrarily) $p_x(y) = 1$ (i.e., 'recognize' the system); but to what extent is such a procedure physically justifiable?

In reality, we should expect that $p_x(y) < 1$ – a fact which cannot but increase the uncertainty of the results. In particular, numerical 'modelling' of the light curves (p. 11) and any results based upon them also cannot guarantee the *uniqueness* of the representation of the observations: such work may furnish particular sets of the elements which can do so, but will not disclose the range of other combinations of the elements by which the light changes can likewise be represented within the limits of observational errors. In point of fact, the smaller the intrinsic determinacy of the case from available evidence (photometric, spectroscopic), the easier it should become to reproduce the observed evidence by a wide range of combinations of the elements (and this range may indeed encompass the trial values adopted to begin with; thus engendering spurious confidence in easy success). In cases bordering on indeterminacy, merely to establish that a certain set of the elements can reproduce the observed light changes within the limits of observational errors may mean still very little from the physical point of view. Their probability can be established only by full-dress analytical methods outlined in this chapter; and to critical investigators who will pursue this path in the future the writer now wishes to bid Godspeed.

Bibliographical Notes

VII-**1**. An analysis of the errors which affect the solutions for the geometrical elements of eclipsing variables based on observational data of finite precision did not receive systematic consideration before the advent of the iterative methods in the time-domain, outlined in Sections 2–3 of Chapter IV. The fundamental papers underlying this iterative approach have already been listed in the Bibliographical Notes to that chapter; while those concerning the respective errors analysis are those by S. L. Piotrowski (1948a) and Z. Kopal (1948a).

The presentation of the subject as given in this section follows largely that contained in Chapter VI of the writer's treatise on *Close Binary Systems* (Kopal, 1959), brought up to date in certain details only.

VII-**2**. An improvement of preliminary elements of eclipsing binary systems by the method of differential corrections, as outlined in this section, constitutes a subject with a long and distinguished pedigree; the method having been invoked (to a different extent) in the early stage of our subject as far back as by G. W. Myers (1898), A. W. Roberts (1908) or H. N. Russell and H. Shapley (1914).

A. Pannekoek and E. van Dien (1937) attempted to extend the method of differential corrections to include a determination of stellar limb darkening simultaneously with other elements of the eclipse; but the first adequate treatment of this problem was not given till by A. B. Wyse (1939). In this study (which constituted a significant landmark in the development of our subject) Wyse formulated, however, the coefficients of the differential corrections in a closed form only for the case of uniformly bright discs; while the case of limb-darkened stars (not tractable in terms of elementary functions) was dealt with by him only by approximate methods.

An analytical theory of the method of differential corrections has been developed since by the author of this book in a number of publications (cf. Z. Kopal, 1943, 1946, 1950 or 1959); but several aspects of its presentation in this section (such as its extension to distorted stars, or to higher than linear coefficients of limb-darkening) are new.

VII-**3**. An outline of the error analysis in the frequency domain, as given in this section, is mostly new. Of earlier papers containing background information cf. Z. Kopal (1976b, 1979).

For fuller details concerning the application of least-squares techniques to our problems cf. Appendix to Chapter VI of the author's *Close Binary Systems* (1959) and references quoted therein. The astronomical reader without much previous acquaintance with such methods can do no better than to consult that old classic on *The Calculus of Observations*, by E. T. Whittaker and G. Robinson (1924), which has since passed through many successive editions and in which a derivation of all theorems used in this section can be found.

EXPLICIT FORMS OF THE ASSOCIATED α_n^0-FUNCTIONS AT INTERNAL TANGENCY

In Section 1 of Chapter III we found that the fractional loss of light α_n^0 of an arbitrarily limb-darkened spherical star at the moment of internal tangency (where $\delta_2 = r_1 - r_2$) can be generally expressed as

$$\alpha_n^0 = \frac{(n+2)!k^{(n+4)/2}}{\Gamma(\frac{1}{2}n+2)\Gamma(\frac{1}{2}n+3)}\,{}_2F_1(-\tfrac{1}{2}n, \tfrac{1}{2}n+2; \tfrac{1}{2}n+3; k), \tag{I.1}$$

where $k \equiv r_2/r_1 \leqslant 1$, and n stands for an integer. Should n be even, the hypergeometric series on the right-hand side of Equation (I.1) will reduce to polynomials in ascending powers of the ratio of the radii k. For odd values of n the series ${}_2F_1$ consist of an infinite number of terms. It is, however, of interest to note that even in such cases the expansions on the r.h.s. of Equation (I.1) for α_n^0 represent combinations of certain elementary functions (i.e., $\sin^{-1}\sqrt{k}$ and $\sqrt{k(1-k)}$), which for $n = 0(1)4$ assume the following forms:

$$\alpha_0^0 = k^2, \tag{I.2}$$

$$\alpha_1^0 = (4/3\pi)\{\sin^{-1}\sqrt{k} + \tfrac{1}{3}(4k-3)(2k+1)\sqrt{k(1-k)}\}, \tag{I.3}$$

$$\alpha_2^0 = \tfrac{1}{2}k^3(4-3k), \tag{I.4}$$

$$\alpha_3^0 = (4/5\pi)\{\sin^{-1}\sqrt{k} + \tfrac{1}{3}(4k-3)(2k+1)\sqrt{k(1-k)} + \tfrac{16}{15}k^2(7k-2)(1-k)\sqrt{k(1-k)}\}, \tag{I.5}$$

$$\alpha_4^0 = \tfrac{1}{3}k^4(15-24k+10k^2), \tag{I.6}$$

etc.

EXPLICIT FORMS OF ASSOCIATED α-FUNCTIONS OF EVEN ORDERS AND INDICES

In Section 1 of Chapter III we have established that, for any phase of the eclipses, the associated α-functions of even orders n can be expressed in a closed form in terms of elementary functions; while those of odd orders call for a resort to incomplete elliptic integrals of first and second kind for their specification. If $n = 0$ and $m = 0, 1$, the respective associated α-functions during partial eclipses have already been found to be given by

$$\pi r_1^2 \alpha_0^0 = r_1^2 \cos^{-1}\frac{s}{r_1} + r_2^2 \cos^{-1}\frac{\delta - s}{r_2} - \delta\sqrt{r_1^2 - s^2} \tag{II.1}$$

and

$$\pi r_1^3 \alpha_0^1 = \delta r_2^2 \cos^{-1}\frac{\delta - s}{r_2} - \delta(\delta - s)\sqrt{r_1^2 - s^2}; \tag{II.2}$$

where s continues to be given by Equation (2.4) of Chapter I; while the respective functions of higher order and indices for $m + n \leqslant 5$ assume the forms

$$\alpha_0^2 = \frac{1}{4}\alpha_0^0 + \frac{1}{4}(5D - 2S)\alpha_0^1 - \frac{5D}{6\pi}(1 - S^2)^{3/2}, \tag{II.3}$$

$$\alpha_0^3 = \quad + \frac{1}{4}(7D^2 - 6DS + 3)\alpha_0^1 - \frac{D}{6\pi}(7D + S)(1 - S^2)^{3/2}, \tag{II.4}$$

$$\begin{aligned}\alpha_0^4 = \ & \frac{1}{8}\alpha_0^0 + \frac{1}{8}\{7D^2(3D - 4S) + 2(7D - 2S) + 4DS^2\}\alpha_0^1 \\ & - \frac{D}{60\pi}\{35D(3D - S) + 6S^2 + 49\}(1 - S^2)^{3/2},\end{aligned} \tag{II.5}$$

$$\alpha_2^0 = \frac{1}{2}\alpha_0^0 - \frac{1}{2}(3D - 2S)\alpha_0^1 + \frac{D}{\pi}(1 - S^2)^{3/2}, \tag{II.6}$$

$$\alpha_2^1 = \quad -2D(D - S)\alpha_0^1 + \frac{4D^2}{3\pi}(1 - S^2)^{3/2}, \tag{II.7}$$

$$\begin{aligned}\alpha_2^2 = \ & \frac{1}{12}\alpha_0^0 - \frac{1}{12}\{5D^2(7D - 10S) + 2(5D - S) + 8DS^2\}\alpha_0^2 \\ & + \frac{D}{18\pi}\{5D(7D - 3S) + 3\}(1 - S^2)^{3/2},\end{aligned} \tag{II.8}$$

$$\begin{aligned}\alpha_2^3 = \ & -\frac{D}{2}\{D^2(9D - 17S) + 3S(2DS - 1) + 5D\}\alpha_0^1 \\ & + \frac{D^2}{15\pi}\{5D(9D - 8S) + 16 - S^2\}(1 - S^2)^{3/2},\end{aligned} \tag{II.9}$$

$$\alpha_4^0 = \frac{1}{3}\alpha_0^0 + \frac{1}{3}\{2D^2(5D - 8S) + D(4S^2 - 1) + 2S\}\alpha_0^1$$

$$- \frac{2D}{9\pi}\{2D(5D - 3S) - 3\}(1 - S^2)^{3/2}, \tag{II.10}$$

$$\alpha_4^1 = -\frac{D^2}{35}\{177(D - S)^2 + 37(1 - S^2) - 2D^2\}\alpha_0^1$$

$$- \frac{10D^2}{3\pi}(D - S)(1 - S^2)^{3/2}; \tag{II.11}$$

where we have abbreviated $\delta/r_1 \equiv D$ and $s/r_1 \equiv S$.

The expressions for the associated α-functions of odd orders n lack symmetry and are too long to be reproduced in this place; for their explicit forms the reader is referred to Kopal (1942).

DEFINITION OF HYPERGEOMETRIC SERIES

Throughout the entire theory of the light curves of eclipsing binary systems we frequently encountered certain types of *hypergeometric series*, in one or two variables, which appear to be basic to a mathematical description of astronomical phenomena with which we have been concerned in this book. A concept of the hypergeometric series has been introduced in mathematical analysis by Gauss (1812); and since his time the literature on this subject has become so large that only a merest outline of the results actually used in the text can be given in this place.

Let, as usual,

$$(\alpha)_n \equiv \alpha(\alpha+1)(\alpha+2)\ldots(\alpha+n-j) \qquad \text{(III.1)}$$

denote the usual Pochhammer symbol (such that $(\alpha)_0 = 1$). If so, a generalized hypergeometric series ${}_pF_q$ of an argument x is defined by

$${}_pF_q = \sum_{n=0}^{\infty} \frac{(\alpha_1)_n(\alpha_2)_n\ldots(\alpha_p)_n}{(\gamma_1)_n(\gamma_2)_n\ldots(\gamma_2)_n}\frac{x^n}{n!}. \qquad \text{(III.2)}$$

When $p \leqslant q$, this series converges for all values of x. If, on the other hand, $p > q+1$, the series converges only for $x = 0$ and is, therefore, significant only if it terminates.

In all cases met in this book $p = q+1$, and x is a positive quantity. If so, the infinite series (III.2) converges when $x < 1$; while if $x = 1$ it does so provided that

$$\gamma_1 + \gamma_2 + \cdots + \gamma_q - \alpha_1 - \alpha_2 - \cdots - \alpha_p > 0. \qquad \text{(III.3)}$$

With the exception of Equation (2.9) of Chapter V where $p = 3$ and $q = 2$, all hypergeometric series of this type correspond to $p = 2$ and $q = 1$, which for (say) $\alpha_1 = -n$ reduces (cf. Equation (3.90) of Chapter I) to the Jacobi polynomial

$${}_2F_1(-n, n+a; \gamma; x) \equiv G_n(a, \gamma, x). \qquad \text{(III.4)}$$

The 'ordinary' hypergeometric series ${}_2F_1(\alpha, \beta; \gamma; x)$ is known to satisfy a second-order differential equation

$$\left\{x(1-x)\frac{d^2}{dx^2} + [\gamma - (\alpha+\beta+1)x]\frac{d}{dx} - \alpha\beta\right\}{}_2F_1 = 0; \qquad \text{(III.5)}$$

and, in addition, a large number of algebraic as well as differential recursion formulae for which the reader is referred to standard sources (e.g., Chapter XIV of Whittaker and Watson, 1920; Bailey, 1935; Erdélyi *et al.*, 1953). Two integral formulae (of which frequent use has been made in the text) should, however, be especially pointed out: namely, that

$$
{}_2F_1(\alpha, \beta; \gamma; x) = \frac{\Gamma(\gamma)}{\Gamma(\beta)\Gamma(\gamma-\beta)} \int_0^1 t^{\beta-1}(1-t)^{\gamma-\beta-1}(1-xt)^{-\alpha}\,\mathrm{d}t, \tag{III.6}
$$

valid for $x < 1$ if $\gamma < \beta > 0$; and

$$
{}_2F_1(\alpha, \beta; \gamma; x) = \frac{\Gamma(\gamma)}{\Gamma(\lambda)\Gamma(\gamma-\lambda)} \int_0^1 t^{\lambda-1}(1-t)^{\gamma-\lambda-1}{}_2F_1(\alpha, \beta; \lambda; xt)\,\mathrm{d}t \tag{III.7}
$$

provided that $\gamma > \lambda > 0$.

In addition to the hypergeometric series of single variables of the form ${}_pF_q$, frequent occurrence of two particular types of Appell's hypergeometric functions of two variables has been encountered in Chapters III and V. In Section III-3 we found that the $J^m_{\beta,\gamma}$-integrals (arising in connection with a distortion of the components of close eclipsing systems) are expressible in terms of Appell's hypergeometric series of the form

$$
F^{(1)}(\alpha; \beta, \beta'; \gamma; \alpha, y) = \sum_{m=0}^{\infty} \sum_{n=0}^{\infty} \frac{(\alpha)_{m+n}(\beta)_m(\beta')_n}{(\gamma)_{m+n}} \frac{x^m y^n}{m!n!}, \tag{III.8}
$$

where $y = 2x$; while in Section V-2 we found that the 'moments of the light-curves' A_{2m} for arbitrary types of eclipses result as series in terms of the functions

$$
F^{(4)}(\alpha, \beta; \gamma, \gamma'; x, y) = \sum_{m=0}^{\infty} \sum_{n=0}^{\infty} \frac{(\alpha)_{m+n}(\beta)_{m+n}}{(\gamma)_m(\gamma')_n} \frac{x^m y^n}{m!n!}. \tag{III.9}
$$

The (partial) differential equations and recursion relation satisfied by Appell's hypergeometric functions of two variables are much less well known than is the case with hypergeometric functions of a single variable; apart from the classical treatise by Appell and Kampé de Fériet (1926) the reader should again consult Bailey (1935) or Erdélyi *et al.* (1953); in addition to the more recent papers by Lanzano (1976a, b, c) already quoted in the text.

EXPLICIT FORMS AND TABLES OF INTEGRALS OF THE TYPE $J^m_{-1,\gamma}$

In our investigation, in Section 2 of Chapter II, of the photometric effects arising from a distortion of the limb of the eclipsing star we have been led to introduce a family of integrals of the form $J^m_{\beta,\gamma}$; whose analytical properties were subsequently investigated in Section III-3. The aim of the present appendix will be to present numerical tabulations of such functions for $m = 0(1)5$, $\beta = -1$ (i.e., corresponding to the effects of tidal distortion of m-th spherical-harmonic symmetry) and $\gamma = 0, 1$ (corresponding to the case of zero or linear limb-darkening).

The explicit forms of the equations defining these functions are given by

$$\pi J^0_{-1,0} = \cos^{-1}\mu, \tag{IV.1}$$

$$\pi J^1_{-1,0} = \sqrt{1-\mu^2}, \tag{IV.2}$$

$$2\pi J^2_{-1,0} = \cos^{-1}\mu + \mu\sqrt{1-\mu^2}, \tag{IV.3}$$

$$3\pi J^3_{-1,0} = (2+\mu^2)\sqrt{1-\mu^2}, \tag{IV.4}$$

$$8\pi J^4_{-1,0} = 3\cos^{-1}\mu + \mu(3+2\mu^2)\sqrt{1-\mu^2}, \tag{IV.5}$$

$$15\pi J^5_{-1,0} = (8+4\mu^2+3\mu^4)\sqrt{1-\mu^2}, \tag{IV.6}$$

etc; and

$$\pi J^0_{-1,1} = 2\{2E-(1+\mu)F\}, \tag{IV.7}$$

$$3\pi J^1_{-1,1} = 2\{(1+\mu)F-2\mu E\}, \tag{IV.8}$$

$$15\pi J^2_{-1,1} = 2\{2(9-2\mu^2)E-(1+\mu)(9-2\mu)F\}, \tag{IV.9}$$

$$105\pi J^3_{-1,1} = 2\{(25-6\mu+8\mu^2)(1+\mu)F-2\mu(19+8\mu^2)E\}, \tag{IV.10}$$

$$315\pi J^4_{-1,1} = 2\{2(147-24\mu^2-16\mu^4)E - (147-36\mu+12\mu^2-16\mu^3)(1+\mu)F\}, \tag{IV.11}$$

$$3465\pi J^5_{-1,1} = 2\{(675-204\mu+252\mu^2-96\mu^3+128\mu^4)(1+\mu)F - 2\mu(471+156\mu^2+128\mu^4)E\}, \tag{IV.12}$$

etc. if the eclipse is partial; and

$$\pi J^0_{-1,0} = 1, \tag{IV.13}$$

$$\pi J^1_{-1,0} = 0, \tag{IV.14}$$

$$\pi J^2_{-1,0} = \tfrac{1}{2}, \tag{IV.15}$$

$$\pi J^3_{-1,0} = 0, \tag{IV.16}$$

$$\pi J^4_{-1,0} = \tfrac{3}{8}, \tag{IV.17}$$

$$\pi J^{5}_{-1,0} = 0; \tag{IV.18}$$

and

$$\pi\kappa J^{0}_{-1,1} = 4E, \tag{IV.19}$$

$$3\pi\kappa J^{1}_{-1,1} = 4\{(1+\mu)F - \mu E\}, \tag{IV.20}$$

$$15\pi\kappa J^{2}_{-1,1} = 4\{(9-2\mu^2)E + 2\mu(1+\mu)F\}, \tag{IV.21}$$

$$105\pi\kappa J^{3}_{-1,1} = 4\{(25+8\mu^2)(1+\mu)F - \mu(19+8\mu^2)E\}, \tag{IV.22}$$

$$315\pi\kappa J^{4}_{-1,1} = 4\{(147-24\mu^2-16\mu^4)E + 4(9+4\mu^2)(1+\mu)F\}, \tag{IV.23}$$

$$3465\pi\kappa J^{5}_{-1,1} = 4\{(675+252\mu^2+128\mu^4)(1+\mu)F - \mu(471+156\mu^2+128\mu^4)E\}, \tag{IV.24}$$

etc. if they are annular.

In these equations, as before,

$$\mu \equiv \frac{\delta - s}{r_2} = 1 - 2\kappa^2 \tag{IV.25}$$

in accordance with Equation (3.27) of Chapter III; while F and E denote the complete elliptic integrals of the first and second kind, respectively of modulus κ^2. It should only be kept in mind that, for annular eclipse, the relation (IV.25) between μ and κ^2 is to be replaced by

$$\kappa^2 = \frac{2}{1-\mu}; \tag{IV.26}$$

i.e., that the 'partial' and 'annular' κ's expressed as functions of μ are mutually reciprocal.

Five-digit numerical values of the functions defined by Equations (IV.1)–(IV.12) valid for partial eclipses, and Equations (IV.19)–(IV.24) for annular eclipses, as given in the accompanying tables, have been taken from a previous work by the present writer (cf. Kopal, 1947); and are listed as functions of the argument

$$\nu = 2\sin^{-1}\kappa, \tag{IV.27}$$

in accordance with Equation (1.52) of Chapter V. Within the domain considered, the angle ν is a single-valued function of μ for either type of eclipse; and if the latter is partial, Equation (IV.27) reduces to $\nu = \cos^{-1}\mu$. At the moment of first contact of the eclipse, $\nu = 0°$. If the eclipse is an occultation (i.e., $r_1 < r_2$), the angle ν increases with diminishing light until a maximum (the location of which depends on the ratio r_1/r_2), and decreases thereafter to zero at the moment of inner contact (i.e., the beginning of totality). If, on the other hand, the eclipse happens to be a transit, ν increases from 0° to 180° at the moment of internal tangency; and decreases again during annular phase to zero at the moment of central eclipse (when $\delta = 0$).

In the tables which follow 5-digit values of $J^{m}_{-1,\gamma}$ are listed for $\nu = 0°(5°)180°$, together with the second 'modified' differences M'' of the respective tabulations to facilitate non-linear interpolation for intermediate entries by means of the Everett interpolation formula

$$(J)_n = (1-n)(J)_0 + n(J)_1 + M''_0 E''_0(n) + M''_1 E''_1(n) + \cdots, \qquad \text{(IV.28)}$$

within each interval of tabulation (in which $0 \leqslant n \leqslant 1$). The use of Everett's formula (IV.28) in the above form is equivalent to that of any other central-difference formula employing tabular differences up to the fourth (cf., e.g., Kopal, 1955; Sections II-K and L); but it possesses the advantage that only one modified difference M'' needs to be tabulated alongside each entry. The second-difference Everett interpolation coefficients $E''_0(n)$ and $E''_1(n)$ have been extensively tabulated (for a bibliography of such tables cf., e.g., Fletcher, Miller and Rosenhead, 1946); and both are numerically smaller than 0.1 for $0 \leqslant n \leqslant 1$.

TABLE IV-1(p)

ν	$J^{0}_{-1,0}$	M″	$J^{1}_{-1,0}$	M″	$J^{2}_{-1,0}$	M″	$J^{3}_{-1,0}$	M″	$J^{4}_{-1,0}$	M″	$J^{5}_{-1,0}$	M″
0°	0.00000	0.0000	0.00000	0.0000	0.00000	0.0000	0.00000	0.0000	0.00000	0.0000	0.00000	0.000
5	0.02778	0.0000	0.02774	-0.0002	0.02771	-0.0004	0.02767	-0.0006	0.02764	-0.0008	0.02760	-0.001
10	0.05556	0.0000	0.05527	-0.0004	0.05500	-0.0008	0.05471	-0.0012	0.05445	-0.0016	0.05417	-0.002
15	0.08333	0.0000	0.08239	-0.0006	0.08146	-0.0012	0.08054	-0.0018	0.07965	-0.0023	0.07878	-0.002
20	0.11111	0.0000	0.10887	-0.0008	0.10671	-0.0016	0.10462	-0.0022	0.10261	-0.0028	0.10068	-0.003
25	0.13889	0.0000	0.13452	-0.0010	0.13041	-0.0019	0.12651	-0.0025	0.12284	-0.0031	0.11937	-0.003
30	0.16667	0.0000	0.15915	-0.0012	0.15225	-0.0021	0.14589	-0.0027	0.14003	-0.0032	0.13462	-0.003
35	0.19444	0.0000	0.18258	-0.0014	0.17200	-0.0023	0.16255	-0.0028	0.15409	-0.0031	0.14648	-0.003
40	0.22222	0.0000	0.20461	-0.0016	0.18948	-0.0024	0.17643	-0.0028	0.16510	-0.0028	0.15523	-0.002
45	0.25000	0.0000	0.22508	-0.0017	0.20458	-0.0024	0.18757	-0.0026	0.17332	-0.0024	0.16131	-0.002
50	0.27778	0.0000	0.24384	-0.0019	0.21726	-0.0024	0.19614	-0.0023	0.17913	-0.0020	0.16524	-0.001
55	0.30556	0.0000	0.26074	-0.0020	0.22756	-0.0023	0.20242	-0.0020	0.18297	-0.0015	0.16758	-0.001
60	0.33333	0.0000	0.27566	-0.0021	0.23559	-0.0021	0.20675	-0.0016	0.18530	-0.0010	0.16884	-0.000
65	0.36111	0.0000	0.28849	-0.0022	0.24152	-0.0019	0.20950	-0.0012	0.18658	-0.0007	0.16944	-0.000
70	0.38889	0.0000	0.29911	-0.0023	0.24560	-0.0016	0.21107	-0.0008	0.18719	-0.0004	0.16968	-0.000
75	0.41667	0.0000	0.30746	-0.0023	0.24812	-0.0012	0.21184	-0.0005	0.18742	-0.0002	0.16975	-0.000
80	0.44444	0.0000	0.31347	-0.0024	0.24944	-0.0008	0.21213	-0.0002	0.18749	-0.0001	0.16976	0.000
85	0.47222	0.0000	0.31710	-0.0024	0.24993	-0.0004	0.21220	0.0000	0.18750	0.0000	0.16977	0.000
90	0.50000	0.0000	0.31831	-0.0024	0.25000	0.0000	0.21221	0.0000	0.18750	0.0000	0.16977	0.000
95	0.52778	0.0000	0.31710	-0.0024	0.25007	0.0004	0.21220	0.0000	0.18750	0.0000	0.16977	0.000
100	0.55556	0.0000	0.31347	-0.0024	0.25056	0.0008	0.21213	-0.0002	0.18751	0.0001	0.16976	0.000
105	0.58333	0.0000	0.30746	-0.0023	0.25188	0.0012	0.21184	-0.0005	0.18758	0.0002	0.16975	-0.000
110	0.61111	0.0000	0.29911	-0.0023	0.25441	0.0016	0.21107	-0.0008	0.18781	0.0004	0.16968	-0.000
115	0.63889	0.0000	0.28849	-0.0022	0.25849	0.0019	0.20950	-0.0012	0.18842	0.0007	0.16944	-0.000
120	0.66667	0.0000	0.27566	-0.0021	0.26442	0.0021	0.20675	-0.0016	0.18970	0.0010	0.16884	-0.000
125	0.69444	0.0000	0.26074	-0.0020	0.27244	0.0023	0.20242	-0.0020	0.19203	0.0015	0.16758	-0.001
130	0.72222	0.0000	0.24384	-0.0019	0.28274	0.0024	0.19614	-0.0023	0.19586	0.0020	0.16524	-0.001
135	0.75000	0.0000	0.22508	-0.0017	0.29542	0.0024	0.18757	-0.0026	0.20167	0.0024	0.16131	-0.002
140	0.77778	0.0000	0.20461	-0.0016	0.31052	0.0024	0.17643	-0.0028	0.20990	0.0028	0.15523	-0.002
145	0.80556	0.0000	0.18258	-0.0014	0.32800	0.0023	0.16255	-0.0028	0.22091	0.0031	0.14648	-0.003
150	0.83333	0.0000	0.15915	-0.0012	0.34775	0.0021	0.14589	-0.0027	0.23497	0.0032	0.13462	-0.0034
155	0.86111	0.0000	0.13452	-0.0010	0.36959	0.0019	0.12651	-0.0025	0.25216	0.0031	0.11937	-0.003
160	0.88889	0.0000	0.10887	-0.0008	0.39329	0.0016	0.10462	-0.0022	0.27239	0.0028	0.10068	-0.003
165	0.91667	0.0000	0.08239	-0.0006	0.41855	0.0012	0.08054	-0.0018	0.29535	0.0023	0.07878	-0.002
170	0.94444	0.0000	0.05527	-0.0004	0.44501	0.0008	0.05471	-0.0012	0.32056	0.0016	0.05417	-0.002
175	0.97222	0.0000	0.02774	-0.0002	0.47229	0.0004	0.02767	-0.0006	0.34736	0.0008	0.02760	-0.001
180°	1.00000	0.0000	0.00000	0.0000	0.50000	0.0000	0.00000	0.0000	0.37500	0.0000	0.00000	0.000

TABLE IV-2(p)

ν	$J^0_{-1,1}$	M″	$J^1_{-1,1}$	M″	$J^2_{-1,1}$	M″	$J^3_{-1,1}$	M″	$J^4_{-1,1}$	M″	$J^5_{-1,1}$	M″
0°	0.00000	0.0038	0.00000	0.0038	0.00000	0.0038	0.00000	0.0038	0.00000	0.0038	0.00000	0.0038
5	0.00190	0.0038	0.00190	0.0038	0.00190	0.0038	0.00190	0.0038	0.00189	0.0038	0.00187	0.0038
10	0.00760	0.0038	0.00757	0.0037	0.00755	0.0036	0.00751	0.0036	0.00749	0.0035	0.00745	0.0036
15	0.01707	0.0037	0.01693	0.0035	0.01679	0.0034	0.01664	0.0032	0.01651	0.0030	0.01637	0.0028
20	0.03027	0.0037	0.02982	0.0033	0.02937	0.0030	0.02894	0.0027	0.02852	0.0025	0.02811	0.0022
25	0.04713	0.0036	0.04602	0.0031	0.04496	0.0026	0.04395	0.0022	0.04299	0.0019	0.04207	0.0015
30	0.06756	0.0035	0.06528	0.0028	0.06315	0.0022	0.06116	0.0017	0.05931	0.0013	0.05756	0.0009
35	0.09148	0.0034	0.08730	0.0024	0.08349	0.0017	0.08003	0.0011	0.07686	0.0007	0.07396	0.0003
40	0.11877	0.0032	0.11172	0.0020	0.10549	0.0012	0.09998	0.0006	0.09507	0.0001	0.09069	-0.0001
45	0.14929	0.0031	0.13814	0.0016	0.12865	0.0006	0.12049	0.0001	0.11344	-0.0003	0.10731	-0.0004
50	0.18289	0.0029	0.16616	0.0012	0.15246	0.0002	0.14109	-0.0003	0.13157	-0.0005	0.12350	-0.0006
55	0.21941	0.0027	0.19533	0.0007	0.17646	-0.0002	0.16140	-0.0006	0.14917	-0.0007	0.13908	-0.0007
60	0.25866	0.0025	0.22519	0.0002	0.20023	-0.0006	0.18111	-0.0008	0.16606	-0.0008	0.15396	-0.0007
65	0.30045	0.0023	0.25528	-0.0002	0.22342	-0.0009	0.20002	-0.0009	0.18217	-0.0008	0.16814	-0.0007
70	0.34455	0.0021	0.28513	-0.0007	0.24574	-0.0011	0.21802	-0.0009	0.19750	-0.0008	0.18166	-0.0006
75	0.39075	0.0018	0.31429	-0.0011	0.26699	-0.0012	0.23508	-0.0009	0.21207	-0.0007	0.19457	-0.0006
80	0.43879	0.0016	0.34233	-0.0015	0.28705	-0.0012	0.25123	-0.0008	0.22593	-0.0007	0.20690	-0.0005
85	0.48841	0.0013	0.36884	-0.0019	0.30590	-0.0012	0.26653	-0.0008	0.23913	-0.0006	0.21869	-0.0005
90	0.53935	0.0010	0.39345	-0.0023	0.32361	-0.0010	0.28103	-0.0007	0.25170	-0.0006	0.22994	-0.0005
95	0.59132	0.0007	0.41581	-0.0025	0.34030	-0.0009	0.29478	-0.0007	0.26365	-0.0006	0.24065	-0.0005
100	0.64401	0.0004	0.43564	-0.0028	0.35615	-0.0006	0.30778	-0.0008	0.27499	-0.0006	0.25082	-0.0005
105	0.69711	0.0001	0.45270	-0.0030	0.37140	-0.0003	0.31997	-0.0009	0.28573	-0.0006	0.26045	-0.0006
110	0.75029	-0.0003	0.46681	-0.0031	0.38631	-0.0001	0.33125	-0.0011	0.29589	-0.0005	0.26949	-0.0006
115	0.80321	-0.0006	0.47786	-0.0031	0.40115	0.0002	0.34144	-0.0013	0.30556	-0.0004	0.27792	-0.0007
120	0.85552	-0.0010	0.48581	-0.0031	0.41615	0.0004	0.35032	-0.0016	0.31486	-0.0002	0.28563	-0.0009
125	0.90684	-0.0014	0.49066	-0.0030	0.43153	0.0005	0.35765	-0.0018	0.32396	0.0000	0.29249	-0.0011
130	0.95678	-0.0018	0.49251	-0.0028	0.44744	0.0006	0.36317	-0.0020	0.33309	0.0003	0.29830	-0.0013
135	1.00495	-0.0022	0.49153	-0.0026	0.46395	0.0006	0.36667	-0.0022	0.34248	0.0005	0.30282	-0.0016
140	1.05093	-0.0027	0.48795	-0.0023	0.48104	0.0005	0.36798	-0.0023	0.35235	0.0006	0.30578	-0.0018
145	1.09425	-0.0031	0.48208	-0.0019	0.49859	0.0002	0.36706	-0.0022	0.36285	0.0007	0.30694	-0.0020
150	1.13443	-0.0036	0.47433	-0.0014	0.51635	-0.0002	0.36398	-0.0020	0.37403	0.0006	0.30614	-0.0020
155	1.17096	-0.0042	0.46517	-0.0009	0.53394	-0.0006	0.35896	-0.0016	0.38578	0.0003	0.30337	-0.0018
160	1.20325	-0.0049	0.45517	-0.0002	0.55086	-0.0013	0.35239	-0.0009	0.39779	-0.0002	0.29881	-0.0014
165	1.23063	-0.0056	0.44503	0.0007	0.56643	-0.0021	0.34486	0.0000	0.40953	-0.0010	0.29290	-0.0007
170	1.25231	-0.0067	0.43561	0.0017	0.57979	-0.0033	0.33724	0.0010	0.42019	-0.0021	0.28641	0.0005
175	1.26716	-0.0085	0.42807	0.0036	0.58972	-0.0052	0.33073	0.0029	0.42855	-0.0042	0.28059	0.0024
180°	1.27324	-0.0134	0.42441	0.0086	0.59418	-0.0102	0.32741	0.0079	0.43250	-0.0088	0.27743	0.0075

TABLE IV-3(a)

ν	$J^0_{-1,1}$	M″	$J^1_{-1,1}$	M″	$J^2_{-1,1}$	M″	$J^3_{-1,1}$	M″	$J^4_{-1,1}$	M″	$J^5_{-1,1}$	M″
0°	∞		0.00000	0.0000	∞		0.00000	0.0000	∞		0.00000	0.000
5	45.8294		0.01093	0.0000	22.91		0.00817	0.0000	17.17		0.00687	0.000
10	22.9038		0.02184	0.0000	11.452		0.01636	0.0000	8.5807		0.01373	0.000
15	15.2571		0.03277	0.0000	7.6289		0.02457	0.0000	5.7168		0.02059	0.000
20	11.4312		0.04374	0.0001	5.7151		0.03280	0.0000	4.2838		0.02746	0.000
25	9.13126		0.05476	0.0001	4.56555		0.04106	0.0000	3.4222		0.03435	0.000
30	7.59632		0.06582	0.0001	3.79802		0.04936	0.0001	2.8476	0.1524	0.04127	0.000
35	6.49802		0.07695	0.0001	3.24878		0.05771	0.0001	2.4361	0.0970	0.04823	0.000
40	5.67265		0.08815	0.0001	2.83598		0.06612	0.0001	2.1267	0.0656	0.05523	0.000
45	5.02931		0.09944	0.0001	2.51416	0.0619	0.07459	0.0001	1.88542	0.0463	0.06228	0.000
50	4.51343	0.0907	0.11081	0.0001	2.25604	0.0453	0.08312	0.0001	1.69184	0.0339	0.06939	0.000
55	4.09033	0.0681	0.12229	0.0001	2.04425	0.0341	0.09173	0.0001	1.53296	0.0256	0.07656	0.000
60	3.73686	0.0528	0.13389	0.0001	1.86724	0.0263	0.10044	0.0001	1.40012	0.0197	0.08380	0.000
65	3.43704	0.0417	0.14561	0.0001	1.71695	0.0209	0.10925	0.0001	1.28733	0.0156	0.09112	0.000
70	3.17944	0.0335	0.15746	0.0001	1.58776	0.0166	0.11815	0.0001	1.19031	0.0125	0.09853	0.000
75	2.95569	0.0273	0.16945	0.0002	1.47541	0.0136	0.12716	0.0001	1.10594	0.0102	0.10604	0.000
80	2.75955	0.0226	0.18159	0.0002	1.37677	0.0112	0.13629	0.0001	1.03184	0.0084	0.11365	0.000
85	2.58623	0.0190	0.19388	0.0002	1.28946	0.0094	0.14555	0.0001	0.96618	0.0070	0.12137	0.000
90	2.43202	0.0161	0.20634	0.0002	1.21161	0.0079	0.15495	0.0002	0.90758	0.0059	0.12921	0.000
95	2.29402	0.0138	0.21897	0.0002	1.14174	0.0068	0.16449	0.0002	0.85496	0.0051	0.13718	0.000
100	2.16993	0.0120	0.23177	0.0002	1.07870	0.0059	0.17417	0.0002	0.80744	0.0044	0.14528	0.000
105	2.05789	0.0105	0.24473	0.0002	1.02156	0.0051	0.18400	0.0002	0.76430	0.0038	0.15352	0.000
110	1.95641	0.0093	0.25787	0.0002	0.96955	0.0045	0.19399	0.0002	0.72496	0.0031	0.16190	0.000
115	1.86428	0.0083	0.27117	0.0001	0.92206	0.0040	0.20413	0.0001	0.68895	0.0029	0.17043	0.000
120	1.78051	0.0075	0.28461	0.0001	0.87857	0.0036	0.21442	0.0001	0.65590	0.0026	0.17910	0.000
125	1.70430	0.0069	0.29817	0.0001	0.83867	0.0033	0.22486	0.0001	0.62549	0.0024	0.18791	0.000
130	1.63502	0.0064	0.31183	0.0000	0.80203	0.0030	0.23542	0.0001	0.59746	0.0022	0.19686	0.000
135	1.57216	0.0060	0.32554	0.0000	0.76840	0.0028	0.24609	0.0001	0.57160	0.0020	0.20593	0.000
140	1.51534	0.0058	0.33924	−0.0001	0.73756	0.0027	0.25685	0.0000	0.54777	0.0019	0.21511	0.000
145	1.46430	0.0056	0.35284	−0.0002	0.70938	0.0026	0.26763	−0.0001	0.52585	0.0018	0.22436	0.000
150	1.41887	0.0055	0.36622	−0.0004	0.68379	0.0026	0.27834	−0.0001	0.50577	0.0018	0.23362	−0.000
155	1.37899	0.0056	0.37923	−0.0006	0.66079	0.0027	0.28889	−0.0003	0.48753	0.0019	0.24280	−0.000
160	1.34474	0.0058	0.39162	−0.0009	0.64046	0.0029	0.29910	−0.0006	0.47121	0.0021	0.25177	−0.0004
165	1.31636	0.0062	0.40306	−0.0014	0.62301	0.0032	0.30869	−0.0010	0.45698	0.0024	0.26030	−0.0007
170	1.29429	0.0069	0.41304	−0.0021	0.60883	0.0038	0.31726	−0.0016	0.44517	0.0029	0.26803	−0.0013
175	1.27933	0.0086	0.42075	−0.0038	0.59866	0.0054	0.32406	−0.0032	0.43647	0.0044	0.27428	−0.002
180°	1.27324	0.0134	0.42441	−0.0085	0.59418	0.0102	0.32740	−0.0079	0.43250	0.0091	0.27743	−0.007

ASSOCIATED α-FUNCTIONS FOR $m > 0$

In Section 2 of Chapter I we were led to introduce a class of associated α_n^0-functions of order $n \geqslant -1$ to describe the photometric effects of the eclipses of limb-darkened stellar discs which are spherical in form; and in Section II-2 generalized these to functions α_n^n of order n and index $m \geqslant 0$, required to account for the photometric effects, within eclipses, of rotational or tidal distortion. The associated α-functions of zero index were, moreover, found in Section I-3 to be expressible as Hankel transforms (3.32) or (3.34) in terms of the geometrical elements of the eclipses; and their Fourier approximations constructed in Sections I-3 and I-4. The aim of the present Appendix will be to extend the same formulation to the associated alpha-functions α_n^m of any (integral) index m in terms of suitable integral transforms.

In order to do so, let us return to Equation (3.1) of Chapter I, and generalize Equation (3.2) to the form

$$f(x, y) \equiv x^m z^n \tag{V.1}$$

where, in accordance with Equations (2.11) and (3.3) of Chapter I, $x = r\cos\theta$ and $z^2 = r_1^2 - x^2 - y^2 = r_1^2 - r^2$. The Fourier transform (3.1) of the aperture of radius r_1 will then be given by

$$F(u, v) = \int_0^\infty f(r)\left\{\int_{-\pi}^{\pi} e^{-2\pi iqr\cos(\theta-\phi)}\cos^m\theta\,\mathrm{d}\theta\right\} r^{m+1}\,\mathrm{d}r, \tag{V.2}$$

representing a generalization of (3.4) of Chapter I, where the function $f(r)$ continues to be given by Equations (3.12) and (3.13) of that chapter.

In order to evaluate the integral in curly brackets on the r.h.s. of Equation (V.2), we resort again to the use of the Jacobi expansion (3.6) of Chapter I. Since, moreover, for integral values of m and $j(>0)$,

$$\int_{-\pi}^{\pi} \cos 2j(\tfrac{1}{2}\pi + \theta - \phi)\cos^m\theta\,\mathrm{d}\theta = \frac{(-1)^j m![1-(-1)^{m+2j+1}]\pi}{2^m\Gamma(1+\frac{1}{2}m+j)\Gamma(1+\frac{1}{2}m-j)}\cos 2j\phi \tag{V.3}$$

and

$$\int_{-\pi}^{\pi} \sin(2j+1)(\tfrac{1}{2}\pi + \theta - \phi)\cos^m\theta\,\mathrm{d}\theta = \frac{(-1)^j m![1-(-1)^{m+2j+2}]\pi}{2^m\Gamma(\frac{3}{2}+\frac{1}{2}m+j)\Gamma(\frac{1}{2}+\frac{1}{2}m-j)}\cos(2j+1)\phi \tag{V.4}$$

as a generalization of Equations (3.7) and (3.8) of Chapter I; while, for $j = 0$,

$$\int_{-\pi}^{\pi} \cos^{2\mu} \theta \, d\theta = \begin{cases} 2\pi & \mu = 0, \\ 2B(\mu + \frac{1}{2}, \frac{1}{2}) & \mu > 0; \end{cases} \tag{V.5}$$

and

$$\int_{-\pi}^{\pi} \cos^{2\mu+1} \theta \, d\theta = 0 \tag{V.6}$$

for $\mu = 0, 1, 2, \ldots$, it follows that for *even* values of $m \equiv 2\mu$, the Fourier transform $F(u, v)$ of the aperture function $f(r)$ will assume the form

$$F(u, v) = 2B(\mu + \tfrac{1}{2}, \tfrac{1}{2}) \int_0^\infty f(r) J_0(2\pi qr) r^{2\mu+1} \, dr$$
$$+ \frac{\pi}{4^\mu(\mu + 1)} \sum_{j=1}^{\mu} \frac{(-1)^j \cos 2j\phi}{B(\mu + j + 1, \mu - j + 1)} \int_0^\infty f(r) J_{2_j}(2\pi qr) r^{2\mu+1} \, dr, \tag{V.7}$$

where $J_{2j}(qr)$ stands for the Bessel function of the first kind (of the respective index and argument), which for $\mu = 0$ reduces to Equation (3.11) of Chapter I. For *odd* values of $m \equiv 2\mu + 1$,

$$F(u, v) = -\frac{\pi i}{4^\mu(\mu + 1)} \sum_{j=0}^{\mu} \frac{(-1)^j \cos (2j + 1)\phi}{B(\mu + j + 2, \mu - j + 1)}$$
$$\times \int_0^\infty f(r) J_{2j+1}(2\pi qr) r^{2\mu+2} \, dr \tag{V.8}$$

which, unlike (V.7), becomes a purely imaginary quantity.

If, lastly, the function $f(r)$ continues to be identical with the law of limb-darkening of the form expressed by Equation (3.12) of Chapter I, the remaining integrations with respect to r on the right-hand sides of Equations (V.7)–(V.8) can be performed (cf., e.g., Erdélyi *et al.*, 1954; p. 26) to yield

$$F(u, v) = \frac{L_1}{4^\mu(2\mu + 1)} \sum_{l=0}^{\Lambda} C^{(l)} \sum_{j=0}^{\mu} \frac{(-1)^j B(\mu + j + 1, \nu)}{(2j)! B(\mu + j + 1, \mu - j + 1)}$$
$$\times (\pi q r_1)^{2j} {}_1F_2\{\mu + j + 1; \quad 2j + 1, \mu + \nu + j + 1; -(\pi q r_1)^2\}$$
$$\times \cos 2j\phi \tag{V.9}$$

if $m \equiv 2\mu$ is zero or an even integer; while if $m \equiv 2\mu + 1$ is odd,

$$F(u, v) = -\frac{iL_1}{4^\mu(2\mu + 2)} \sum_{l=0}^{\Lambda} C^{(l)} \sum_{j=0}^{\mu} \frac{(-1)^j B(\mu + j + 2, \nu)}{(2j + 1)! B(\mu + j + 2, \mu - j + 1)}$$
$$\times (\pi q r_1)^{2j} {}_1F_2\{\mu + j + 2; \quad 2j + 2, \quad \mu + \nu + j + 2; -(\pi q r_1)^2\}$$
$$\times \cos (2j + 1)\phi, \tag{V.10}$$

where L_1 stands for the fractional luminosity of the 'aperture' of radius r_1; the $C^{(l)}$'s are limb-darkening coefficients as defined by Equations (2.8)–(2.9) of Chapter I; and ${}_1F_2$ represents a generalized hypergeometric series defined by Equation (III.2) of Appendix III for $p = 1$ and $q = 2$.

On the other hand, the Fourier transform $G(u, v)$ of the (opaque) occulting disc with centre displaced by δ continues to be given by Equation (3.25) of Chapter I. Since, moreover, by a combination of Jacobi's expansion (3.6) of that chapter with Equations (V.3)–(V.4) of this Appendix,

$$\int_{-\pi}^{\pi} e^{-2\pi i\delta q\cos\phi}\cos 2j\phi\,\mathrm{d}\phi = 2\pi(-1)^j J_{2j}(2\pi q\delta) \tag{V.11}$$

and

$$\int_{-\pi}^{\pi} e^{-2\pi i\delta q\cos\phi}\cos(2j+1)\phi\,\mathrm{d}\phi = 2\pi i(-1)^j J_{2j+1}(2\pi q\delta), \tag{V.12}$$

Equations (3.30) and (3.31) of Chapter I disclose that, for *even* values of $m \equiv 2\mu$,

$$\begin{aligned}\alpha_l^{2\mu} = \frac{b}{2\mu+1}\sum_{j=0}^{\infty}\frac{1}{(2j)!4^{\mu+j}}\frac{B(\mu+j+1,\nu)}{B(\mu+j+1,\mu-j+1)}\\ \times\int_0^{\infty}(ay)^{2j}{}_1F_2\{\mu+j+1;\quad 2j+1,\mu+\nu+j+1;\quad -(ay/2)^2\}\\ \times J_1(by)J_{2j}(cy)\,\mathrm{d}y;\end{aligned} \tag{V.13}$$

where if $m \equiv 2\mu + 1$ is odd,

$$\begin{aligned}\alpha_l^{2\mu+1} = \frac{b}{\mu+1}\sum_{j=0}^{\mu}\frac{1}{(2j+1)!4^{\mu+\nu+1}}\frac{B(\mu+j+2,\nu)}{B(\mu+j+2,\mu-j+1)}\\ \times\int_0^{\infty}(ay)^{2j+1}{}_1F_2\{\mu+j+2;\quad 2j+2,\mu+\nu+j+2;\quad -(ay/2)^2\}\\ \times J_1(by)J_{2j+1}(cy)\,\mathrm{d}y,\end{aligned} \tag{V.14}$$

where a, b, c are constants characterizing the geometry of the eclipse and defined by Equations (3.56)–(3.58) of Chapter I, and $\nu = \frac{1}{2}(l+2)$.

From the foregoing Equations (V.13) and (V.14) it therefore transpires that *all associated α_n^m-functions of integral index $m \geqslant 0$ can be expressed as linear combinations of Hankel transforms of the products* ${}_1F_2J_{2j-}$. Since, moreover, for arbitrary values of β and ν,

$${}_1F_2\{\beta;\beta,\nu+1;-(ay/2)^2\} \equiv \Gamma(\nu+1)(2/ay)^{\nu}J_{\nu}(ay), \tag{V.15}$$

Equations (V.13) and (V.14) can be alternatively rewritten as

$$\alpha_l^{2\mu} = (b/4^\mu)2^\nu\Gamma(\nu)\int_0^\infty (ay)^{-\nu}J_{2\mu+\nu}(ay)J_1(by)J_{2\mu}(cy)\,\mathrm{d}y$$

$$+\frac{b}{2\mu+1}\sum_{j=0}^{\mu-1}\frac{1}{(2j)!4^{\mu+j}}\frac{B(\mu+j+1,\nu)}{B(\mu+j+1,\mu-j+1)}\int_0^\infty (ay)^{2j}\times$$

$$\times\,{}_1F_2\{\mu+j+1;\quad 2j+1,\mu+\nu+j+1;\quad -(ay/2)^2\}$$

$$\times\,J_1(by)J_{2j}(cy)\,\mathrm{d}y \tag{V.16}$$

and

$$\alpha_l^{2\mu+1} = (b/4^\mu)2^\nu\Gamma(\nu)\int_0^\infty (ay)^{-\nu}J_{2\mu+\nu+1}(ay)J_1(by)J_{2\mu+1}(cy)\,\mathrm{d}y$$

$$+\frac{b}{\mu+1}\sum_{j=0}^{\mu-1}\frac{1}{(2j+1)!4^{\mu+\nu+1}}\frac{B(\mu+j+2,\nu)}{B(\mu+j+2,\mu-j+1)}\int_0^\infty (ay)^{2j+1}\times$$

$$\times\,{}_1F_2\{\mu+j+2;\quad 2j+2,\mu+\nu+j+2;\quad -(ay/2)^2\}$$

$$\times\,J_1(by)J_{2j+1}(cy)\,\mathrm{d}y. \tag{V.17}$$

For $\mu = 0$, the sums of the right-hand sides of the preceding equations vanish; and the rest reduce to Equations (3.76) of Chapter III.

APPENDIX VI

TABLES OF $J^0_{1,n}$ INTEGRALS

The aim of the present Appendix will be to provide 5-digit tables of the functions $J^0_{-1,\gamma}$ for $\gamma = 2(1)7$, $J^0_{1,\gamma}$ $(\gamma = -1(1)6)$ and $J^0_{3,\gamma}$ $(\gamma = -1(1)3)$, valid for partial as well as annular eclipses, the need of which arises (cf. Section II-2) in connection with nonlinear limb-darkening and rotational distortion of the eclipsing star. Analytical expressions defining such functions can be obtained from Equations (IV.1)–(IV.24) given in the preceding Appendix, with the aid of the recursion formulae deduced in Section III-3; but they are long and their derivation may be left as an exercise for the interested reader.

Explanatory notes on the use of these tables would not differ from those which prefaced tables given in Appendix IV; and need not be repeated in this place; the numerical data given here have likewise been taken from Kopal (1947).

TABLE VI-1(p)

α	$J^{o}_{-1,2}$	M''	$J^{o}_{-1,3}$	M''	$J^{o}_{-1,4}$	M''	$J^{o}_{-1,5}$	M''	$J^{o}_{-1,6}$	M''	$J^{o}_{-1,7}$	M''
0°	0.00000	0.0000	0.00000	0.0000	0.00000	0.0000	0.00000	0.0000	0.00000	0.0000	0.00000	0.0000
5	0.00014	0.0008	0.00001	0.0001	0.00000	0.0000	0.00000	0.0000	0.00000	0.0000	0.00000	0.0000
10	0.00112	0.0017	0.00017	0.0005	0.00002	0.0001	0.00000	0.0000	0.00000	0.0000	0.00000	0.0000
15	0.00378	0.0025	0.00087	0.0011	0.00020	0.0004	0.00005	0.0001	0.00001	0.0000	0.00000	0.0000
20	0.00892	0.0032	0.00274	0.0020	0.00086	0.0010	0.00028	0.0005	0.00008	0.0002	0.00002	0.0001
25	0.01730	0.0040	0.00661	0.0031	0.00259	0.0020	0.00103	0.0012	0.00040	0.0006	0.00016	0.0003
30	0.02964	0.0046	0.01354	0.0042	0.00634	0.0033	0.00302	0.0023	0.00144	0.0015	0.00069	0.0010
35	0.04659	0.0052	0.02472	0.0056	0.01345	0.0050	0.00745	0.0041	0.00415	0.0032	0.00233	0.0023
40	0.06875	0.0057	0.04147	0.0069	0.02564	0.0072	0.01614	0.0067	0.01025	0.0060	0.00656	0.0051
45	0.09661	0.0061	0.06516	0.0084	0.04507	0.0097	0.03171	0.0102	0.02256	0.0101	0.01620	0.0096
50	0.13058	0.0064	0.09720	0.0097	0.07422	0.0125	0.05764	0.0146	0.04530	0.0160	0.03594	0.0168
55	0.17097	0.0066	0.13896	0.0110	0.11588	0.0155	0.09828	0.0198	0.08438	0.0238	0.07314	0.0275
60	0.21800	0.0067	0.19172	0.0121	0.17301	0.0186	0.15884	0.0258	0.14764	0.0337	0.13853	0.0421
65	0.27175	0.0067	0.25660	0.0131	0.24869	0.0216	0.24526	0.0323	0.24490	0.0454	0.24682	0.0613
70	0.33221	0.0066	0.33456	0.0139	0.34593	0.0244	0.36402	0.0391	0.38789	0.0589	0.41722	0.0850
75	0.39924	0.0064	0.42632	0.0142	0.46752	0.0269	0.52188	0.0459	0.58998	0.0736	0.67334	0.1129
80	0.47259	0.0060	0.53228	0.0144	0.61589	0.0288	0.72555	0.0522	0.86576	0.0887	1.0430	0.144
85	0.55188	0.0055	0.65255	0.0141	0.79297	0.0301	0.98129	0.0577	1.2301	0.103	1.5570	0.177
90	0.63662	0.0049	0.78689	0.0135	1.00000	0.0306	1.29445	0.0618	1.6977	0.117	2.2483	0.210
95	0.72620	0.0041	0.93469	0.0125	1.23742	0.0301	1.66907	0.0641	2.2813	0.127	3.1490	0.240
100	0.81989	0.0030	1.09493	0.0111	1.50473	0.0286	2.10744	0.0644	2.9914	0.134	4.2887	0.264
105	0.91688	0.0024	1.26624	0.0094	1.80043	0.0261	2.60971	0.0621	3.8345	0.136	5.6913	0.280
110	1.01625	0.0014	1.44685	0.0072	2.12198	0.0224	3.17358	0.0571	4.8122	0.132	7.3718	0.284
115	1.11699	0.0003	1.63464	0.0048	2.46574	0.0177	3.79406	0.0493	5.9202	0.121	9.3337	0.274
120	1.21800	-0.0009	1.82714	0.0020	2.82700	0.0120	4.46336	0.0388	7.1477	0.103	11.5668	0.248
125	1.31812	-0.0021	2.02161	-0.0011	3.20006	0.0050	5.17095	0.0256	8.4769	0.079	14.0448	0.205
130	1.41615	-0.0034	2.21506	-0.0042	3.57849	-0.0019	5.90373	0.0101	9.8835	0.047	16.7248	0.145
135	1.51082	-0.0047	2.40430	-0.0075	3.95493	-0.0097	6.64626	-0.0072	11.3363	0.011	19.5476	0.071
140	1.60084	-0.0060	2.58602	-0.0109	4.32165	-0.0178	7.38134	-0.0260	12.7991	-0.030	22.4394	-0.016
145	1.68490	-0.0073	2.75688	-0.0142	4.67060	-0.0259	8.09056	-0.0448	14.2313	-0.073	25.3139	-0.111
150	1.76169	-0.0086	2.91356	-0.0174	4.99367	-0.0338	8.75494	-0.0638	15.5899	-0.117	28.0765	-0.210
155	1.82991	-0.0098	3.05284	-0.0205	5.28299	-0.0412	9.35572	-0.0817	16.8316	-0.159	30.6293	-0.308
160	1.88830	-0.0111	3.17171	-0.0232	5.53123	-0.0478	9.87519	-0.0976	17.9146	-0.198	32.8760	-0.397
165	1.93563	-0.0122	3.26746	-0.0255	5.73185	-0.0534	10.2975	-0.111	18.8009	-0.230	34.7279	-0.473
170	1.97074	-0.0133	3.33778	-0.0274	5.87936	-0.0575	10.6093	-0.121	19.4586	-0.255	36.1096	-0.533
175	1.99253	-0.0144	3.38081	-0.0287	5.96962	-0.0602	10.8005	-0.127	19.8634	-0.270	36.9628	-0.569
180°	2.00000	-0.0152	3.39531	-0.0292	6.00000	-0.0611	10.8650	-0.130	20.0000	-0.275	37.2514	-0.582

TABLE VI-2(a)

α	$J^{o}_{-1,2}$	M″	$J^{o}_{-1,3}$	M″	$J^{o}_{-1,4}$	M″	$J^{o}_{-1,5}$	M″	$J^{o}_{-1,6}$	M″	$J^{o}_{-1,7}$	M″
0°	∞		∞		∞		∞		∞		∞	
5	2100.3		96256		44114×10^{2}		20217×10^{4}		92654×10^{5}		42463×10^{7}	
10	524.58		12015		27519×10		63030×10^{2}		14436×10^{4}		33065×10^{5}	
15	232.78		3551.6		54190		82681×10		12615×10^{3}		19248×10^{4}	
20	130.65		1493.5		17072		19518×10		22311×10^{2}		25509×10^{3}	
25	83.386		761.53		6955.2		63528		58030×10		53012×10^{2}	
30	57.713		438.54		3332.8		25332		192574		14642×10^{2}	
35	42.236		274.60		1785.9		11618		75598		49206×10	
40	32.195	2.96	182.81		1038.5		5902.3		33562		19094×10	
45	25.314	1.87	127.51		642.78		3242.8		16373		82726	
50	20.396	1.23	92.276		417.98		1895.6		8606.6		39122	
55	16.761	0.85	68.801	6.91	282.92		1165.4		4809.0		19877	
60	14.000	0.60	52.584	4.48	198.00		747.39		2828.0		10726	
65	11.856	0.44	41.040	3.01	142.56	17.2	496.87		1737.5		6095.4	
70	10.158	0.33	32.613	2.08	105.19	11.0	340.85		1109.2		3625.0	
75	8.7936	0.247	26.330	1.48	79.327	7.26	240.42	32.2	732.75		2245.1	
80	7.6811	0.192	21.560	1.07	60.999	4.92	173.90	20.4	499.27		1442.9	
85	6.7638	0.152	17.881	0.79	47.749	3.40	128.72	13.3	350.02		959.33	
90	6.0000	0.121	15.005	0.59	38.000	2.40	97.352	8.84	252.00	30.6	658.36	
95	5.3587	0.098	12.731	0.45	30.715	1.73	75.134	6.01	186.03	19.8	465.45	
100	4.8164	0.080	10.915	0.35	25.197	1.26	59.116	4.17	140.63	13.0	338.44	39.5
105	4.3552	0.067	9.4519	0.273	20.967	0.94	47.383	2.94	108.74	8.8	252.71	25.5
110	3.9612	0.056	8.2654	0.216	17.691	0.70	38.662	2.11	85.921	6.03	193.51	16.9
115	3.6234	0.047	7.2974	0.173	15.129	0.54	32.095	1.53	69.314	4.22	151.78	11.4
120	3.3333	0.041	6.5044	0.140	13.111	0.41	27.093	1.13	57.037	3.00	121.78	7.8
125	3.0840	0.035	5.8523	0.115	11.511	0.32	23.242	0.85	47.835	2.17	99.835	5.52
130	2.8698	0.031	5.3158	0.094	10.236	0.25	20.253	0.64	40.853	1.60	83.554	3.94
135	2.6863	0.027	4.8744	0.079	9.2162	0.202	17.917	0.49	35.503	1.20	71.309	2.89
140	2.5299	0.024	4.5122	0.066	8.4004	0.162	16.082	0.39	31.372	0.91	62.011	2.15
145	2.3977	0.022	4.2166	0.056	7.7487	0.133	14.639	0.31	28.169	0.71	54.905	1.64
150	2.2872	0.020	3.9777	0.048	7.2312	0.110	13.508	0.25	25.688	0.56	49.465	1.28
155	2.1966	0.018	3.7875	0.042	6.8250	0.093	12.629	0.21	23.778	0.46	45.323	1.02
160	2.1244	0.017	3.6399	0.038	6.5129	0.081	11.958	0.18	22.333	0.38	42.216	0.84
165	2.0693	0.016	3.5301	0.034	6.2821	0.072	11.465	0.16	21.277	0.33	39.961	0.72
170	2.0306	0.016	3.4543	0.031	6.1234	0.065	11.127	0.14	20.557	0.30	38.430	0.64
175	2.0076	0.015	3.4099	0.029	6.0306	0.062	10.930	0.13	20.138	0.28	37.542	0.59
180°	2.0000	0.015	3.3953	0.029	6.0000	0.061	10.865	0.12	20.000	0.27	37.251	0.58

TABLE VI-3(p)

α	$J^{\circ}_{1,-1}$	M″	$J^{\circ}_{1,0}$	M″	$J^{\circ}_{1,1}$	M″	$J^{\circ}_{1,2}$	M″	$J^{\circ}_{1,3}$	M″	$J^{\circ}_{1,4}$	M″	$J^{\circ}_{1,5}$	M″	$J^{\circ}_{1,6}$	M″
0°	0.00000	0.0038	0.00000	0.0000	0.00000	0.0000	0.00000	0.0000	0.00000	0.0000	0.00000	0.0000	0.00000	0.0000	0.00000	0.0000
5	0.00190	0.0038	0.00007	0.0004	0.00000	0.0001	0.00000	0.0000	0.00000	0.0000	0.00000	0.0000	0.00000	0.0000	0.00000	0.0000
10	0.00757	0.0037	0.00056	0.0008	0.00006	0.0002	0.00001	0.0000	0.00000	0.0000	0.00000	0.0000	0.00000	0.0000	0.00000	0.0000
15	0.01693	0.0035	0.00188	0.0012	0.00029	0.0004	0.00005	0.0001	0.00001	0.0000	0.00000	0.0000	0.00000	0.0000	0.00000	0.0000
20	0.02982	0.0033	0.00440	0.0016	0.00090	0.0007	0.00021	0.0003	0.00005	0.0001	0.00001	0.0001	0.00000	0.0000	0.00000	0.0000
25	0.04607	0.0031	0.00848	0.0019	0.00217	0.0010	0.00064	0.0005	0.00020	0.0002	0.00007	0.0001	0.00003	0.0000	0.00001	0.0000
30	0.06528	0.0028	0.01441	0.0021	0.00441	0.0013	0.00156	0.0008	0.00059	0.0004	0.00024	0.0003	0.00010	0.0001	0.00005	0.0001
35	0.08730	0.0024	0.02244	0.0023	0.00799	0.0017	0.00328	0.0012	0.00145	0.0008	0.00068	0.0005	0.00033	0.0003	0.00018	0.0002
40	0.11172	0.0020	0.03274	0.0024	0.01328	0.0021	0.00620	0.0017	0.00313	0.0013	0.00167	0.0009	0.00092	0.0007	0.00054	0.0005
45	0.13814	0.0016	0.04542	0.0024	0.02064	0.0024	0.01079	0.0022	0.00611	0.0019	0.00364	0.0016	0.00225	0.0013	0.00144	0.0011
50	0.16616	0.0012	0.06052	0.0024	0.03043	0.0027	0.01759	0.0028	0.01101	0.0027	0.00725	0.0025	0.00495	0.0023	0.00347	0.0021
55	0.19533	0.0007	0.07800	0.0023	0.04295	0.0030	0.02717	0.0033	0.01859	0.0036	0.01339	0.0036	0.00999	0.0037	0.00766	0.0036
60	0.22519	0.0002	0.09775	0.0021	0.05843	0.0031	0.04009	0.0039	0.02974	0.0045	0.02321	0.0051	0.01878	0.0055	0.01560	0.0059
65	0.25528	-0.0002	0.11959	0.0019	0.07703	0.0032	0.05689	0.0044	0.04541	0.0055	0.03813	0.0067	0.03318	0.0079	0.02964	0.0092
70	0.28513	-0.0007	0.14329	0.0016	0.09881	0.0032	0.07806	0.0048	0.06662	0.0065	0.05978	0.0085	0.05558	0.0108	0.05304	0.0134
75	0.31429	-0.0011	0.16854	0.0012	0.12376	0.0031	0.10400	0.0051	0.09435	0.0074	0.08996	0.0104	0.08884	0.0141	0.09006	0.0187
80	0.34233	-0.0015	0.19500	0.0008	0.15174	0.0028	0.13496	0.0052	0.12950	0.0082	0.13054	0.0123	0.13625	0.0177	0.14595	0.0249
85	0.36884	-0.0019	0.22229	0.0004	0.18251	0.0025	0.17105	0.0051	0.17281	0.0088	0.18334	0.0139	0.20133	0.0213	0.22685	0.0317
90	0.39345	-0.0023	0.25000	0.0000	0.21574	0.0021	0.21221	0.0049	0.22483	0.0090	0.25000	0.0153	0.28765	0.0247	0.33953	0.0388
95	0.41581	-0.0025	0.27771	-0.0004	0.25102	0.0016	0.25820	0.0045	0.28581	0.0089	0.33186	0.0162	0.39858	0.0276	0.49097	0.0456
100	0.43564	-0.0028	0.30500	-0.0008	0.28786	0.0010	0.30861	0.0038	0.35568	0.0086	0.42977	0.0165	0.53695	0.0297	0.68784	0.0515
105	0.45270	-0.0030	0.33146	-0.0012	0.32571	0.0004	0.36282	0.0031	0.43404	0.0078	0.54401	0.0161	0.70476	0.0306	0.93586	0.0557
110	0.46681	-0.0031	0.35671	-0.0016	0.36398	-0.0002	0.42008	0.0022	0.52009	0.0066	0.67417	0.0149	0.90289	0.0301	1.23913	0.0576
115	0.47786	-0.0031	0.38041	-0.0019	0.40207	-0.0008	0.47951	0.0012	0.61269	0.0051	0.81908	0.0129	1.13083	0.0280	1.59942	0.0565
120	0.48581	-0.0031	0.40225	-0.0021	0.43937	-0.0014	0.54008	0.0001	0.71034	0.0033	0.97679	0.0102	1.38649	0.0243	2.01560	0.0521
125	0.49066	-0.0030	0.42200	-0.0023	0.47531	-0.0019	0.60074	-0.0010	0.81128	0.0013	1.14459	0.0068	1.66614	0.0189	2.48319	0.0441
130	0.49251	-0.0028	0.43948	-0.0024	0.50935	-0.0024	0.66038	-0.0021	0.91351	-0.0009	1.31910	0.0028	1.96437	0.0119	2.99417	0.0324
135	0.49153	-0.0026	0.45458	-0.0024	0.54101	-0.0028	0.71790	-0.0032	1.01485	-0.0031	1.49636	-0.0016	2.27428	0.0037	3.53696	0.0175
140	0.48795	-0.0023	0.46726	-0.0024	0.56989	-0.0032	0.77224	-0.0042	1.11306	-0.0054	1.67198	-0.0063	2.58769	-0.0055	4.09679	0.0000
145	0.48208	-0.0019	0.47756	-0.0023	0.59566	-0.0034	0.82243	-0.0051	1.20591	-0.0075	1.84134	-0.0110	2.89549	-0.0152	4.65626	-0.0194
150	0.47433	-0.0014	0.48559	-0.0021	0.61809	-0.0035	0.86759	-0.0058	1.29125	-0.0095	1.99974	-0.0155	3.18805	-0.0250	5.19620	-0.0396
155	0.46517	-0.0009	0.49152	-0.0019	0.63702	-0.0036	0.90695	-0.0065	1.36709	-0.0113	2.14273	-0.0198	3.45573	-0.0343	5.69667	-0.0593
160	0.45517	-0.0002	0.49560	-0.0016	0.65239	-0.0036	0.93990	-0.0069	1.43167	-0.0129	2.26603	-0.0234	3.68930	-0.0427	6.13815	-0.0774
165	0.44503	0.0007	0.49812	-0.0011	0.66420	-0.0035	0.96598	-0.0073	1.48348	-0.0140	2.36602	-0.0265	3.88047	-0.0496	6.50267	-0.0930
170	0.43561	0.0017	0.49944	-0.0008	0.67252	-0.0034	0.98482	-0.0075	1.52134	-0.0149	2.43969	-0.0287	4.02231	-0.0549	6.77488	-0.1048
175	0.42807	0.0036	0.49993	-0.0004	0.67744	-0.0033	0.99619	-0.0076	1.54440	-0.0154	2.48481	-0.0301	4.10958	-0.0582	6.94309	-0.1122
180°	0.42441	0.0086	0.50000	0.0000	0.67906	-0.0032	1.00000	-0.0076	1.55214	-0.0155	2.50000	-0.0306	4.13904	[illegible]	7.00000	[illegible]

α	$J^{\circ}_{1,-1}$	M''	$J^{\circ}_{1,0}$	M''	$J^{\circ}_{1,1}$	M''	$J^{\circ}_{1,2}$	M''	$J^{\circ}_{1,3}$	M''	$J^{\circ}_{1,4}$	M''	$J^{\circ}_{1,5}$	M''	$J^{\circ}_{1,6}$	M''
0°	0.00000	0.0000	0.50000	0.0000	∞		∞		∞		∞		∞		∞	
5	0.01093	0.0000	0.50000	0.0000	22.923		1050.16		48137		22057×10^{2}		10110×10^{4}		46327×10^{5}	
10	0.02184	0.0000	0.50000	0.0000	11.452		262.29		6006.0		13759×10		31514×10^{2}		72181×10^{3}	
15	0.03277	0.0000	0.50000	0.0000	7.6286		116.39		1775.8		27094		41339×10		63073×10^{2}	
20	0.04374	0.0001	0.50000	0.0000	5.7151		65.327		746.73		8535.7		97571		11154×10^{2}	
25	0.05476	0.0001	0.50000	0.0000	4.5657		41.693		380.74		3477.1		31756		29003×10	
30	0.06582	0.0001	0.50000	0.0000	3.7983	0.203	28.856		219.24		1665.9		12659		96201	
35	0.07695	0.0001	0.50000	0.0000	3.2492	0.130	21.118		137.27		892.44		5802.8		37735	
40	0.08815	0.0001	0.50000	0.0000	2.8367	0.088	16.097	1.48	91.369		518.74		2945.8		16733	
45	0.09944	0.0001	0.50000	0.0000	2.5152	0.062	12.657	0.93	63.717		320.89		1616.7		8148.3	
50	0.11081	0.0001	0.50000	0.0000	2.2574	0.046	10.198	0.62	46.096		208.49		943.55		4272.7	
55	0.12229	0.0001	0.50000	0.0000	2.0461	0.034	8.3803	0.424	34.355	3.45	140.96		578.88		2379.4	
60	0.13389	0.0001	0.50000	0.0000	1.8697	0.026	7.0000	0.300	26.242	2.24	98.500		370.19		1393.0	
65	0.14561	0.0001	0.50000	0.0000	1.7201	0.021	5.9278	0.219	20.465	1.51	70.778		245.21		850.98	
70	0.15746	0.0001	0.50000	0.0000	1.5917	0.017	5.0792	0.163	16.248	1.04	52.097		167.44		539.38	
75	0.16945	0.0002	0.50000	0.0000	1.4803	0.014	4.3968	0.124	13.102	0.74	39.164	3.63	117.43		353.18	
80	0.18159	0.0002	0.50000	0.0000	1.3828	0.011	3.8406	0.096	10.712	0.53	30.000	2.46	84.354		238.11	
85	0.19388	0.0002	0.50000	0.0000	1.2968	0.009	3.3819	0.076	8.8677	0.394	23.375	1.70	61.927		164.87	
90	0.20634	0.0002	0.50000	0.0000	1.2204	0.008	3.0000	0.061	7.4253	0.295	18.500	1.20	46.385	4.40	117.00	
95	0.21897	0.0002	0.50000	0.0000	1.1523	0.007	2.6793	0.049	6.2837	0.225	14.858	0.86	35.403	2.99	84.975	
100	0.23177	0.0002	0.50000	0.0000	1.0912	0.006	2.4082	0.040	5.3710	0.174	12.099	0.63	27.508	2.07	63.088	
105	0.24473	0.0002	0.50000	0.0000	1.0363	0.006	2.1776	0.033	4.6349	0.136	9.9837	0.468	21.744	1.46	47.836	4.29
110	0.25787	0.0002	0.50000	0.0000	0.98686	0.0048	1.9806	0.029	4.0369	0.108	8.3456	0.351	17.475	1.04	37.019	2.93
115	0.27117	0.0001	0.50000	0.0000	0.94222	0.0043	1.8117	0.024	3.5481	0.087	7.0646	0.268	14.274	0.76	29.222	2.04
120	0.28461	0.0001	0.50000	0.0000	0.90194	0.0040	1.6667	0.020	3.1468	0.070	6.0556	0.206	11.848	0.56	23.519	1.44
125	0.29817	0.0001	0.50000	0.0000	0.86562	0.0037	1.5420	0.017	2.8161	0.057	5.2554	0.161	9.9699	0.416	19.291	1.04
130	0.31183	0.0000	0.50000	0.0000	0.83298	0.0034	1.4349	0.015	2.5433	0.047	4.6178	0.126	8.5555	0.314	16.122	0.75
135	0.32554	0.0000	0.50000	0.0000	0.80377	0.0032	1.3431	0.014	2.3180	0.040	4.1081	0.101	7.4408	0.241	13.722	0.56
140	0.33924	-0.0001	0.50000	0.0000	0.77779	0.0031	1.2649	0.012	2.1325	0.033	3.7002	0.081	6.5710	0.187	11.891	0.42
145	0.35284	-0.0002	0.50000	0.0000	0.75492	0.0030	1.1988	0.011	1.9805	0.028	3.3744	0.066	5.8912	0.148	10.488	0.32
150	0.36622	-0.0004	0.50000	0.0000	0.73507	0.0030	1.1436	0.010	1.8571	0.024	3.1156	0.055	5.3614	0.119	9.4132	0.252
155	0.37923	-0.0006	0.50000	0.0000	0.71820	0.0030	1.0983	0.009	1.7583	0.022	2.9126	0.047	4.9521	0.098	8.5942	0.202
160	0.39162	-0.0009	0.50000	0.0000	0.70429	0.0030	1.0622	0.009	1.6812	0.020	2.7565	0.040	4.6416	0.082	7.9801	0.167
165	0.40306	-0.0014	0.50000	0.0000	0.69336	0.0030	1.0347	0.008	1.6235	0.018	2.6411	0.036	4.4142	0.072	7.5346	0.142
170	0.41304	-0.0021	0.50000	0.0000	0.68546	0.0031	1.0154	0.008	1.5835	0.017	2.5617	0.033	4.2590	0.064	7.2324	0.126
175	0.42075	-0.0038	0.50000	0.0000	0.68067	0.0032	1.0039	0.008	1.5599	0.016	2.5153	0.031	4.1687	0.060	7.0574	0.116
180°	0.42441	-0.0085	0.50000	0.0000	0.67906	0.0032	1.0000	0.008	1.5521	0.016	2.5000	0.030	4.1390	0.059	7.0000	0.114

TABLE VI-5(p)

α	$J^{o}_{3,-1}$	M″	$J^{o}_{3,0}$	M″	$J^{o}_{3,1}$	M″	$J^{o}_{3,2}$	M″	$J^{o}_{3,3}$	M″
0°	0.00000	0.0000	0.00000	0.0000	0.00000	0.0000	0.00000	0.0000	0.00000	0.000
5	0.00001	0.0001	0.00000	0.0000	0.00000	0.0000	0.00000	0.0000	0.00000	0.000
10	0.00017	0.0004	0.00001	0.0000	0.00000	0.0000	0.00000	0.0000	0.00000	0.000
15	0.00085	0.0011	0.00008	0.0001	0.00001	0.0000	0.00000	0.0000	0.00000	0.000
20	0.00262	0.0018	0.00031	0.0004	0.00005	0.0001	0.00001	0.0000	0.00000	0.000
25	0.00620	0.0026	0.00092	0.0007	0.00020	0.0002	0.00005	0.0001	0.00001	0.000
30	0.01235	0.0033	0.00220	0.0011	0.00057	0.0004	0.00017	0.0002	0.00006	0.000
35	0.02181	0.0040	0.00453	0.0015	0.00136	0.0007	0.00048	0.0004	0.00019	0.000
40	0.03521	0.0044	0.00837	0.0020	0.00286	0.0011	0.00116	0.0007	0.00051	0.000
45	0.05295	0.0046	0.01418	0.0024	0.00544	0.0015	0.00247	0.0010	0.00123	0.000
50	0.07519	0.0044	0.02240	0.0028	0.00955	0.0020	0.00480	0.0015	0.00266	0.001
55	0.10179	0.0039	0.03341	0.0031	0.01566	0.0025	0.00863	0.0021	0.00524	0.001
60	0.13224	0.0031	0.04747	0.0032	0.02426	0.0029	0.01455	0.0027	0.00960	0.002
65	0.16577	0.0021	0.06466	0.0031	0.03578	0.0033	0.02320	0.0034	0.01652	0.003
70	0.20132	0.0008	0.08489	0.0028	0.05057	0.0035	0.03523	0.0040	0.02691	0.004
75	0.23762	-0.0006	0.10786	0.0023	0.06884	0.0035	0.05124	0.0045	0.04175	0.005
80	0.27329	-0.0020	0.13306	0.0016	0.09061	0.0034	0.07173	0.0049	0.06205	0.006
85	0.30693	-0.0034	0.15986	0.0008	0.11573	0.0030	0.09705	0.0050	0.08876	0.007
90	0.33724	-0.0045	0.18750	0.0000	0.14383	0.0025	0.12732	0.0049	0.12263	0.007
95	0.36308	-0.0054	0.21514	-0.0008	0.17438	0.0018	0.16242	0.0045	0.16418	0.007
100	0.38357	-0.0060	0.24195	-0.0016	0.20671	0.0010	0.20197	0.0039	0.21359	0.007
105	0.39816	-0.0061	0.26715	-0.0023	0.24004	0.0002	0.24535	0.0030	0.27064	0.007
110	0.40667	-0.0060	0.29011	-0.0028	0.27356	-0.0006	0.29174	0.0021	0.33472	0.0061
115	0.40926	-0.0055	0.31034	-0.0031	0.30648	-0.0013	0.34017	0.0010	0.40484	0.0048
120	0.40645	-0.0046	0.32753	-0.0032	0.33808	-0.0020	0.38956	-0.0001	0.47966	0.0031
125	0.39904	-0.0037	0.34159	-0.0031	0.36774	-0.0025	0.43882	-0.0012	0.55757	0.0013
130	0.38805	-0.0025	0.35260	-0.0028	0.39500	-0.0028	0.48689	-0.0022	0.63677	-0.0006
135	0.37461	-0.0013	0.36082	-0.0024	0.41954	-0.0029	0.53279	-0.0030	0.71536	-0.0025
140	0.35990	-0.0001	0.36663	-0.0020	0.44119	-0.0029	0.57568	-0.0038	0.79147	-0.0043
145	0.34506	0.0009	0.37047	-0.0015	0.45992	-0.0029	0.61484	-0.0043	0.86327	-0.0061
150	0.33105	0.0016	0.37280	-0.0010	0.47577	-0.0028	0.64969	-0.0048	0.92906	-0.0075
155	0.31864	0.0022	0.37408	-0.0007	0.48885	-0.0026	0.67978	-0.0051	0.98735	-0.0088
160	0.30837	0.0025	0.37469	-0.0004	0.49932	-0.0025	0.70478	-0.0053	1.03684	-0.0099
165	0.30051	0.0025	0.37492	-0.0001	0.50731	-0.0024	0.72445	-0.0055	1.07645	-0.0108
170	0.29510	0.0023	0.37499	0.0000	0.51293	-0.0023	0.73861	-0.0056	1.10535	-0.0114
175	0.29201	0.0021	0.37500	0.0000	0.51627	-0.0022	0.74715	-0.0057	1.12293	-0.0117
180°	0.29103	0.0019	0.37500	0.0000	0.51738	-0.0022	0.75000	-0.0057	1.12883	-0.0119

TABLE VI-6(a)

α	$J^o_{3,-1}$	M″	$J^o_{3,0}$	M″	$J^o_{3,1}$	M″	$J^o_{3,2}$	M″	$J^o_{3,3}$	M″
0°	0.00000	0.0000	0.37500	0.0000	∞		∞		∞	
5	0.00813	0.0000	0.37500	0.0000	17.3		787.62		11029	
10	0.01630	0.0000	0.37500	0.0000	8.613		196.72		4946.4	
15	0.02450	0.0000	0.37500	0.0000	5.731		87.293		1325.8	
20	0.03273	0.0000	0.37500	0.0000	4.290	0.47	48.995		559.95	
25	0.04100	0.0000	0.37500	0.0000	3.426	0.26	31.270		285.54	
30	0.04931	0.0001	0.37500	0.0000	2.8490	0.153	21.642		164.43	
35	0.05767	0.0001	0.37500	0.0000	2.4367	0.098	15.839		102.95	
40	0.06608	0.0001	0.37500	0.0000	2.1274	0.066	12.073	1.01	68.519	
45	0.07455	0.0001	0.37500	0.0000	1.8864	0.046	9.4926	0.662	47.778	
50	0.08308	0.0001	0.37500	0.0000	1.6932	0.034	7.6484	0.445	34.562	
55	0.09168	0.0001	0.37500	0.0000	1.5348	0.026	6.2853	0.309	25.755	2.59
60	0.10035	0.0001	0.37500	0.0000	1.4025	0.020	5.2500	0.221	19.669	1.68
65	0.10910	0.0001	0.37500	0.0000	1.2904	0.016	4.4459	0.162	15.335	1.08
70	0.11793	0.0001	0.37500	0.0000	1.1942	0.013	3.8094	0.121	12.171	0.78
75	0.12686	0.0001	0.37500	0.0000	1.1108	0.010	3.2976	0.092	9.8103	0.553
80	0.13589	0.0001	0.37500	0.0000	1.0379	0.008	2.8804	0.072	8.0169	0.400
85	0.14500	0.0001	0.37500	0.0000	0.97348	0.0073	2.5364	0.057	6.6327	0.295
90	0.15419	0.0001	0.37500	0.0000	0.91640	0.0062	2.2500	0.045	5.5497	0.222
95	0.16346	0.0001	0.37500	0.0000	0.86552	0.0053	2.0095	0.037	4.6924	0.167
100	0.17281	0.0001	0.37500	0.0000	0.81996	0.0046	1.8061	0.030	4.0067	0.131
105	0.18222	0.0000	0.37500	0.0000	0.77906	0.0041	1.6332	0.025	3.4535	0.102
110	0.19166	0.0000	0.37500	0.0000	0.74226	0.0037	1.4855	0.021	3.0038	0.081
115	0.20111	0.0000	0.37500	0.0000	0.70912	0.0033	1.3588	0.018	2.6361	0.065
120	0.21055	-0.0001	0.37500	0.0000	0.67928	0.0030	1.2500	0.015	2.3340	0.053
125	0.21994	-0.0001	0.37500	0.0000	0.65245	0.0028	1.1565	0.013	2.0849	0.043
130	0.22922	-0.0002	0.37500	0.0000	0.62841	0.0026	1.0762	0.012	1.8791	0.036
135	0.23832	-0.0003	0.37500	0.0000	0.60697	0.0025	1.0074	0.010	1.7091	0.030
140	0.24716	-0.0004	0.37500	0.0000	0.58800	0.0024	0.94871	0.0091	1.5689	0.025
145	0.25565	-0.0005	0.37500	0.0000	0.57139	0.0023	0.89912	0.0081	1.4539	0.021
150	0.26365	-0.0006	0.37500	0.0000	0.55705	0.0022	0.85770	0.0074	1.3604	0.019
155	0.27101	-0.0008	0.37500	0.0000	0.54494	0.0022	0.82372	0.0069	1.2855	0.016
160	0.27756	-0.0010	0.37500	0.0000	0.53504	0.0022	0.79664	0.0064	1.2270	0.015
165	0.28309	-0.0013	0.37500	0.0000	0.52733	0.0022	0.77600	0.0061	1.1832	0.013
170	0.28734	-0.0015	0.37500	0.0000	0.52181	0.0022	0.76148	0.0059	1.1528	0.012
175	0.29007	-0.0018	0.37500	0.0000	0.51849	0.0022	0.75286	0.0057	1.1348	0.012
180°	0.29103	-0.0020	0.37500	0.0000	0.51738	0.0022	0.75000	0.0057	1.1288	0.012

REFERENCES

Aitken, R. G.: 1935, *The Binary Stars*, McGraw Hill, New York (Dover Reprint, 1964).

Alkan, H.: 1978, *Astrophys. Space Sci.* **58**, 453.

Alkan, H.: 1979, *Astrophys. Space Sci.* **60**, 233.

Alkan, H. and Edalati, M. T.: 1978, *Astrophys. Space Sci.* **59**, 431.

Al-Naimiy, H. M. K.: 1978, *Astrophys. Space Sci.* **59**, 3.

André, Ch.: 1900, *Traité d'Astronomie Stellaire*, (Vol. 2), Gauthier Villars, Paris.

Appell, P.: 1880, *Ann. Ecole Normale Paris* (2) **9**, 119.

Appell, P. and Kampé de Fériet, J.: 1926, *Fonctions Hypergéometriques et Hypersphériques*, Gauthier Villars, Paris.

Bailey, W. N.: 1935, *Generalized Hypergeometric Series* (Cambridge Tracts in Mathematics and Math. Physics, No. 32), Cambr. Univ. Press.

Bailey, W. N.: 1936, *Proc. London Math. Soc.* **40**, 37.

Bateman, H.: 1905, *Proc. London Math. Soc.* **3**, 111.

Binnendijk, L.: 1966, *Astron. J.* **71**, 340.

Blazhko, S.: 1911, *Ann. de l'Observ. Astr. de Moscou (2e sér)* **5**, 76.

Brown, O. E.: 1938, *Astron. J.* **47**, 93.

Brown, W. E. and Shook, C. A.: 1933, *Planetary Theory*, Cambr. Univ. Press.

Budding, E.: 1973, *Astrophys. Space Sci.* **22**, 87.

Burchnall, J. L.: 1942, *Quart. Journ. Math. (Oxford)* **13**, 90.

Chanan, G. A., Middleditch, J., and Nelson, J. E.: 1976, *Astrophys. J.* **208**, 512.

Chaundy, T. W.: 1943, *Quart. Journ. Math. (Oxford)* **14**, 55.

Crawford, R. T.: 1930, *Determination of Orbits of Comets and Asteroids*, McGraw Hill, New York.

Demircan, O.: 1977, *Astrophys. Space Sci.* **52**, 189.

Demircan, O.: 1978a, *Astrophys. Space Sci.* **53**, 257.

Demircan, O.: 1978b, *Astrophys. Space Sci.* **56**, 389.

Demircan, O.: 1978c, *Astrophys. Space Sci.* **56**, 453.

Demircan, O.: 1978d, *Astrophys. Space Sci.* **59**, 313.

Demircan, O.: 1979a, *Astrophys. Space Sci.* **61**, 499.

Demircan, O.: 1979b, *Astrophys. Space Sci.* **61**, 507.

Demircan, O.: 1979c, *Astrophys. Space Sci.* **62**, 189.

Demircan, O.: 1979d, *Astrophys. Space Sci.* **62**, 235.

Dugan, R. S.: 1911, Princeton Contr. No. 1.

Dunér, N. C.: 1900, *Astrophys. J.* **11**, 175.

Edalati, M. T.: 1978a, *Astrophys. Space Sci.* **58**, 3.

Edalati, M. T.: 1978b, *Astrophys. Space Sci.* **59**, 333.

Edalati, M. T.: 1978c, *Astrophys. Space Sci.* **59**, 443.

Edalati, M. T. and Budding.: Astrophys. Space Sci. **57**, 181.

Ellsworth, J.: 1936, *Publ. Obs. Lyon* (1) **2**, Fasc. 1.

Erdélyi, A., Magnus, W., Oberhettinger, F., and Tricomi, F. G.: 1953, *Higher Transcendental Functions*, McGraw Hill Publ. Co., New York; Vol. 1.

Erdélyi, A., Magnus, W., Oberhettinger, F., and Tricomi, F. G.: 1954, *Tables of Integral Transforms*, McGraw Hill Publ. Co., New York; Vol. 2.

Ferrari, K.: 1938, *Sitzungsber d. Akad. Wiss. Wien (IIa)* **147**, 497.

Ferrari, K.: 1939, *Sitzungsber. d. Akad. Wiss. Wien (IIa)* **148**, 217.

Fetlaar, J.: 1923, *Utrecht Recherches* **9**, Part 1.

Fletcher, A., Miller, J. C. P., and Rosenhead, L.: 1946, *Index of Mathematical Tables*, Sci. Comp. Service, London.

Fracastoro, M. G.: 1972, *Atlas of the Light Curves of Eclipsing Variables*, Torino, Italy.

Fredricks, L. W.: 1960, *Astron. J.* **65**, 628

Gauss, C. F.: 1809, *Theoria Motus Corporum Coelestium*, Hamburg; Section 177.

Goodricke, J.: 1783, *Phil. Trans. Roy. Soc. London* **73**, 474.

Hagen, J. G.: 1921, *Die Veränderlichen Sterne*, Freiburg; pp. 609–671.

Harting, J.: 1889, Diss. München.
Hartwig, E.: 1900, *Astron. Nachr.* **152**, 309.
Harzer, P.: 1927, Kiel Publ., No. 16.
Henroteau, F. C.: 1928, in *Handb. der Astrophysik*, Julius Springer, Berlin, Vol. 6, Pt. 2.
Herschel, W.: 1783, 'Observations upon Algol', read before the Royal Society of London on 8 May 1783; but not printed till in *The Scientific Papers of Sir William Herschel*, London 1912; Vol. I, p. CVII.
Heun, K.: 1889, *Math. Ann.* **33**, 161.
Huffer, C. M. and Collins, G. W.: 1962, *Astrophys. J. Suppl.* **7**, 351.
Huffer, C.M. and Kopal, Z.: 1951, *Astrophys. J.* **114**, 297.
Jurkevich, I.: 1970, in *Vistas in Astronomy*, (ed. A. Beer), Pergamon Press, Vol. 12, pp. 63ff.
Jurkevich, I. and Heard, W. B.: 1977, *Astrophys. Space Sci.* **52**, 327.
Jurkevich, I., Willman, W. W. and Petty, A. F.: 1976, *Astrophys. Space Sci.* **44**, 63.
Kamp, P. van de.: 1978, *Sky and Telescope* **56**, 397.
Kitamura, M.: 1965, in *Advances in Astronomy and Astrophysics*, (ed. Z. Kopal), Acad. Press, New York; Vol. 3, pp. 27–87.
Kitamura, M.: 1967, *Tables of the Characteristic Functions of the Eclipse for the Solution of Light Curves of Eclipsing Binary Systems*, Tokyo Univ. Press, Tokyo, Japan.
Kopal, Z.: 1932, *Astron. Nachr.* **245**, 335; **247**, 117.
Kopal, Z.: 1941, *Astrophys. J.* **94**, 145.
Kopal, Z.: 1942, *Proc. Amer. Phil. Soc.* **85**, 399.
Kopal, Z.: 1945, *Proc. Amer. Phil. Soc.* **89**, 517.
Kopal, Z.: 1946, *An Introduction to the Study of Eclipsing Variables* (Harvard Obs. Mono. No. 6), Harvard Univ. Press, Cambridge, Mass.
Kopal, Z.: 1947, Harvard Obs. Circular No. 450.
Kopal, Z.: 1948a, *Astrophys. J.* **108**, 46.
Kopal, Z.: 1948b, *Journ. of Math. Tables and Other Aids to Computation*, **3**, 191.
Kopal, Z.: 1949, Harvard Obs. Circular No. 454.
Kopal, Z.: 1950, *Computation of the Elements of Eclipsing Variables*, Harvard Obs. Monograph No. 8.
Kopal, Z.: 1954, *Jodrell Bank Annals* **1**, 37–57.
Kopal, Z.: 1955, *Numerical Analysis*, Chapman-Hall and John Wiley, London and New York.
Kopal, Z.: 1959, *Close Binary Systems*, Chapman-Hall and John Wiley, London and New York.
Kopal, Z.: 1960, in *Atti del Convegno per le Celebrazioni di G. V. Schiaparelli* (ed. F. Zagar), Milano, Italy; pp. 156–161.
Kopal, Z.: 1962, in *Transactions Internat. Astron. Union XI B* (ed. D. H. Sadler), Acad. Press, New York, and London; p. 369.
Kopal, Z.: 1972, in *Advances in Astronomy and Astrophysics* (Acad. Press, New York), Vol. 9, pp. 1–65.
Kopal, Z.: 1975a, *Astrophys. Space Sci.* **34**, 431.
Kopal, Z.: 1975b, *Astrophys. Space Sci.* **35**, 159.
Kopal, Z.: 1975c, *Astrophys. Space Sci.* **35**, 171.
Kopal, Z.: 1975d, *Astrophys. Space Sci.* **36**, 227.
Kopal, Z.: 1975e, *Astrophys. Space Sci.* **38**, 191.
Kopal, Z.: 1976a, *Astrophys. Space Sci.* **40**, 461.
Kopal, Z.: 1976b, *Astrophys. Space Sci.* **45**, 269.
Kopal, Z.: 1977a, *Astrophys. Space Sci.* **46**, 87.
Kopal, Z.: 1977b, *Astrophys. Space Sci.* **50**, 225.
Kopal, Z.: 1977c, *Astrophys. Space Sci.* **51**, 439.
Kopal, Z.: 1978a, *Astrophys. Space Sci.* **57**, 439.
Kopal, Z.: 1978b, *Dynamics of Close Binary Systems* (D. Reidel Publ. Co., Dordrecht).
Kopal, Z.: 1979, *Astrophys. Space Sci.* **66**, 91.
Kopal, Z. and Al-Naimiy, H. M.: 1978, *Astrophys. Space Sci.* **57**, 479.
Kopal, Z. and Demircan, O.: 1978, *Astrophys. Space Sci.* **55**, 241.
Kopal, Z. and Kitamura, M.: 1968, in *Advances in Astronomy and Astrophysics* (Acad. Press, New York), Vol. 6, pp. 125–172.

Kopal, Z., Markellos, V., and Niarchos, P.: 1976, *Astrophys. Space Sci.* **40**, 183.
Kopal, Z. and Shapley, M. B.: 1946, *Astrophys. J.* **104**, 160.
Kopal, Z. and Shapley, M. B.: 1956, *Jodrell Bank Annals* **1**, 141–221.
Krat, V. A.: 1934, *Astron. Zhurnal* **11**, 407.
Krat, V. A.: 1935, *Astron. Zhurnal* **12**, 21.
Krat, V. A.: 1936, *Astron. Zhurnal* **13**, 521.
Kron, G. E.: 1939, *Lick Observ. Bull.*, No. 499.
Kron, G. E. and Gordon, K. C.: 1943, *Astrophys. Journ.* **97**, 311.
Kurutac, M.: 1978, *Astrophys. Space Sci.* **57**, 71.
Lanczos, C.: 1956, *Applied Analysis* (Prentice-Hall, New York), Chapter IV.
Lanczos, C.: 1966, *Discourse on Fourier Series* (Oliver and Boyd, Edinburgh).
Lanzano, P.: 1976a, *Astrophys. Space Sci.* **42**, 425.
Lanzano, P.: 1976b, *Astrophys. Space Sci.* **45**, 419.
Lanzano, P.: 1976c, *Astrophys. Space Sci.* **45**, 483.
Lighthill, M. J.: 1958, *Fourier Analysis and Generalized Functions*, Cambr. Univ. Press.
Linnell, A. P. and Proctor, D. D.: 1970, *Astrophys. J.* **161**, 1045; **162**, 683.
Linnell, A. P. and Proctor, D. D.: 1971, *Astrophys. J.* **164**, 131.
Livaniou, H.: 1977, *Astrophys. Space Sci.* **51**, 77 (see also Rovithis–Livaniou).
Look, K. H., Chen, J. S. and Zou, Z. L.: 1978, *Scientia Sinica* **21**, 613.
Mauder, H.: 1966, Kleine Veröff. Sternw. Bamberg, Ser. 3, Nr. 38 and 39.
Merrill, J. E.: 1950–53, Contr. Princeton, Nos. 23–24.
Montanari, G.: 1670, 'Sopra la sparizione d'Alcune Stelle e altre Novità celesti' in Prose di Academici Gelati di Bologna.
Myers, G. N.: 1898, *Astrophys. J.* **7**, 1; *Illinois Obs. Bull.*, No. 1.
Niarchos, P.: 1978, *Astrophys. Space Sci.* **58**, 301.
Oberhettinger, F.: 1973, *Fourier Expansions*, Academic Press, New York.
Otrebski, A.: 1948, *Acta Astron.* (a) **4**, 139.
Pannekoek, A.: 1902, 'Untersuchung über den Lichtwechsel Algols' (Diss. Leiden).
Pannekoek, A. and van Dien, E.: 1937, *Bull. Astr. Inst. Netherlands* **8**, 141.
Pickering, E. C.: 1880, *Proc. Amer. Acad. Arts and Sci.* **16**, 1.
Piotrowski, S. L.: 1937, *Acta Astron.* (a) **4**, 1.
Piotrowski, S. L.: 1947, *Astrophys. J.* **106**, 472.
Piotrowski, S. L.: 1948a, *Astrophys. J.* **108**, 36, 510.
Piotrowski, S. L.: 1948b, *Proc. U.S. Nat. Acad. Sci.* **34**, 23.
Plassmann, J.: 1888, *Die Veränderlichen Sterne*, Köln.
Rice, S. O.: 1935, *Quart. Journ. Math. (Oxford)* **6**, 52.
Roberts, A. W.: 1903, *Mon. Not. Roy. Astr. Soc.* **63**, 527.
Roberts, A. W.: 1908, *Mon. Not. Roy. Astr. Soc.* **68**, 490.
Rödiger, C.: 1902, 'Untersuchungen über das Doppelstersystem Algol' (Diss. Königsberg).
Rovithis-Livaniou, H.: 1977, *Astrophys. Space Sci.* **52**, 271.
Rovithis-Livaniou, H.: 1978, *Astrophys. Space Sci.* **59**, 463.
Russell, H. N.: 1912, *Astrophys. J.* **35**, 315; **36**, 54.
Russell, H. N.: 1942, *Astrophys. J.* **95**, 345.
Russell, H. N.: 1945, *Astrophys. J.* **102**, 1.
Russell, H. N. and Merrill, J. E.: 1952, Princeton Contr. No. 26.
Russell, H. N. and Shapley, H.: 1912, *Astrophys. J.* **36**, 239, 285.
Scharbe, S.: 1925, *Pulkovo Observ. Bull.*, No. 94.
Schiller, K.: 1923, *Einführung in das Studium der Veränderlichen Sterne* (J. A. Barth, Leipzig); Chapter 5.
Sievert, C. E. and Burniston, E. E.: 1972, *Celestial Mechanics* **6**, 294.
Sitterly, B. W.: 1930, Princeton Contr. No. 11.
Sneddon, I. N.: 1951, *Fourier Transforms*, McGraw Hill, New York.
Söderhjelm, S.: 1974, *Astron. Astrophys.* **34**, 59.
Smith, S. A. H.: 1976, *Astrophys. Space Sci.* **40**, 315.
Stein, J. G.: 1924a, *Bull. Astr. Inst. Netherlands* **2**, 123.

Stein, J. G.: 1924b, *Die Veränderlichen Sterne*, Herder und Co., Freiburg.
Strand, K. Aa.: 1959, *Astron. J.* **64**, 346.
Szegö, G.: 1939, *Orthogonal Polynomials* (Amer. Math. Soc. Colloquium Publ. Vol. 23), New York.
Takeda, S.: 1934, *Mem. Coll. Sci. Kyoto Univ.* (*Ser. A*) **17**, 197.
Takeda, S.: 1937, *Mem. Coll. Sci. Kyoto Univ.* (*Ser. A*) **20**, 47.
Truesdell, C.: 1948, *A Unified Theory of Special Functions* (Ann. Math. Studies No. 18), Princeton Univ. Press.
Tsesevich, V. P.: 1936, *Publ. Univ. Obs. Leningrad* **6**, 48.
Tsesevich, V. P.: 1939, *Bull. Astr. Inst. USSR Acad. Sci.*, No. 45.
Tsesevich, V. P.: 1940, *Bull. Astr. Inst. USSR Acad. Sci.*, No. 50.
Tsesevich, V. P.: 1971, *Zatmennye Peremennye Zvjozdy*, Izd. Nauka, Moscow.
Vogel, H. C.: 1890, *Astron. Nachr.* **123**, 289.
Vogt, H.: 1919, *Heidelberg Veröff.* **7**, 183.
Walter, K.: 1931, Königsberg Veröff. Nr. 2.
Watson, G. N.: 1945, *Treatise on the Theory of Bessel Functions* (2nd ed.), Cambr. Univ. Press.
Whittaker, E. T. and Robinson, G.: 1924, *The Calculus of Observations* (Blackie and Sons, London and Glasgow), Chapter IX.
Whittaker, E. T. and Watson, G. N.: 1920, *Modern Analysis*, Cambr. Univ. Press.
Wyse, A. B.: 1939, Lick Observ. Bull., No. 494.

NAME INDEX

SUBJECT INDEX

ASTROPHYSICS AND SPACE SCIENCE LIBRARY

Edited by

J. E. Blamont, R. L. F. Boyd, L. Goldberg, C. de Jager, Z. Kopal, G. H. Ludwig, R. Lüst, B. M. McCormac, H. E. Newell, L. I. Sedov, Z. Švestka, and W. de Graaff

1. C. de Jager (ed.), *The Solar Spectrum, Proceedings of the Symposium held at the University of Utrecht, 26–31 August, 1963.* 1965, XIV + 417 pp.
2. J. Orthner and H. Maseland (eds.), *Introduction to Solar Terrestrial Relations, Proceedings of the Summer School in Space Physics held in Alpbach, Austria, July 15–August 10, 1963 and Organized by the European Preparatory Commission for Space Research.* 1965, IX + 506 pp.
3. C. C. Chang and S. S. Huang (eds.), *Proceedings of the Plasma Space Science Symposium, held at the Catholic University of America, Washington, D.C., June 11–14, 1963.* 1965, IX + 377 pp.
4. Zdeněk Kopal, *An Introduction to the Study of the Moon.* 1966, XII + 464 pp.
5. B. M. McCormac (ed.), *Radiation Trapped in the Earth's Magnetic Field. Proceedings of the Advanced Study Institute, held at the Chr. Michelsen Institute, Bergen, Norway, August 16–September 3, 1965.* 1966, XII + 901 pp.
6. A. B. Underhill, *The Early Type Stars.* 1966, XII + 282 pp.
7. Jean Kovalevsky, *Introduction to Celestial Mechanics.* 1967, VIII + 427 pp.
8. Zdeněk Kopal and Constantine L. Goudas (eds.), *Measure of the Moon. Proceedings of the 2nd International Conference on Selenodesy and Lunar Topography, held in the University of Manchester, England, May 30–June 4, 1966.* 1967, XVIII + 479 pp.
9. J. G. Emming (ed.), *Electromagnetic Radiation in Space. Proceedings of the 3rd ESRO Summer School in Space Physics, held in Alpbach, Austria, from 19 July to 13 August, 1965.* 1968, VIII + 307 pp.
10. R. L. Carovillano, John F. McClay, and Henry R. Radoski (eds.), *Physics of the Magnetosphere, Based upon the Proceedings of the Conference held at Boston College, June 19–28, 1967.* 1968, X + 686 pp.
11. Syun-Ichi Akasofu, *Polar and Magnetospheric Substorms.* 1968, XVIII + 280 pp.
12. Peter M. Millman (ed.), *Meteorite Research. Proceedings of a Symposium on Meteorite Research, held in Vienna, Austria, 7–13 August, 1968.* 1969, XV + 941 pp.
13. Margherita Hack (ed.), *Mass Loss from Stars. Proceedings of the 2nd Trieste Colloquium on Astrophysics, 12–17 September, 1968.* 1969, XII + 345 pp.
14. N. D'Angelo (ed.), *Low-Frequency Waves and Irregularities in the Ionosphere. Proceedings of the 2nd ESRIN-ESLAB Symposium, held in Frascati, Italy, 23–27 September, 1968.* 1969, VII + 218 pp.
15. G. A. Partel (ed.), *Space Engineering. Proceedings of the 2nd International Conference on Space Engineering, held at the Fondazione Giorgio Cini, Isola di San Giorgio, Venice, Italy, May 7–10, 1969.* 1970, XI + 728 pp.
16. S. Fred Singer (ed.), *Manned Laboratories in Space. Second International Orbital Laboratory Symposium.* 1969, XIII + 133 pp.
17. B. M. McCormac (ed.), *Particles and Fields in the Magnetosphere. Symposium Organized by the Summer Advanced Study Institute, held at the University of California, Santa Barbara, Calif., August 4–15, 1969.* 1970, XI + 450 pp.
18. Jean-Claude Pecker, *Experimental Astronomy.* 1970, X + 105 pp.
19. V. Manno and D. E. Page (eds.), *Intercorrelated Satellite Observations related to Solar Events. Proceedings of the 3rd ESLAB/ESRIN Symposium held in Noordwijk, The Netherlands, September 16–19, 1969.* 1970, XVI + 627 pp.
20. L. Mansinha, D. E. Smylie, and A. E. Beck, *Earthquake Displacement Fields and the Rotation of the Earth, A NATO Advanced Study Institute Conference Organized by the Department of Geophysics, University of Western Ontario, London, Canada, June 22–28, 1969.* 1970, XI + 308 pp.
21. Jean-Claude Pecker, *Space Observatories.* 1970, XI + 120 pp.
22. L. N. Mavridis (ed.), *Structure and Evolution of the Galaxy. Proceedings of the NATO Advanced Study Institute, held in Athens, September 8–19, 1969.* 1971, VII + 312 pp.

23. A. Muller (ed.), *The Magellanic Clouds. A European Southern Observatory Presentation: Principal Prospects, Current Observational and Theoretical Approaches, and Prospects for Future Research, Based on the Symposium on the Magellanic Clouds, held in Santiago de Chile, March 1969, on the Occasion of the Dedication of the European Southern Observatory.* 1971, XII + 189 pp.
24. B. M. McCormac (ed.), *The Radiating Atmosphere. Proceedings of a Symposium Organized by the Summer Advanced Study Institute, held at Queen's University, Kingston, Ontario, August 3–14, 1970.* 1971, XI + 455 pp.
25. G. Fiocco (ed.), *Mesospheric Models and Related Experiments. Proceedings of the 4th ESRIN-ESLAB Symposium, held at Frascati, Italy, July 6–10, 1970.* 1971, VIII + 298 pp.
26. I. Atanasijević, *Selected Exercises in Galactic Astronomy.* 1971, XII + 144 pp.
27. C. J. Macris (ed.), *Physics of the Solar Corona. Proceedings of the NATO Advanced Study Institute on Physics of the Solar Corona, held at Cavouri-Vouliagmeni, Athens, Greece, 6–17 September 1970.* 1971, XII + 345 pp.
28. F. Delobeau, *The Environment of the Earth.* 1971, IX + 113 pp.
29. E. R. Dyer (general ed.), *Solar-Terrestrial Physics/1970. Proceedings of the International Symposium on Solar-Terrestrial Physics, held in Leningrad, U.S.S.R., 12–19 May 1970.* 1972, VIII + 938 pp.
30. V. Manno and J. Ring (eds.), *Infrared Detection Techniques for Space Research. Proceedings of the 5th ESLAB-ESRIN Symposium, held in Noordwijk, The Netherlands, June 8–11, 1971.* 1972, XII + 344 pp.
31. M. Lecar (ed.), *Gravitational N-Body Problem. Proceedings of IAU Colloquium No. 10, held in Cambridge, England, August 12–15, 1970.* 1972, XI + 441 pp.
32. B. M. McCormac (ed.), *Earth's Magnetospheric Processes. Proceedings of a Symposium Organized by the Summer Advanced Study Institute and Ninth ESRO Summer School, held in Cortina, Italy, August 30–September 10, 1971.* 1972, VIII + 417 pp.
33. Antonin Rükl, *Maps of Lunar Hemispheres.* 1972, V + 24 pp.
34. V. Kourganoff, *Introduction to the Physics of Stellar Interiors.* 1973, XI + 115 pp.
35. B. M. McCormac (ed.), *Physics and Chemistry of Upper Atmospheres. Proceedings of a Symposium Organized by the Summer Advanced Study Institute, held at the University of Orléans, France, July 31–August 11, 1972.* 1973, VIII + 389 pp.
36. J. D. Fernie (ed.), *Variable Stars in Globular Clusters and in Related Systems. Proceedings of the IAU Colloquium No. 21, held at the University of Toronto, Toronto, Canada, August 29–31, 1972.* 1973, IX + 234 pp.
37. R. J. L. Grard (ed.), *Photon and Particle Interaction with Surfaces in Space. Proceedings of the 6th ESLAB Symposium, held at Noordwijk, The Netherlands, 26–29 September, 1972.* 1973, XV + 577 pp.
38. Werner Israel (ed.), *Relativity, Astrophysics and Cosmology. Proceedings of the Summer School, held 14–26 August, 1972, at the BANFF Centre, BANFF, Alberta, Canada.* 1973, IX + 323 pp.
39. B. D. Tapley and V. Szebehely (eds.), *Recent Advances in Dynamical Astronomy. Proceedings of the NATO Advanced Study Institute in Dynamical Astronomy, held in Cortina d'Ampezzo, Italy, August 9–12, 1972.* 1973, XIII + 468 pp.
40. A. G. W. Cameron (ed.), *Cosmochemistry. Proceedings of the Symposium on Cosmochemistry, held at the Smithsonian Astrophysical Observatory, Cambridge, Mass., August 14–16, 1972.* 1973, X + 173 pp.
41. M. Golay, *Introduction to Astronomical Photometry.* 1974, IX + 364 pp.
42. D. E. Page (ed.), *Correlated Interplanetary and Magnetospheric Observations. Proceedings of the 7th ESLAB Symposium, held at Saulgau, W. Germany, 22–25 May, 1973.* 1974, XIV + 662 pp.
43. Riccardo Giacconi and Herbert Gursky (eds.), *X-Ray Astronomy.* 1974, X + 450 pp.
44. B. M. McCormac (ed.), *Magnetospheric Physics. Proceedings of the Advanced Summer Institute, held in Sheffield, U.K., August 1973.* 1974, VII + 399 pp.
45. C. B. Cosmovici (ed.), *Supernovae and Supernova Remnants. Proceedings of the International Conference on Supernovae, held in Lecce, Italy, May 7–11, 1973.* 1974, XVII + 387 pp.
46. A. P. Mitra, *Ionospheric Effects of Solar Flares.* 1974, XI + 294 pp.
47. S.-I. Akasofu, *Physics of Magnetospheric Substorms.* 1977, XVIII + 599 pp.

48. H. Gursky and R. Ruffini (eds.), *Neutron Stars, Black Holes and Binary X-Ray Sources.* 1975, XII + 441 pp.
49. Z. Švestka and P. Simon (eds.), *Catalog of Solar Particle Events 1955–1969. Prepared under the Auspices of Working Group 2 of the Inter-Union Commission on Solar-Terrestrial Physics.* 1975, IX + 428 pp.
50. Zdeněk Kopal and Robert W. Carder, *Mapping of the Moon.* 1974, VIII + 237 pp.
51. B. M. McCormac (ed.), *Atmospheres of Earth and the Planets. Proceedings of the Summer Advanced Study Institute, held at the University of Liège, Belgium, July 29–August 8, 1974.* 1975, VII + 454 pp.
52. V. Formisano (ed.), *The Magnetospheres of the Earth and Jupiter. Proceedings of the Neil Brice Memorial Symposium, held in Frascati, May 28–June 1, 1974.* 1975, XI + 485 pp.
53. R. Grant Athay, *The Solar Chromosphere and Corona: Quiet Sun.* 1976, XI + 504 pp.
54. C. de Jager and H. Nieuwenhuijzen (eds.), *Image Processing Techniques in Astronomy. Proceedings of a Conference, held in Utrecht on March 25–27, 1975.* XI + 418 pp.
55. N. C. Wickramasinghe and D. J. Morgan (eds.), *Solid State Astrophysics. Proceedings of a Symposium, held at the University College, Cardiff, Wales, 9–12 July 1974.* 1976, XII + 314 pp.
56. John Meaburn, *Detection and Spectrometry of Faint Light.* 1976, IX + 270 pp.
57. K. Knott and B. Battrick (eds.), *The Scientific Satellite Programme during the International Magnetospheric Study. Proceedings of the 10th ESLAB Symposium, held at Vienna, Austria, 10–13 June 1975.* 1976, XV + 464 pp.
58. B. M. McCormac (ed.), *Magnetospheric Particles and Fields. Proceedings of the Summer Advanced Study School, held in Graz, Austria, August 4–15, 1975.* 1976, VII + 331 pp.
59. B. S. P. Shen and M. Merker (eds.), *Spallation Nuclear Reactions and Their Applications.* 1976, VIII + 235 pp.
60. Walter S. Fitch (ed.), *Multiple Periodic Variable Stars. Proceedings of the International Astronomical Union Colloquium No. 29, held at Budapest, Hungary, 1–5 September 1976.* 1976, XIV + 348 pp.
61. J. J. Burger, A. Pedersen, and B. Battrick (eds.), *Atmospheric Physics from Spacelab. Proceedings of the 11th ESLAB Symposium, Organized by the Space Science Department of the European Space Agency, held at Frascati, Italy, 11–14 May 1976.* 1976, XX + 409 pp.
62. J. Derral Mulholland (ed.), *Scientific Applications of Lunar Laser Ranging. Proceedings of a Symposium held in Austin, Tex., U.S.A., 8–10 June, 1976.* 1977, XVII + 302 pp.
63. Giovanni G. Fazio (ed.), *Infrared and Submillimeter Astronomy. Proceedings of a Symposium held in Philadelphia, Penn., U.S.A., 8–10 June, 1976.* 1977, X + 226 pp.
64. C. Jaschek and G. A. Wilkins (eds.), *Compilation, Critical Evaluation and Distribution of Stellar Data. Proceedings of the International Astronomical Union Colloquium No. 35, held at Strasbourg, France, 19–21 August, 1976.* 1977, XIV + 316 pp.
65. M. Friedjung (ed.), *Novae and Related Stars. Proceedings of an International Conference held by the Institut d'Astrophysique, Paris, France, 7–9 September, 1976.* 1977, XIV + 228 pp.
66. David N. Schramm (ed.), *Supernovae. Proceedings of a Special IAU-Session on Supernovae held in Grenoble, France, 1 September, 1976.* 1977, X + 192 pp.
67. Jean Audouze (ed.), *CNO Isotopes in Astrophysics. Proceedings of a Special IAU Session held in Grenoble, France, 30 August, 1976.* 1977, XIII + 195 pp.
68. Z. Kopal, *Dynamics of Close Binary Systems,* XIII + 510 pp.
69. A. Bruzek and C. J. Durrant (eds.), *Illustrated Glossary for Solar and Solar-Terrestrial Physics.* 1977, XVIII + 204 pp.
70. H. van Woerden (ed.), *Topics in Interstellar Matter.* 1977, VIII + 295 pp.
71. M. A. Shea, D. F. Smart, and T. S. Wu (eds.), *Study of Travelling Interplanetary Phenomena.* 1977, XII + 439 pp.
72. V. Szebehely (ed.), *Dynamics of Planets and Satellites and Theories of Their Motion. Proceedings of IAU Colloquium No. 41, held in Cambridge, England, 17–19 August 1976.* 1978, XII + 375 pp.
73. James R. Wertz (ed.), *Spacecraft Attitude Determination and Control.* 1978, XVI + 858 pp.

74. Peter J. Palmadesso and K. Papadopoulos (eds.), *Wave Instabilities in Space Plasmas. Proceedings of a Symposium Organized Within the XIX URSI General Assembly held in Helsinki, Finland, July 31–August 8, 1978.* 1979, VII + 309 pp.
75. Bengt E. Westerlund (ed.), *Stars and Star Systems. Proceedings of the Fourth European Regional Meeting in Astronomy held in Uppsala, Sweden, 7–12 August, 1978.* 1979, XVIII + 264 pp.
76. Cornelis van Schooneveld (ed.), *Image Formation from Coherence Functions in Astronomy. Proceedings of IAU Colloquium No. 49 on the Formation of Images from Spatial Coherence Functions in Astronomy, held at Groningen, The Netherlands, 10–12 August 1978.* 1979, XII + 338 pp.